MW00452417

Maintaining Mission Critical Systems in a 24/7 Environment

IEEE Press
445 Hoes Lane
Piscataway, NJ 08854

Books in the IEEE Press Series on Power Engineering

Rating of Electric Power Cables in Unfavorable Thermal Environments
George J. Anders

Power System Protection
P. M. Anderson

Understanding Power Quality Problems: Voltage Sags and Interruptions
Math H. J. Bollen

Signal Processing of Power Quality Disturbances
Math H. J. Bollen and Irene Yu-Hua Gu

Modeling and High-Performance Control of Electric Machines
John Chiasson

Electric Power Applications of Fuzzy Systems
Edited by M. E. El-Hawary

Principles of Electric Machines with Power Electronic Applications, Second Edition
M. E. El-Hawary

Pulse Width Modulation for Power Converters: Principles and Practice
D. Grahame Holmes and Thomas Lipo

Analysis of Electric Machinery and Drive Systems, Second Edition
Paul C. Krause, Oleg Wasynczuk, and Scott D. Sudhoff

Risk Assessment for Power Systems: Models, Methods, and Applications
Wenyuan Li

Optimization Principles: Practical Applications to the Operations and Markets of the Electric Power Industry
Narayan S. Rau

Electric Economics: Regulation and Deregulation
Geoffrey Rothwell and Tomas Gomez

Electric Power Systems: Analysis and Control
Fabio Saccomanno

Electrical Insulation for Rotating Machines: Design, Evaluation, Aging, Testing, and Repair
Greg Stone, Edward A. Boulter, Ian Culbert, and Hussein Dhirani

MAINTAINING MISSION CRITICAL SYSTEMS IN A 24/7 ENVIRONMENT

Peter M. Curtis

WILEY-INTERSCIENCE
A JOHN WILEY & SONS, INC., PUBLICATION

Published by John Wiley & Sons, Inc. Published simultaneously in Canada.

For general information on our other products and services or for technical support, please contact our Customer Care Department within the United States at (800) 762-2974, outside the United States at (317) 572-3993 or fax (317) 572-4002.

Wiley also publishes its books in a variety of electronic formats. Some content that appears in print may not be available in electronic formats. For more information about Wiley products, visit our web site at www.wiley.com.

Library of Congress Cataloging-in-Publication Data is available.

ISBN 978-0-471-68374-2

Printed in the United States of America

10 9 8 7 6 5 4 3 2

Contents

Foreword

Our lives, livelihoods, and way of life are increasingly dependent on computers and data communication. And this dependence more and more relies on critical facilities or data centers where servers, mainframes, storage devices and communication gear are brought. In short, we are becoming a *data centric* or data center society.

Events like 9/11 and the 2003 Northeast power blackout highlight our incredible dependence on information and data communication. These and similar events have resulted in a slew of government regulations. Likewise, technological advances in a variety of industries place ever increasing demands on our information infrastructure, especially our data centers, changing the way we design, build, use, and maintain these facilities. However, the industry experts have been very slow to document and communicate the vital processes, tools, and techniques.

Not only is ours a dynamic environment, but it also is complex and requires an understanding of electrical, mechanical, fire protection and security systems, of reliability concepts, operating processes, and much more. I realized the great benefit Peter Curtis' book will bring to our mission critical community soon after I started reviewing the manuscript. I believe this is the first attempt to provide a comprehensive overview of all of the interrelated systems, components, and processes that define the data center space. The results are remarkable!

Data center facilities are shaped by a paradox, where critical infrastructure support systems and the facility housing them are designed to last 15+ years while the IT equipment typically has a life of about 3 years. Thus, every few years we are faced with major IT changes that dramatically alter the computer technology and invariably impact the demand for power, heat dissipation, and the physical characteristics of the facility design and operation.

It's no secret that one of the most difficult challenges facing our industry is our ability to objectively assess risk and critical facility robustness. In general, we lack the metrics needed to quantify reliability and availability perspective—the ability to identify and align the function or business mission of each building with its performance expectation. Other industries, particularly aircraft maintenance and nuclear power plants, have spent years developing analytical tools to assess systems resiliency and the work has yielded substantial performance improvements. In addition, the concept of reliability is sometimes misunderstood by professionals serving the data center industry. Peter's efforts to define and explain reliability concepts will help to raise the performance of mission critical space.

Furthermore, the process of integrating all of the interrelated components—programming space allocation, design, redundancy level planning, engineered systems quality, construction, commissioning, operation, documentation, contingency planning, personnel training, etc.—to achieving reliability objectives, is clear and well reasoned. The book plainly demonstrates how and why each element must be addressed to achieve reliability goals. Although this concept appears obvious, it often is not fully understood or accepted.

The comprehensive review of essential electrical and mechanical systems populating these facilities, from uninterruptible power supplies to chillers and generators, has great benefits not only from a functional standpoint but also because it provides the necessary maintenance and testing data needed for effective system operation. And, maybe most importantly, Mr. Curtis recognizes and deals with the vital human factor, "... perhaps the most poorly understood aspect of process safety and reliability management."

I am confident that time will validate the approach and ideas covered here. Meanwhile, we will all profit from this admirable effort to bring a better understanding to this complex and fast changing environment.

PETER GROSS

Los Angeles, California
June 2006

Preface

Maintaining mission critical systems in a 24/7 environment was written for the purpose of building the skills of our future younger work force. As we enter the next phase of business resiliency, our country needs trained people armed with the essentials of critical infrastructure. It is imperative to have right mind-set to make critical decisions during both natural and man-made disasters. This book provides that foundation.

The goal of the book is to provide the fundamentals of the critical infrastructure that protect mission critical environments or the digital lifestyle we live today. This is achieved by understanding mission critical facilities engineering, electrical systems maintenance, power quality, uninterruptible power supplies, automatic transfer switches, static transfer switches, accessed raised floors, fire alarm systems, and HVAC systems. There is also a chapter on policies and regulations that are beginning to drive operational risk assessment.

A special part of the book is the appendix section, which includes the Guide to Enterprise Power and Critical Power Whitepapers as well as an alternate way of measuring mission critical criticality levels.

Maintaining mission critical systems in a 24/7 environment also gathered data from various companies as a preliminary step to bring ideas together that allows for greater industry collaboration resulting in the best practices for business resiliency.

PETER M. CURTIS

Woodbury, New York
October 2006

Acknowledgments

Creating this book could not be possible through the effort of only one person. I've attended various conferences throughout my career including 7/24 Exchange, IFMA, AFE, Data Center Dynamics, AFCOM, and BOMI and harvested insight offered by many mission critical professionals from all walks of the industry. I'm grateful for the professional relationships that were built at these conferences, which allow the sharing of knowledge, know-how, information, and experiences upon which this book is based. I'm also grateful to IEEE/Wiley for taking on this project almost 9 years ago. The format that initially began as material for online educational classes transcended into an entire manuscript passionately assembled by all who came across it.

Professionals in the mission critical field have witnessed its evolution from a fledgling 40-hour-a-week operation, into the 24/7 environment that business and people demand today. The people responsible for the growth and maintenance of the industry have amassed an invaluable cache of knowledge and experience along the way. Compiling this information into a book provides a way for those new to this industry to tap into the years of experience that have emerged since the industry's humble beginnings only a couple of decades ago.

This book's intended audience includes every business that understands the consequences of downtime and seeks to improve its own resiliency. Written by members of senior management, technicians, vendors, manufacturers, and contractors alike, this book gives a comprehensive, 360-degree perspective on the mission critical industry as it stands today. Its importance lies in its use as a foundation toward a seamless transition to the next stages of business resiliency.

I am thankful to the following people and organizations for their help, support, and contributions that have enabled this information to be shared with the next generation of mission critical engineers and business continuity professionals.

CHAPTER CONTRIBUTORS

- Don Beaty, P.E., DLB Associates (Chapter 11—Data Center Cooling Systems and Components)
- Dan Catalfu, Tate Access Floors (Chapter 12—Raised Access Floors)
- Howard L. Chesneau, Fuel Quality Services (Chapter 6—Fuel Systems and Design)
- Edward English III, Fuel Quality Services (Chapter 6—Fuel Systems and Design)

- Brian K. Fabel, P.E., ORR Protection Systems (Chapter 13—Fire Protection in Mission Critical Infrastructures)
- Ron Ritorto, P.E., Mission Critical Fuel Systems (Chapter 6—Fuel Systems and Design)

TECHNICAL REVIEWERS AND EDITORS

- Jerry Burkhardt, Syska Hennessy Group (Appendix C—Criticality Levels)
- Bill Campbell, Emerson Network Power (Chapter 10—An Overview of UPS Systems)
- Charles Cottitta, ADDA (Chapter 5—Standby Generators)
- John C. Day, PDI Corp. (Chapter 8—Static Transfer Switches)
- John Diamond, Strategic Facilities Inc. (Chapter 5—Standby Generators and Chapter 10—An Overview of UPS Systems)
- John DeAngelo, Power Service Concepts, Inc. (Chapter 10—An Overview of UPS Systems)
- Michael Fluegeman, P.E., Power Management Concepts, LLC, (Chapter 7—Automatic Transfer Switches, Chapter 10—An Overview of UPS Systems, and Appendix C—Criticality Levels)
- Richard Greco, P.E., California Data Center Design Group (Chapter 3—Mission Critical Facilities Engineering)
- Ross M. Ignall, Dranetz-BMI (Chapter 9—Fundamentals of Power Quality)
- Cyrus Izzo, P.E., Syska Hennessy Group (Appendix C—Criticality Levels)
- John Kammeter, Chief Technology Officer, PDI Corp. (Chapter 8—Static Transfer Switches)
- Ellen Leinfuss, Dranetz-BMI (Chapter 9—Fundamentals of Power Quality)
- Teresa Lindsey, BITS (Appendix B—BITS Guide to Business-Critical Power)
- Wai-Lin Litzke, Brookhaven National Labs, (Chapter 2—Policies and Regulations)
- Byran Magnum, ADDA (Chapter 5—Standby Generators)
- John Menoche, American Power Conversion (Chapter 10—An Overview of UPS Systems)
- Joseph McPartland III, American Power Conversion, (Chapter 10—An Overview of UPS Systems)
- Mark Mills, Digital Power Group, (Appendix A—Critical Power)
- Gary Olsen, P.E., Cummins (Chapter 5—Standby Generators)
- Ted Pappas, Keyspan Engineering (Chapter 3—Mission Critical Facilities Engineering)

- Robert Perry, American Power Conversion (Chapter 10—An Overview of UPS Systems)
- Dan Sabino, Power Management Concepts, (Chapter 7—Automatic Transfer Switches)
- Douglas H. Sandberg, ASCO Service Inc. (Chapter 7—Automatic Transfer Switches)
- Ron Shapiro, P.E., EYP Mission Critical Facilities, Inc. (Chapter 4—Mission Critical Electrical Systems Maintenance)
- Reza Tajali, P.E., Square D (Chapter 4—Mission Critical Electrical Systems Maintenance)
- Kenneth Uhlman, P.E., Eaton/Cutler Hammer, technical discussions
- Steve Vechy, Enersys (Chapter 10—An Overview of UPS Systems)

Thank you Dr. Robert Amundsen, Director of the Energy Management Graduate Program at New York Institute of Technology, who gave me my first teaching opportunity in 1994. I am grateful that it has allowed me to continually develop professionally, learn, and pollinate many groups with the information presented in this book.

I'd like to thank two early pioneers of this industry for defining what "mission critical" really means to me and the industry. I am appreciative for the knowledge they have imparted to me. Borio Gatto for sharing his engineering wisdom, guidance and advice and Peter Gross, P.E., for his special message in the Foreword as well as expanding my views of the mission critical world.

Thank you to my good friends and colleagues for their continued support, technical dialogue, feedback, advice, over the years: Thomas Weingarten, P.E., Power Management Concepts, LLC, Vecas Gray, P.E., Mark Keller, Esq., Abramson and Keller.

I would also like to thank Joseph F. McPartland and Al Baker who were the early industry educators providing the essential building blocks in electrical design *Electrical Design. These fine gentlemen* assisted many young engineers early in their careers, and I am very happy to have been one of them.

Thank you Lois Hutchinson, for assisting in organizing and editing the initial material, acting as my sounding board and supporting me while I was assimilating my ideas. Also for keeping me on track and focused when necessary.

I would like to express gratitude to all the early contributors, students, mentors, interns, and organizations that I have been working with and learning from over the years for their assistance, guidance, advice, research, and organization of the information presented in this book: John Altherr, P.E., Tala Awad, Anna Benson, Ralph Ciardulli, Ralph Gunther, P.E., Al Law, John Nadon, John Mezic, David Nowak, Arnold Peterson, P.E., Kenneth Davis, MSEM, CEM, Richard Realmuto, P.E., Edward Rosavitch, P.E., Jack Willis, P.E., Shahnaz Karim, and my special friends at AFE Region 4, 7×24 Exchange, and Data Center Dynamics.

Special thanks to my partners Bill Leone and Eduardo Browne for keeping the business flowing and sharing my initial vision and building upon it and also for keeping me focused on the big picture. They have been pillars of strength for me.

Most of all I would also like to express gratitude to my loving and understanding wife Elizabeth for her continual support, encouragement, and acceptance of my creative process. I am also grateful to Elizabeth for her editorial assistance and feedback when my ideas were flowing 24/7.

Lastly, my deepest apologies for anyone I have forgotten.

P.M.C

Chapter 1

An Overview of Reliability and Resilience in Today's Mission Critical Facilities

1.1 INTRODUCTION

Continuous, clean, and uninterrupted power is the lifeblood of any data center, especially one that operates 24 hours a day, 7 days a week. Critical enterprise power is the power without which an organization would quickly be unable to achieve its business objectives. Today more than ever, enterprises of all types and sizes are demanding 24-hour system availability. This means that enterprises must have 24-hour power day after day, year after year. One such example is the banking and financial services industry. Business practices mandate continuous uptime for all computer and network equipment to facilitate round-the-clock trading and banking activities anywhere and everywhere in the world. Banking and financial service firms are completely intolerant of unscheduled downtime, given the guaranteed loss of business that invariably results. However, providing the best equipment is not enough to ensure 24-hour operation throughout the year. The goal is to achieve reliable 24-hour power supply at all times, regardless of the technological sophistication of the equipment or the demands placed upon that equipment by the end user, be it business or municipality.

The banking and financial services industry is constantly expanding to meet the needs of the growing global digital economy. The industry as a whole has been innovative in the design and use of the latest technologies, driving its businesses to become increasingly digitized in this highly competitive business environment. The industry is progressively more dependent on continuous operation of its data centers in reaction to the competitive realities of a world economy. To achieve

Maintaining Mission Critical Systems in a 24/7 Environment By Peter M. Curtis

optimum reliability when the supply and availability of power are becoming less certain is challenging to say the least. The data center of the past required only the installation of stand-alone protective electrical and mechanical equipment mainly for computer rooms. Data centers today operate on a much larger scale—that is, 24/7. The proliferation of distributed systems using hundreds of desktop PCs and workstations connected through LANs and WANs that simultaneously use dozens of software business applications and reporting tools makes each building a "computer room." If we were to multiply the total amount of locations utilized by each bank and financial service firm that are tied together all over the would through the internet, then we would see the necessity of uninterrupted power supply, uptime, and reliability.

The face of Corporate America was severely scarred in the last few years by a number of historically significant events: the collapse of the dot.com bubble and high-profile corporate scandals like Enron and WorldCom. These events have taken a significant toll on financial markets and have served to deflate the faith and confidence of investors. In response, governments and other global organizations enacted new or revised existing laws, policies, and regulations. In the United States, laws such as the Sarbanes–Oxley Act of 2002 (SOX), Basel II, and the U.S. Patriot Act were created. In addition to management accountability, another embedded component of SOX makes it imperative that companies not risk losing data or even risk downtime that could jeopardize accessing information in a timely fashion. Basel II recognizes that infrastructure implementation involves identifying operational risk and emphasizes the allocation of adequate capital to cover potential loss. These laws can actually improve business productivity and processes.

Many companies thoughtlessly fail to consider installing backup equipment or the proper redundancy based on their risk profile. Then when the lights go out due to a major power outage, these same companies suddenly wake up, and they end up taking a huge hit operationally and financially. During the months following the Blackout of 2003, there was a marked increase in the installation of UPS systems and standby generators. Small and large businesses alike learned how susceptible they are to power disturbances and the associated costs of not being prepared. Some businesses that were not typically considered mission critical learned that they could not afford to be unprotected during a power outage. The Blackout of 2003 emphasized the interdependencies across the critical infrastructure, as well as the cascading impacts that occur when one component falters. Most ATMs in the affected areas stopped working, although many had backup systems that enabled them to function for a short period. Soon after the power went out, the Comptroller of the Currency signed an order authorizing national banks to close at their discretion. Governors in a number of affected states made similar proclamations for state-chartered depository institutions. The end result was a loss of revenue and profits and almost the loss of confidence in our financial system. More prudent planning and the proper level of investment in mission critical infrastructure for electric, water, and telecommunications utilities coupled with proactive building infrastructure preparation could have saved the banking and financial services industry millions.

At the present time, the risks associated with cascading power supply interruptions from the public electrical grid in the United States have increased due to the ever-increasing reliance on computer and related technologies. As the number of computers and related technologies continue to multiply in this ever-increasing digital world, the demand for reliable quality power increases as well. Businesses not only compete in the marketplace to deliver whatever goods and services are produced for consumption, but now must compete to hire the best engineers from a dwindling pool of talent who can design the best infrastructures needed to obtain and deliver reliable power to keep mission critical manufacturing and technology centers up and running to produce the very goods and services that are up for sale. The idea that businesses today must compete for the best talent to obtain reliable power is not new, as are the consequences of failing to meet this competition. Without reliable power, there are no goods and services for sale, no revenues, and no profits—only losses while the power is out. Hiring and keeping the best-trained engineers employing the very best analyses, making the best strategic choices, and following the best operational plans to keep ahead of the power supply curve are essential for any technologically sophisticated business to thrive and prosper. A key to success is to provide proper training and educational resources to engineers so they may increase their knowledge and keep current on the latest mission critical technologies available the world over. In addition, all companies need to develop a farm system of young mission critical engineers to help combat the continuing diluted work force for the growing mission critical industry.

It is also necessary for critical industries to constantly and systematically evaluate their mission critical systems, assess and reassess their level of risk tolerance versus the cost of downtime, and plan for future upgrades in equipment and services designed to ensure uninterrupted power supply in the years ahead. Simply put, minimizing unplanned downtime reduces risk. Unfortunately, the most common approach is reactive—that is, spending time and resources to repair a faulty piece of equipment after it fails as opposed to identifying when the equipment is likely to fail and repairing or replacing it without power interruption. If the utility goes down, install a generator. If a ground-fault trips critical loads, redesign the distribution system. If a lightning strike burns power supplies, install a new lightning protection system. Such measures certainly make sense, because they address real risks associated with the critical infrastructure; however, they are always performed after the harm has occurred. Strategic planning can identify internal risks and provide a prioritized plan for reliability improvements that identify the root causes of failure <u>before</u> they occur.

In the world of high-powered business, owners of real estate have come to learn that they, too, must meet the demands for reliable power supply to their tenants. As more and more buildings are required to deliver service guarantees, management must decide what performance is required from each facility in the building. Availability levels of 99.999% (5.25 minutes of downtime per year) allow virtually no facility downtime for maintenance or for other planned or unplanned events. Moving toward high reliability is imperative. Moreover, avoiding the landmines that can cause outages and unscheduled downtime never ends. Even planning and impact assessments are tasks that are never completed; they should be viewed afresh at least once every budget cycle.

The evolution of data center design and function has been driven by the need for uninterrupted power. Data centers now employ many unique designs developed specifically to achieve the goal of uninterrupted power within defined project constraints based on technological need, budget limitations, and the specific tasks each center must achieve to function usefully and efficiently. Providing continuous operation under all foreseeable risks of failure such as power outages, equipment breakdown, internal fires, and so on, requires use of modern design techniques to enhance reliability. These include redundant systems and components, standby power generation, fuel systems, automatic transfer and static switches, pure power quality, UPS systems, cooling systems, raised access floors, and fire protection, as well as the use of Probability Risk Analysis modeling software (each will be discussed in detail later in this book) to predict potential future outages and develop maintenance and upgrade action plans for all major systems.

Also vital to the facilities life cycle is two-way communication between upper management and facilities management. Only when both ends fully understand the three pillars of power reliability—design, maintenance, and operation of the critical infrastructure (including the potential risk of downtime and recovery time)—can they fund and implement an effective plan. Because the costs associated with reliability enhancements are significant, sound decisions can only be made by quantifying performance benefits against downtime cost estimates for each upgrade option to determine the best course of action. Planning and careful implementation will minimize disruptions while making the business case to fund necessary capital improvements and implement comprehensive maintenance strategies. When the business case for additional redundancy, specialized consultants, documentation, and ongoing training reaches the boardroom, the entire organization can be galvanized to prevent catastrophic data losses, damage to capital equipment, and danger to life and limb.

1.2 RISK ASSESSMENT

Critical industries require an extraordinary degree of planning and assessing. It is important to identify the best strategies to reach the targeted level of reliability. In order to design a critical building with the appropriate level of reliability, the cost of downtime and the associated risks need to be assessed. It is important to understand that downtime occurs due to more than one type of failure: design failure, catastrophic failures, compounding failures, or human error failures. Each type of failure will require a different approach on prevention. A solid and realistic approach to disaster recovery must be a priority, especially because the present critical area is inevitably designed with all the eggs located in one basket.

Planning the critical area in scope of banking and financial services places considerable pressure to design an infrastructure that will change over time in an effort to support the continuous business growth. Routine maintenance and upgrading equipment alone does not ensure continuous power. The 24/7 operations of such service mean an absence of scheduled interruptions for any reason including routine maintenance, modification, or upgrades. The main question is how and why infrastructure failures occur. Employing new methods of distributing critical power, understanding capital constraints, and developing processes that minimize human error are some key factors in improving recovery time in the event critical systems are impacted by base-building failures.

The infrastructure reliability can be enhanced by conducting a formal Risk Management Assessment (RMA) and a gap analysis and by following the guidelines of the Critical Area Program (CAP). The RMA and the CAP are used in other industries and customized specifically for needs of Data Center environments. The RMA is an exercise that produces a system of detailed, documented processes, procedures, checks, and balances designed to minimize operator and service provider errors. The practice CAP ensures that only trained and qualified people are associated and authorized to have access to critical sites. These programs coupled with Probability Risk Assessment (PRA) address the hazards of data center uptime. The PRA looks at the probability of failure of each type of electrical power equipment. Performing a PRA can be used to predict availability, number of failures per year, and annual downtime. The PRA, RMA, and CAP are facilitating agents when assessing each step listed below.

- Engineering and design
- Project management
- Testing and commissioning
- Documentation
- Education and training
- Operation and maintenance
- Employee certification
- Risk indicators related to ignoring the Facility Life Cycle Process
- Standard and benchmarking

Industry regulations and policies are more stringent than ever. They are heavily influenced by Basel II, Sarbanes–Oxley Act (SOX), NFPA 1600, and U.S. Securities and Exchange Commission (SEC). Basel II recommends "three pillars"—risk appraisal and control, supervision of the assets, and monitoring of the financial market—to bring stability to the financial system and other critical industries. Basel II implementation involves identifying operational risk and then allocating adequate capital to cover potential loss. As a response to corporate scandals such as Enron and WorldCom, SOX came into force in 2002 and passed the following act: The financial statement published by issuers is required to be accurate (Sec 401); issuers are required to publish information in their annual reports (Sec 404); and issuers are required to disclose to the public, on an urgent basis, information on material changes in their financial condition or operations (Sec 409) and impose penalties of fines and/or imprisonment for not complying (Sec 802).

The purpose of the NFPA 1600 Standard is to help the disaster management, emergency management, and business continuity communities to cope with disasters and emergencies. Keeping up with the rapid changes in technology has been a longstanding priority. The constant dilemma of meeting the required changes within an already constrained budget can become a limiting factor in achieving optimum reliability.

1.3 CAPITAL COSTS VERSUS OPERATION COSTS

Businesses rest at the mercy of the mission critical facilities sustaining them. Each year, $20.6 billion is spent on the electrical and mechanical infrastructure that supports IT in the United States. Business losses due to downtime alone total $46 billion per year globally. An estimated 94% of all businesses that suffer a large data loss go out of business within two years regardless of the size of the business. The daily operations of our economic system depend on the critical infrastructure of the banking and finance services.

Critical industries are operating continuously, 365 days. Because conducting daily operations necessitate the use of new technology, more and more servers are being packed into a single rack. The growing numbers of servers operating 24/7 increases the need for power, cooling, and airflow. When a disaster causes the facility to experience lengthy downtime, a prepared organization is able to quickly resume normal business operations by using a predetermined recovery strategy. Strategy selection involves focusing on key risk areas and selecting a strategy for each one. Also, in an effort to boost reliability and security, the potential impacts and probabilities of these risks as well as the costs to prevent or mitigate damages and the time to recover should be established.

Many organizations associate disaster recovery and business continuity only with IT and communications functions and miss other critical areas that can seriously impact their business. One major area that necessitates strategy development is the banking and financial service industry. The absence of strategy that guarantees recovery has an impact on employees, facilities, power, customer service, billing, and customer and public relations. All areas require a clear, well-thought-out strategy based on recovery time objectives, cost, and profitability impact. The strategic decision is based on the following factors:

- The maximum allowable delay time prior to the initiation of the recovery process
- The time frame required to execute the recovery process once it begins
- The minimum computer configuration required to process critical applications
- The minimum communication device and backup circuits required for critical applications
- The minimum space requirements for essential staff members and equipment
- The total cost involved in the recovery process and the total loss as a result of downtime.

Developing strategies with implementation steps means that no time is wasted in a recovery scenario. The focus is to implement the plan quickly and successfully. The right strategies implemented will effectively mitigate damages and minimize the disruption and cost of downtime.

1.4 CHANGE MANAGEMENT

To provide an uncompromising level of reliability and knowledge of the critical operations contributing to the success of maintaining 100% uptime of the U.S. critical infrastructure, it is crucial to define Mission Critical based on today's industry. The Mission Critical Industry today has an infrastructure primarily composed of silicon and information. Business continuity relies on the mission critical facilities sustaining them.

The Blackout of 2003 resulted in an economic loss estimated at between $700 million and $1 billion to New York City alone. Many city offices and private sector functions did not have sufficient backup power in place, including key agencies such as the Department of Health and Mental Hygiene, Department of Sanitation, and Department of Transportation, and certain priority areas of hospitals. In some cases the backup power failed to operate, failed to initiate power generation, and experienced mechanical failure or exhaustion of fuel supply. Whatever the nature of the problem, it is important to understand the underlying process by which an organization should handle problems and changes. A thorough problem tracking system and a strong change management system is essential to an Emergency Preparedness plan.

Change management crosses departments and must be coordinated and used by all participants to work effectively. Management needs to plan for the future and to make decisions on how to support the anticipated needs of the organization, especially under emergency situations. It is also imperative to understand that we cannot manage today's critical infrastructure the way we did in the early 1980s. Our digital-society needs are very different today.

1.5 TESTING AND COMMISSIONING

Before the facility goes on-line, it is crucial to resolve all potential equipment problems. This is the construction team's sole opportunity to integrate and commission all the systems, due to the facility's 24/7 mission critical status. At this point in the project, all systems installed were tested at the factory and witnessed by a competent independent test engineer familiar with the equipment.

Once the equipment is delivered, set in place, and wired, it is time for the second phase of certified testing and integration. The importance of this phase is to verify and certify that all components work together and to fine-tune, calibrate, and integrate the systems. There is a tremendous amount of preparation in this phase. The facilities engineer must work with the factory, field engineers, and independent test consultants to coordinate testing and calibration. Critical circuit breakers must be tested and calibrated prior to placing any critical electrical load on them. When all the tests are completed, the facilities engineer must compile the certified test reports, which will establish a benchmark for all-future testing. The last phase is to train the staff on each major piece of equipment. This phase is an ongoing process and actually begins during construction.

Many decisions regarding how and when to service a facility's mission critical electrical power distribution equipment are going to be subjective. The objective is easy: a high level of safety and reliability from the equipment, components, and systems. But discovering the most cost-effective and practical methods required to accomplish this can be challenging. Network with colleagues, consult knowledgeable sources, and review industry and professional standards before choosing the approach best suited to your maintenance goals. Also, keep in mind that the individuals performing the testing and service should have the best education, skills, training, and experience available. You depend on their conscientiousness and decision-making ability to avoid potential problems with perhaps the most crucial equipment in your building. Most importantly, learn from your experiences, and those of others. Maintenance programs should be continuously improving. If a task has historically not identified a problem at the scheduled interval, consider adjusting the schedule respectively. Examine your maintenance programs on a regular basis and make appropriate adjustments.

Acceptance and maintenance testing are pointless unless the test results are evaluated and compared to standards, as well as to previous test reports that have established benchmarks. It is imperative to recognize failing equipment and to take appropriate action as soon as possible. Common practice in this industry is for maintenance personnel to perform maintenance without reviewing prior maintenance records. This approach defeats the value of benchmarking and trending and must be avoided. The mission critical facilities engineer can then keep objectives in perspective and depend upon his options when faced with a real emergency.

The importance of taking every opportunity to perform preventive maintenance thoroughly and completely—especially in mission critical facilities—cannot be stressed enough. If not, the next opportunity will come at a much higher price: downtime, lost business, and lost potential clients, not to mention the safety issues that arise when technicians rush to fix a maintenance problem. So do it correctly ahead of time, and avoid shortcuts.

1.6 DOCUMENTATION AND HUMAN FACTOR

The mission critical industry's focus on physical enhancements descends from the early stages of the trade, when all efforts were placed solely in design and construction techniques to enhance mission critical equipment. At the time, technology was primordial and there was a significant need to refine and perfect mission critical electrical and mechanical systems.

Twenty-five years ago the technology supporting mission critical loads was simple. There was little sophistication in the electrical load profile; at that time the Mission Critical Facility Engineering Industry was in its infancy. Over time the data centers have grown from a few mainframes supporting minimal software applications to LAN farms that can occupy 25,000 SF or more.

As more computer hardware occupied the data center, the design of the electrical and mechanical systems supporting the electrical load became more complicated as did the business applications. With businesses relying on this infrastructure, more capital dollars were invested to improve the uptime of the business's lines. Today billions of dollars are invested on an enterprise level into the infrastructure that supports the business 24/7 applications; the major investments are normally in design, equipment procurement, and project management. Few capital dollars are invested in documentation, a significant step in achieving optimum level of reliability.

Years ago, most organizations heavily relied on their workforce to retain much of the information regarding the mission critical systems. A large body of personnel had a similar level of expertise. They remained with their company for decades. Therefore, little emphasis was placed on maintaining a living document for a critical infrastructure.

The mission critical industry can no longer manage their critical system as they did 25 years ago. Today the sophistication of the data center infrastructure necessitates perpetual documentation refreshing. One way to achieve this is to include a living document system that provides the level of granularity necessary to operate a mission critical infrastructure into a capital project. This will assist in keeping the living document current each time a project is completed or milestone is reached. Accurate information is the first level of support that provides first responders the intelligence they need to make informed decisions during critical events. It also acts like a succession plan as employees retire and new employees are hired thus reducing risk and improving their learning curve.

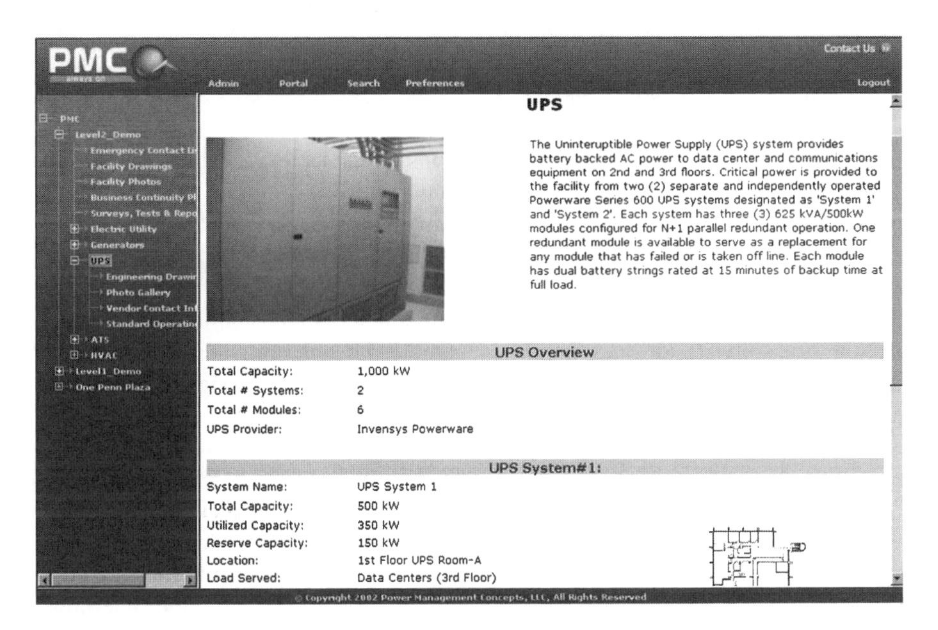

Figure 1.1 Typical screenshot of M. C. Access. (Courtesy of Power Management Concepts, LLC.)

Figure 1.2 M. C. Access. (Courtesy of Power Management Concepts, LLC.)

Human error as a cause of hazard scenarios must be identified, and the factors that influence human errors must be considered. Human error is a given and will arise in all stages of the process. It is vital that the factors influencing the likelihood of errors be identified and assessed to determine if improvements in the human factors design of a process are needed. Surprisingly, human factors are perhaps the most poorly understood aspect of process safety and reliability management.

Balancing system design and training operating staff in a cost-effective manner is essential to critical infrastructure planning. When designing a mission critical facility, the level of complexity and ease of maintainability is a major concern. When there is a problem, the Facilities Manager (FM) is under enormous amounts of pressure to isolate the faulty system while maintaining data center loads and other critical loads. The FM does not have the time to go through complex switching procedures during a critical event. A recipe for human error exists when systems are complex, especially if key system operators and documentation of Emergency Action Procedures (EAP) and Standard Operating Procedures (SOP) are not immediately available. A rather simplistic electrical system design will allow for quicker and easier troubleshooting during this critical time.

To further complicate the problem, equipment manufacturers and service providers are challenged to find and retain the industry's top technicians within their own company. As 24/7 operations become more prevalent, the talent pool available will continue to diminish. Imagine an FM responsible for 20 mission critical facilities around the globe. This would indicate that response times could increase from the current standard of four hours to a much higher and less tolerable timeframe. The need for a simplified, easily accessible, and well-documented design is only further substantiated by the growing imbalance of supply and demand of highly qualified mission critical technicians.

When designing a mission critical facility, a budgeting and auditing plan should be established. Over 60% of downtime is due to human error. Each year, substantial amounts of money are spent on building infrastructure, but inadequate capital is allocated to sustain that critical environment through the use of proper documentation, education and training.

1.7 EDUCATION AND TRAINING

Technology has been progressing faster than Moore's Law. Despite attaining high levels of technological standards in the mission critical industry, most of today's financial resources remain allocated for planning, engineering, equipment procurement, project management, and continued research and development. Unfortunately, little attention is given to the actual management of these systems. As equipment reliability increases, a larger percentage of downtime results from actions by personnel that were not properly trained or do not have access to accurate data during crisis events. Currently, there is a great need to educate and train facility engineers and operators in a swift manner, allowing a more efficient employee learning curve and in the process reducing risk because the employee can respond with situational awareness. Informed decision-making comes with familiar emergency action and standard operating procedures. The diversity among mission critical systems severely hinders people's ability to fully understand and master all necessary equipment and relevant information.

1.8 OPERATION AND MAINTENANCE

What can facility managers do to ensure that their electrical system is as reliable as possible? The seven steps to improved reliability and maintainability are:

- Planning and impact assessment
- Engineering and design
- Project management
- Testing and commissioning
- Documentation
- Education and training
- Operations and maintenance

Hire competent professionals to advise each step of the way. When building a data processing center in an existing building, you do not have the luxury of designing the electrical system from scratch. A competent electrical engineer will design a system that makes the most of the existing electrical system. Use electrical contractors who are experienced in data processing installations. Do not attempt to save money using the full 40% capacity for a conduit, because as quickly as new, state-of-the-art equipment is installed, it is de-installed. Those same number 12 wires will need to come out of the conduit without disturbing the working computer hardware.

Have an experienced electrical testing firm inspect the electrical system, perform tests on circuit breakers, and use thermal-scan equipment to find "hot spots" due to improper connections or faulty equipment. Finally, you should plan for routine shutdowns of your facility so that you can perform preventive maintenance on electrical equipment. Facility managers as well as senior management must not underestimate the cost-effectiveness of a thorough preventive maintenance program. Electrical maintenance is not a luxury; it is a necessity. Again, do you want electrical outages to be scheduled or unscheduled?

Integrating the ideal infrastructure is just about impossible. Therefore, seek out the best possible industry authorities to solve your problems. Competent consultants will have the knowledge, tools, testing equipment, training, and experience necessary to understand the risk tolerance of your company, as well as recommend and implement the proper and most-advanced proven designs.

Whichever firms you choose, always ask for sample reports, testing procedures, and references. Your decisions will determine the system's ultimate reliability, as well as the ease of system maintenance. Seek experienced professionals from within and without your own company: information systems, property, and operations managers, space planners, and the best consultants in the industry for all engineering disciplines. The bottom line is to have proven organizations working on your project.

1.9 EMPLOYEE CERTIFICATION

Empowering employees to function as communication allies can be achieved through a well-planned certification program. Employees have a vested interest in working with management to prevail over the crisis, and many are eager to actively promote the company's positions internally as well as externally. Empowering employees to take charge in times of crisis creates valuable communication allies who not only reinforce core messages internally, but also carry them into the community. The internal crisis communication should be conducted using established communication channels and venues in addition to those that may have been developed to manage specific crisis scenarios. Whichever method of internal crisis communication a company may choose, the more upfront management is about what is happening, the better-informed and more confident employees feel.

In this way, security can be placed on an operation or a task requiring that an employee be certified to perform that action. Certification terms should be defined by the factory and should include training or even periodic recertification as desired. Another way of evaluating certified behavior is to examine employee performance including times and yields. Should these evaluations fall too far below standard over a period of time, the system could recommend decertification.

Technology is driving itself faster than ever. Large investments are made in new technologies to keep up to date with advancements, yet industries are still faced with operational challenges. One possible reason is the limited training provided to employees operating the mission critical equipments. Employee certification is crucial not only to keep up with advanced technology, but also to promote quick emergency response. In the last few years, technologies have been developed to solve the technical problem of linkage and interaction of equipment, but without well-trained personnel. How can we confirm that the employee meets the complex requirements of the facility to ensure high levels of reliability?

1.10 STANDARD AND BENCHMARKING

The past decade has seen wrenching change for many organizations. As firms and institutions have looked for ways to survive and remain profitable, a simple but powerful change strategy called "benchmarking" has become popular. The underlying rationale for the benchmarking process is that learning by example and from best-practice cases is the most effective means of understanding the principles and the specifics of effective practices. Recovery and redundancy together cannot provide sufficient resiliency if they can be disrupted by a single unpredictable event. A mission critical data center designed and developed as mentioned above must be able to endure hazards of nature, such as earthquakes, tornados, floods, and other natural disasters, as well as human made hazards. Great care should be taken to ensure critical functions that will minimize downtime. Standards should be established with guidelines and mandatory requirements for power continuity. Procedures should be developed for the systematic sharing of safety- and performance-related material, best practices, and standards. Supervisory control and data acquisition for networks should be engineered for secure exchange of information between public and private grids.

Preventive maintenance and testing are crucial. The key is to benchmark the facility on a routine basis with the goal of identifying performance deviations from the original design specifications. Done properly, this will provide an early warning mechanism to allow potential failure to be addressed and corrected before it occurs. Once deficiencies are identified, and before any corrective action can be taken, a Method of Operation (MOP) must be written. The MOP will clearly stipulate step-by-step procedures and conditions, including who is to be present, the documentation required, phasing of work, and the state in which the system is to be placed after the work is completed. The MOP will greatly minimize errors and potential system downtime by identifying responsibility of vendors, contractors, the owner, the testing entity, and anyone else involved. In addition, a program of ongoing operational staff training and procedures is important to deal with emergencies outside of the regular maintenance program.

The most important aspect of benchmarking is that it is a process driven by the participants whose goal is to improve their organization. It is a process through which participants learn about successful practices in other organizations and then draw on those cases to develop solutions most suitable for their own organizations. True process benchmarking identifies the "hows" and "whys" for performance gaps and helps organizations learn and understand how to perform with higher standards of practice.

Chapter 2

Policies and Regulations

2.1 EXECUTIVE SUMMARY

Globalization, telecommunications, and advances in technology have dramatically improved today's business operations in terms of speed and sophistication. With all the advantages, it has also added to its complexity and vulnerability, especially as it relates to the IT infrastructure and other interdependent physical infrastructures. This chapter discusses business continuity and how current regulations and policies impact the way companies and government must operate today with regard to data integrity, availability, and disaster recovery.

The *U.S. Patriot Act, Basel II Accord, and the Sarbanes–Oxley Act* are among the most recent laws that have significantly impacted the financial industry and government agencies, particularly the Securities and Exchange Commission. In some cases the rules build on existing policies of business operations, and in others they have tightened the practices. The main objectives of these policies are to:

a. Assure public safety, confidence, and services
b. Establish responsibility and accountability
c. Encourage and facilitate partnering among all levels of government and between government and industry
d. Encourage market solutions whenever possible; compensation for market failure with focused government intervention
e. Facilitate meaningful information sharing
f. Foster international security cooperation
g. Develop technologies and expertise to combat terrorist threats
h. Safeguard privacy and constitutional freedoms

Maintaining Mission Critical Systems in a 24/7 Environment By Peter M. Curtis

Guidelines to attain the above objectives include: *The National Strategy for the Physical Protection of Critical Infrastructures & Key Assets*, detailing the roles of government departments and agencies in the protection of the nation's assets; *Sound Practices to Strengthen the Resilience of the U.S. Financial System*; planning for business continuity for critical financial institutions; and other standards such as *NFPA 1600*. Unfortunately, most of these documents can only provide general requirements for compliance without necessarily having specific details on implementation and technical controls. Regardless, the cost of noncompliance can be severe with personal liabilities attached. Essentially, the specific strategies used are left up to the industry based on their own risk assessment and management objectives.

2.2 INTRODUCTION

Business continuity and disaster recovery are important components to a company's risk-management strategy. In today's regulatory environment, inadequate planning or short-falls can be costly in terms of business operations; noncompliance can mean stiff penalties. On the positive side, the more stringent standards along with new technologies that address security and continuity issues can significantly improve business processes and productivity. Croy (2004) clearly states that business continuity is beyond just IT maintenance. Business decisions are often dependent on the accuracy of financial reports as they are generated in real-time. Plans must be in place to ensure that information is stored and transmitted accurately, and secured should an unexpected event occur. Events do not have to be major disasters, power outages can be even bigger risks. Business continuity must be "woven" into business processes throughout the company.

2.3 INDUSTRY REGULATIONS AND POLICIES

2.3.1 U.S. Patriot Act

Shortly following the September 11th 2001 terrorist attack, Congress passed the U.S. Patriot Act in order to strengthen national security (Provide Appropriate Tools Required to Intercept and Obstruct Terrorism) (Act of 2001). This Act contains many provisions, most of which are to support activities related to counterterrorism, threat assessment, and risk mitigation, to support critical infrastructure protection and continuity, to enhance law enforcement investigative tools, and to promote partnerships between government and industry.

"Critical infrastructure" means systems and assets, whether physical or virtual, so vital to the United States that the incapacity or destruction of such systems and assets would have a debilitating impact on security, national economic security, national public health or safety, or any combination of these. In today's information age, a national effort is necessary for the protection of cyberspace as well as our physical infrastructure, which is relied upon for national defense, continuity of government, economic prosperity, and quality of life in the United States. Private business, government, and the national security depend on an interdependent network of critical physical and information infrastructures, including telecommunications, energy, financial services, water, and transportation sectors. The policy requires that critical infrastructure disruptions be minimized, rare, brief, geographically limited in effect, manageable, and minimally detrimental to the economy, human and government services, and national security.

Through public–private cooperation, comprehensive and effective program are necessary to ensure continuity of businesses and essential U.S. government functions. Although specific methods are not provided in the Act, each institution must determine and assess its own particular risks in developing a program. Programs should include security policies, procedures and practices, reporting requirements, line of responsibility, training, and emergency response.

An essential element for compliance under the Patriot Act is identification of customers and their activities. For both financial and nonfinancial institutions, compliance means that you need to know your customer and acquire knowledge of their business operations. This is being implemented through a Customer Identification Program (CIP) and through periodic review of the Treasury Department's list of Specially Designated Nationals and Blocked Persons (SDN) to ensure that financial transactions are not done with terrorists. This is accomplished by implementing vigilant programs that consist of identity checks, securing and monitoring personal data, and reporting of suspicious activities. In addition, institutions should periodically cross-reference lists published by government agencies listing banned individuals and foreign nations.

Another provision of the Patriot Act is Title III, entitled "International Money Laundering Abatement and Anti-Terrorism Financing Act of 2001," which requires that a financial institution take reasonable steps to safeguard against money laundering activities. The purpose of these requirements is to thwart any possible financial transactions or money laundering activities by terrorists. Some steps required for compliance include: (1) designating a compliance offer, (2) implementing an ongoing training program, (3) adopting an independent audit function, and (4) developing internal policies, procedures, and controls.

Financial institutions include domestic and foreign entities and include banks, brokers, futures merchants, commodities traders, and investment bankers, and insurance companies must take added measures in their business operations. Casinos, real estate brokers, pawnbrokers, travel agencies, jewelers, automobile and boat retailers, and nontraditional financial institutions also need to comply, and the requirements even extend to nonfinancial institutions.

There are two specific federal entities, both of which are housed under the Treasury Department, that support these programs: (1) The Financial Crimes Enforcement Network supports domestic and international law enforcement efforts to combat money laundering and other financial crimes and is the entity that monitors CIP compliance, and (2) the Office of Foreign Asset Control (OFAC) maintains the list of Specially Designated Nationals and Blocked Persons (SDN) and also monitors compliance that prohibits U.S. entities from entering into business transactions with

SDN entities (Executive Order 13224). OFAC has established industry compliance guidelines specifically for insurance, import/export businesses, tourism, securities and banks. Each institution must determine and evaluate its own particular risks in developing a program, which is likely to include an analysis of its customer base, industry, and location.

Business Continuity Plan (BCP)

The BCP allows an organization to quickly recover from a disaster (i.e., natural disaster, terrorism, cyber-terrorism) and to reestablish normal operations, given the scope and severity of the business disruption. The purpose is to safeguard employees and property, make financial and operational assessments, protect records, and resume business operations as quickly as possible.

In financial institutions a business continuity plan addresses the following: data backup and recovery; all mission critical systems; financial and operational assessments; alternative communications with customers, employees, and regulators; alternate physical location of employees; critical supplier, contractor, bank, and counterparty impact; regulatory reporting; and ensuring customers prompt access to their funds and securities.

For federal departments and bureaus business, continuity is developed under an Enterprise Architecture framework, which addresses how a department, agency, or bureau should maintain operations following a major disaster, incident, or disruption. It includes government standards as well as commercial best practices. Every federal agency is responsible for determining which processes and systems are mission critical, and which business processes must be restored in order of priority should a disaster occur. For federal facilities, a business continuity plan consists of the following hierarchy of elements (from higher to lower priority):

- Continuity of operations
- Continuity of government
- Critical infrastructure protection (computer and physical infrastructure)
- Essential processes/systems
- Nonessential processes/systems

To properly plan and administer a business continuity program, the following 10 subject areas should be addressed:

1. *Program Initiation and Management.* This first step establishes policy, budget management, and reporting and assesses new projects.
2. *Risk Evaluation and Control.* This assesses vulnerabilities, identifies mitigation opportunities, and defines backup and restoral procedures.
3. *Business Impact Analysis.* This identifies functions, criticality, impact, and interdependencies and defines recovery time objectives.
4. *Business Continuity (BC) Strategies.* These assess continuity options and conduct business cost analysis.

5. *Emergency Response and Operations.* This step defines incident command structure and establishes emergency response procedures.
6. *Develop and Implement BC Plans.* Define BC process, procedures, and implementation capability.
7. *Awareness and Training Program.* Set up BC training.
8. *Maintain and Exercise BC Plans.* Conduct exercises and maintain current BC plans.
9. *Public Relations and Crisis Coordination.* Establish proactive public relations, media handling, and grief counseling.
10. *Coordination with Public Authorities.* Plan development and test exercises, assess new laws and regulations, assess information sharing.

2.3.2 The National Strategy for the Physical Protection of Critical Infrastructures and Key Assets

This document released in 2003 details a major part of Bush's overall homeland security strategy. It defines "critical infrastructure" sectors and "key assets" as follows:

Critical Infrastructure. This includes agriculture, food, water, public health, emergency services, government, defense industrial base, information and telecommunications, energy, transportation, banking and finance, chemical industry, postage, and shipping.

The National Strategy also discusses "cyber infrastructure" as closely connected to, but distinct from, physical infrastructure.

Key Assets. These targets are not considered vital, but could create local disaster or damage our nation's morale and confidence. These assets include:

- National symbols, monuments, centers of government or commerce, icons, and perhaps high-profile events
- Facilities and structures that represent national economic power and technology
- Prominent commercial centers, office buildings, and sports stadiums, where large numbers of people congregate

National security, business, and the quality of life in the United States rely on the continuous, reliable operation of a complex set of interdependent infrastructures. With the telecommunications and technology available today, infrastructure systems have become more linked, automated, and interdependent. Any disruptions to one system could easily cascade into multiple failures of remaining linked systems, thereby leaving both physical and cyber systems more vulnerable. Ensuring the protection of these assets requires strong collaboration between government agencies and between the public and private sectors. "Currently, there are no tools that allow understanding of the operation of this complex, interdependent system. This makes it difficult to identify critical nodes, determine the consequences of outages, and develop optimized mitigation strategies." (S. G. Varnado, U.S. House of Representatives Committee on Energy and Commerce, 2002).

Clearly, communication of accurate information is an important link to these interdependent infrastructures. Surveillance, critical capabilities (facilities and services), and data systems must be protected. The Department of Homeland Security is working with officials in each sector to ensure adequate redundant communication networks and to improve communications availability and reliability, especially during major disruptions.

2.3.3 U.S. Security and Exchange Commission (SEC)

The primary mission of the U.S. Securities and Exchange Commission (SEC) is to protect investors and maintain the integrity of the securities markets. The SEC works closely with Congress, other federal departments and agencies, the stock exchanges, state securities regulators, and the private sector to oversee the securities markets. It has the power to enforce securities laws to prevent insider trading and accounting fraud, providing false or misleading information to investors about the securities and the companies that issue them. The latest addition to the list of laws that govern this industry is the Sarbanes–Oxley Act 2002, which has made a major shakeup to the industry and has a huge impact in the way "normal" business operations are conducted today. This is discussed in greater detail in the following pages.

2.3.4 Sound Practices to Strengthen the Resilience of the U.S. Financial System

This paper is being drafted by interagency agreement between the Federal Reserve Board (Board), the Office of the Comptroller of the Currency (OCC), and the Securities and Exchange Commission (SEC). The paper provides guidance on business continuity planning for critical financial institutions, particularly important in the post-9/11 environment, and contains sound practices to ensure the resilience of the U.S. financial system. The following highlights the specific relevant areas:

Business Continuity Objectives

- Rapid recovery/timely resumption of critical operations following a wide-scale disruption
- Rapid recovery/timely resumption of critical operations following the loss or inaccessibility of staff in at least one major operating location
- High level of confidence, through ongoing use or robust testing, that critical internal and external continuity arrangements are effective and compatible

Sound practices to consider that cover the following technical areas are:

- Focus on establishing robust backup facilities
- Definition of wide-scale regional disruption and parameters of probable events (e.g., power disruption, natural disaster)
- Duration of outage and achievable recovery times
- Impact on type of data storage technology and telecommunications infrastructure
- "Diversification of risk" and establishing backup sites for operations and data centers that do not rely on the same infrastructure and other risk elements as primary sites
- Geographic distance between primary and backup facilities
- Sufficient critical staffing to recover from wide-scale disruption
- Routine use and testing of backup facilities
- Costs versus risks

The "sound practices" can only provide overall general guidelines with no specific details with respect to recovery times, prescribed technology, backup site locations, testing schedules, or the level of resilience across firms. Each firm must determine their own risk profile based on their existing infrastructure, planned enhancements, resources, and risk management strategies. In the end, it is the role of the Board of Directors and senior management to ensure that their business continuity plans reflect the firm's overall business objectives. It is the role of the regulatory agencies to monitor the implementation of those plans.

2.3.5 Federal Real Property Council (FRPC)

This is an interagency created under Federal Real Property Asset Management (Executive Order 13327) and is responsible for the development of a single and descriptive database of federal real properties. The Executive Order is a requirement for improved real property asset management and complete accountability that integrates into an organization's financial and mission objectives. It is consistent with other federal financial policies such as SOX and the new approach toward business operations. This implication is that facilities and infrastructure will need to be optimized in order to ensure that disclosure of financial data is complete, verifiable, and authenticated to allow for sharing of data across functional business areas.

2.3.6 Basel II Accord

With globalization of financial services firms and sophisticated information technology, banking has become more diverse and complex. The Basel Capital Accord was first introduced in 1988 by the Bank for International Settlements (BIS) and was then updated in 2004 as the Basel II Accord. This provides a regulatory framework that requires all internationally active banks to adopt similar or consistent risk management practices for tracking and publicly reporting their operational, credit, and market risks.

Basel II impacts information technology because of the need for banks to collect, store and process data. Financial organizations also must minimize operational risk by having operational controls such as power protection strategies to ensure reliability, availability and security of their data and business systems.

International policymakers intend Basel II to be a platform for better business management where financial institutions systematically incorporate risk as a key driver in business decisions.

Basel II compliance has the following benefits:

- Improve risk-management practices by integrating, protecting and auditing data.
- Rapidly and effectively track exposure to operation, credit and market risks.
- Reduce capital charges and other costs resulting from noncompliance.
- Control data from the source through to end reporting for improved data integrity.
- Provide secure and reliable access to applications and relational databases 24/7.
- Maintain a historical audit trail to rapidly detect data discrepancies or irregularities.

2.3.7 Sarbanes–Oxley (SOX)

Passed by Congress in 2002, the Sarbanes–Oxley Act was enacted in response to major incidences of corporate fraud, abuse of power, and mismanagement, such as the Enron Scandal, Worldcom (MCI) debacle, and others. The Act, which defines new industry "best practices," mainly targets publicly traded companies, although some parts are legally binding for non-publicly listed and not-for-profit companies. The SEC, the NASDAQ, and the New York Stock Exchange are preparing the regulations and rulings. Its mandates include the following:

- Increased *personal* liability for senior-level managers, board, and audit committees
- New reporting and accounting requirements
- Code of ethics and culture, and new definition of conflict of interest
- Restrictions on services that auditors can provide
- Created the "Public Company Accounting Oversight Board" to oversee the activities of the auditing profession

SOX requires major changes in securities law, reviews of policies and procedures with attention to data standards and IT processes. A key part of the review process is how vital corporate data are stored, managed, and protected. From a power protection standpoint the focus is on availability, integrity, and accountability.

Under Section 404 of SOX, there is an extensive role of IT infrastructure and applications in today's financial reporting and accounting processes. Although SOX itself does not specify how the technical controls should be implemented, there are technical solutions available to help public companies comply with its objectives. An example that was found through a search on the web is SSH Communications Security (http://www.ssh.com/). The system incorporates confidentiality, integrity, and authentications as security services within the network that prevents illegitimate modification of financial data and ensures authorization decisions that are based on true user identities.

The financial controls aimed at preventing corporate fraud have brought about "mixed feelings," with some CIOs indicating that it has proved costly. Some experts estimate that SOX will cost corporate America an additional $35 billion this year, 20 times more than the SEC had predicted (Russo, 2005).

With an optimized and flexible approach, and cooperation between the SEC and corporations the figures will likely land somewhere in between.

2.3.8 NFPA 1600

Congress adopts private sector preparedness standard developed by the American National Standards Institute and based on the National Fire Protection Association 1600 Standard on Disaster/Emergency Management and Business Continuity Programs. NFPA 1600 requires an "all hazards" approach, and there has been significant emphasis over the past three years on preparedness for terrorist attacks. A risk assessment is required to determine all hazards that might impact people, property, operations, and the environment. This should quantify the probability of occurrence and the severity of their consequences. The risk assessment must also take into account the business impact a disaster will have on the organizations' mission, as well as direct and indirect financial consequences. This enables an organization's to evaluate the cost-effectiveness of mitigation efforts and determine how much to invest in preparedness, response, and recovery plans.

A recovery plan addresses continuity of business and recovery of buildings, systems, and equipment. The business continuity plan identifies critical functions, how quickly they must be recovered (recovery time objective), and operating strategies to maintain these critical functions until normal functionality is restored. The required impact analysis can help to identify recovery time objectives and provide justification for investment in mitigation or business continuity efforts.

Chapter 3

Mission Critical Facilities Engineering

3.1 INTRODUCTION

Businesses that are motivated to plug into the Information Age and travel the globe at Internet speeds 24 hours a day, indefinitely, require reliability and flexibility, regardless of whether the companies are large Fortune 1000 corporations or small companies serving global customers. This is the reality of conducting business today. Whatever type of business you are in, many organizations have realized that a 24/7 operation is imperative. An hour of downtime can wreak havoc on project schedules and lost person-hours in re-keying electronic data, not to mention the potential for losing millions of dollars.

Twenty-five years ago, the facilities manager was responsible for the integrity of the building. As long as the electrical equipment worked 95% of the time, the FM was doing a good job. When there was a problem with downtime, it usually was a computer fault. As technology improved on both the hardware and software fronts, information technology began to design their hardware and software systems with redundancy. As a result of IT's efforts, computer systems have become so reliable that they're only down during scheduled upgrades.

Today the major reasons for downtime are human-error utility failure: poor power quality, power distribution failures, and environmental system failures (although the percentage remains small). So the facilities manager is usually in the hot seat when a problem occurs. Problems are not limited just to power quality; it could also be that the staff has not been properly trained in certain situations. Recruiting of inside staff and outside consultants can be difficult, because facilities management, protection equipment manufacturers, and consulting firms all compete for the same personnel pool.

Maintaining Mission Critical Systems in a 24/7 Environment By Peter M. Curtis

Minimizing unplanned downtime reduces risk. But unfortunately, the most common approach is reactive—that is, spending time and resources to repair a faulty piece of equipment. However, strategic planning can identify internal risks and provide a prioritized plan for reliability improvements. Also, only when both ends fully understand the potential risk of outages, including recovery time, can they fund and implement an effective plan. Because the costs associated with reliability enhancement are significant, sound decisions can only be made by quantifying the performance benefits and weighing the options against their respective risks.

Planning and careful implementation will minimize disruptions while making the business case to fund capital improvements and maintenance strategies. When the business case for additional redundancies, consultants, and ongoing training reaches the boardroom, the entire organization can be galvanized to prevent catastrophic data losses, damage to capital equipment, and even danger to life safety.

Figure 3.1 "Seven steps" is a continuous cycle of evaluation, implementation, preparation, and maintenance. (Courtesy of Power Management Concepts, LLC.)

3.2 COMPANIES' EXPECTATIONS: RISK TOLERANCE AND RELIABILITY

In order to design a building with the appropriate level of reliability, a company first must assess the cost of downtime and determine its associated risk tolerance. Because recovery time is now a significant component of downtime, downtime can no longer be equated to simple power availability, measured in terms of *one nine* (90%) or *six nines* (99.9999%). Today, recovery time is typically many times longer than utility outages, since operations have become much more complex. Is a 32-second outage really only 32 seconds? Is it perhaps 2 hours or 2 days? The real question is, How long does it take to fully recover from the 32 second outage and return to normal operational status? Although measuring in terms of nines has its limitations, it remains a useful measurement we need to identify. For a 24/7 facility:

- A 99% uptime or reliability level would be equivalent to 87.6 hours of downtime per year.
- A 99.9% reliability level would be equivalent to 8.76 hours of downtime per year.
- A 99.99% reliability level would be equivalent to 0.876 hours or 52 minutes of downtime per year.
- A 99.999% reliability level would be equivalent to 5.25 minutes of downtime per year.
- A 99.9999% reliability level would be equivalent to 32 seconds of downtime per year.

In new 24/7 facilities, it is imperative to not only design and integrate the most reliable systems, but also to keep them simple. When there is a problem, the facilities manager is under enormous pressure to isolate the faulty system and in the process, not disrupt critical electrical loads. The FM does not have the luxury of time to go through complex switching procedures during critical events. If systems are overly complex and key people who understand the system functionality are not available, this is a recipe for failure via human error. When designing a critical facility, it is important that the building design not outsmart the FM. Companies also maximize profits and minimize cost by using the simplest design approach possible.

In older buildings, facility engineers and senior management need to evaluate the cost of operating with obsolete electrical distribution systems and the associated risk of an outage. Where a high potential for losses exists, serious capital expenditures to upgrade the electrical distribution system are monetarily justified by senior management. The cost of downtime across a spectrum of industries has exploded in recent years, because business has become completely computer-dependent and systems have become increasingly complex (Table 3.1).

Imagine that you are the manager responsible for a major data center that provides approval of checks and other on-line electronic transactions for American

Table 3.1 Cost of Downtime[a]

Industry	Average Cost per Hour
Brokerage	$6,400,000
Energy	$2,800,000
Credit card operations	$2,600,000
Telecommunications	$2,000,000
Manufacturing	$1,600,000
Retail	$1,100,000
Health care	$640,000
Media	$340,000
Human life	"Priceless"

[a] Prepared by a disaster-planning consultant of Contingency Planning Research.

Express, MasterCard, and Visa. On the biggest shopping day of the year, the day after Thanksgiving, you find out that the data center has lost its utility service. Your first reaction is that the data center has a UPS and standby generator, so there is no problem. Right? However, the standby generator has not started due to a fuel problem and the data center will shutdown in 15 minutes, which is the amount of time the UPS system has in battery power at full load. The penalty for not being proactive is loss of revenue and potential loss of major clients, and if the problem is large enough, your business could be at risk of financial collapse. Table 3.1 shows the average cost per hour of downtime.

You, the manager, could have avoided this nightmare scenario by exercising the standby generator every week for 30 minutes–the proverbial ounce of prevention.

There are about three times as many UPS systems in use today than there were 10 years ago, and many more companies are still discovering their worth after losing data during a power-line disturbance. Do you want electrical outages to be scheduled or unscheduled? Serious facilities engineers use comprehensive preventive maintenance procedures to avoid being caught off-guard.

Many companies do not consider installing backup equipment until after an incident has already occurred. During the months following the Blackout of 2003, the industry experienced a boom in the installation of UPS systems and standby generators. Small and large businesses alike learned how susceptible they are to power disturbances and the associated costs of not being prepared. Some businesses that are not typically considered mission critical learned that they could not afford to be unprotected during a power outage. For example, the Blackout of 2003 destroyed $250 million of perishable food in New York City alone.[1] Businesses of every type, everywhere, are reassessing their level of risk tolerance and cost of downtime.

[1] New York City Comptroller William Thompson.

3.3 IDENTIFYING THE APPROPRIATE REDUNDANCY IN A MISSION CRITICAL FACILITY

Mission critical facilities cannot be susceptible at any time to an outage, including maintenance of the subsystems. Therefore, careful consideration must be given in evaluating and implementing redundancy in systems design. Examples of redundancy are classified as $N+1$ and $N+2$ configurations and are normally applied to the systems below:

- Utilities service
- Power distribution
- UPS
- Emergency generator
- Fuel system supplying emergency generator

A standard system $(N+1)$ is a combination of two basic schemes that meet the criteria of furnishing an essential component plus one additional component for backup. This design provides the best of both worlds at a steep price with no economies of scale considered. A standard system protects critical equipment and provides long-term protection to critical operations. In a true $N+1$ design, each subsystem is configured in a parallel redundant arrangement such that full load may be served even if one system is off-line due to scheduled maintenance or a system failure.

The next level of reliability is a premium system. The premium system meets the criteria of an $N+2$ design by providing the essential component plus two components for backup. It also utilizes dual electric service from two different utility substations. Under this configuration, any one of the backup components can be taken off-line for maintenance and still retain $N+1$ reliability. It is also recommended that the electric services be installed underground as opposed to aerial.

3.4 IMPROVING RELIABILITY, MAINTAINABILITY, AND PROACTIVE PREVENTATIVE MAINTENANCE

The average human heart beats approximately 70 times a minute, or a bit more than once per second. Imagine if a heart missed three beats in a minute.

This would be considered a major power-line disturbance if we were to compare it to an electrical distribution system. Take the electrical distribution system that feeds your facility, or, better yet, the output of the UPS system, and interrupt it for 3 seconds. This is an eternity for computer hardware. The critical load is disrupted, your computers crash, and your business loses two days worth of labor or, worse, is fined $10 to $20 million by the federal government because they did not receive the quota, $500 billion dollars of transactions, by the end of the day.

All this could have been prevented if electrical maintenance and testing were performed on a routine basis and if the failed electrical connections were detected and repaired. Repairs could have been quickly implemented during the biannual infrared scanning program that takes place before building maintenance shutdowns. It takes an enormous effort to get to that point, but it is that simple.

What can the data processing or facility manager do to ensure that their electrical system is as reliable as possible? The seven steps to improved reliability and maintainability are:

- Planning and impact assessment
- Engineering and design
- Project management
- Testing and commissioning
- Documentation
- Education and training
- Operations and maintenance

Hire competent professionals to advise each step of the way. When installing a data processing center in an existing building, you do not have the luxury of designing the electrical system from scratch. A competent electrical engineer will design a system that makes the most out of the existing electrical system. Use electrical contractors who are experienced in data processing installations. Do not attempt to save money using the full 40% capacity for a conduit, because as quickly as new, state-of-art equipment is installed, it is deinstalled. Those same number 12 wires will need to come out of the conduit without disturbing the working computer hardware.

Have an experienced electrical testing firm inspect the electrical system, perform tests on circuit breakers, and use thermal-scan equipment to find "hot spots" due to improper connections or faulty equipment. Finally, plan for routine shutdowns of your facility to perform preventive maintenance of electrical equipment. Facility managers must not underestimate the cost-effectiveness of a thorough preventative maintenance program, nor must they let senior management do so. Electrical maintenance is not a luxury; it is a necessity. Again, do you want electrical outages to be scheduled or unscheduled?

Integrating the ideal electrical infrastructure is just about impossible. Therefore, seek out the best possible industry authorities to solve your problems. Competent consultants will have the knowledge, tools, testing equipment, training, and experience necessary to understand the risk tolerance of your company, as well as recommend and implement the proper and most-advanced proven designs.

Equipment manufacturers and service providers are challenged to find and retain the industry's top technicians. As 24/7 operations become more prevalent, the talent pool available will diminish. This would indicate that response times could increase from the current industry standard of 4 hours. Therefore, the human element has a significant impact in risk and reliability.

No matter which firms you choose, always ask for sample reports, testing procedures, and references. Your decisions will determine the system's ultimate reliability, as well as how easy the system is to maintain. Seek experienced professionals from within and without your own company: information systems, property, and operations managers, space planners, and the best consultants in the industry for all engineering disciplines. The bottom line is, you want proven organizations working on your project.

3.5 THE MISSION CRITICAL FACILITIES MANAGER AND THE IMPORTANCE OF THE BOARDROOM

To date, the mission critical facilities manager has not achieved high levels of prestige within the corporate world. This means that if the requirements are 24/7, forever, the mission critical facilities manager must work hard to have a voice in the boardroom. The board can then become a powerful voice that supports the facilities manager. The FM can use the board to establish a standard for managing the risks associated with older equipment or maintenance cuts—for instance, relying on a UPS system that has reached the end of its useful life but is still deployed due to cost constraints. The FM is in a unique position to advise and paint vivid scenarios. Imagine incurring losses due to downtime, plus damage to capital equipment that keeps the Fortune 1000 company in business.

Board members understand this language: It is comparable to managing and analyzing risk in other avenues, such as whether to invest in emerging markets in unstable economies. The risk is one and the same; the loss is measured in the bottom line.

The facilities engineering department should be run and evaluated just like any other business line; it should show a profit. But instead of increased revenue, the business line shows increased uptime, which can be equated monetarily, plus far less risk. It is imperative that the facilities engineering department be given the tools and the human resources necessary to implement the correct preventative maintenance training, and document management requirements, with the support of all company business lines.

3.6 QUANTIFYING RELIABILITY AND AVAILABILITY

Datacenter reliability ultimately depends on the organization as a whole weighing the dangers of outages against available enhancement measures. Reliability modeling is an essential tool for designing and evaluating mission critical facilities. The conceptual phase, or programming, of the design should include full PRA (Probabilistic Risk Assessment) methodology. The design team must quantify performance (reliability and availability) against cost in order to take fundamental design decisions through the approval process.

Reliability predictions are only as good as the ability to model the actual system. In past reliability studies, major insight was gained into various electrical distribution configurations using IEEE Standard 493-1997 *Recommended Practice for the Design of Reliable Industrial and Commercial Power Systems*. It is also the major source of data on failure and repair rates for electrical equipment. There are, however, aspects of the electrical distribution system for a critical facility that differ from other industrial and commercial facilities. Therefore, internal data accumulated from the engineer's practical experience is needed to complement the Gold Book information.

Reliability analysis with PRA software provides a number of significant improvements over earlier, conventional reliability methods.

3.6.1 Review of Reliability Versus Availability

Reliability (R) is the probability that a product or service will operate properly for a specified period under design operating conditions without failure. That is, the likelihood that equipment will function properly over the next, say, year.

Failure rate (λ) is defined as the probability that a failure per unit time occurs in the interval, given that no failure has occurred prior to the beginning of the interval. Look at the other side of the coin: What is the likelihood that equipment will fail over that same year?

For a constant failure rate, λ, reliability as a function of time is:

$$R(t) = 1 - \lambda(t)$$

Mean time between failures (MTBF), as its name implies, is the mean of probability distribution function of failure. For a statistically large sample, it is the average time the equipment performed its intended function between failures. For the example of a constant failure rate, we have:

$$\mathrm{MTBF} = 1/\lambda$$

Mean time to repair (MTTR) is the average time it takes to repair the failure and get the equipment back into service. (Remember, just because a component is back in service does not mean that a computer, bank, or factory goes on-line instantaneously and magically recovers lost data or feedstock.)

Availability (A) is the long-term average fraction of time that a component or system is in service and satisfactorily performing its intended function. This is also known as *steady-state availability*:

$$A = \mathrm{MTBF}/(\mathrm{MTBF} + \mathrm{MTTR})$$

High reliability means that there is a high probability of good performance in a given time interval. *High availability* is a function of a component's or system's failure frequency and repair times, and is the more accurate indication of performance.

3.7 DESIGN CONSIDERATIONS FOR THE MISSION CRITICAL DATA CENTER

In most mission critical facilities, the data center constitutes the critical load in the company's or institution's daily operations. The costs of hardware and software could run up to $3000 or $4000 per square foot, often resulting in an investment of several million dollars. In a datacenter of only 5000 square feet, you could be responsible for a $20 million capital investment, without considering the cost of downtime. Combined, the cost of downtime and damage to equipment could be catastrophic.

Proper data center design will protect the investment and minimize downtime. Early in the planning process, an array of experienced professionals must review all the factors that affect operations. This is no time to be "jack of all trades, master of none." Here are ten basic steps critical to designing and developing a successful mission critical data center:

1. Determine the needs of the client and the mission critical data center.
2. Develop the configuration for the hardware.
3. Calculate the air, water, and power requirements.
4. Determine your total space requirements.
5. Validate the specific site: Be sure that the site is well-located and away from natural disasters and that electric, telecommunications, and water utilities can provide the high level of reliability for your company.
6. Develop a layout after all parties agree.
7. Design the mission critical infrastructure to $N+1$, $N+2$, or higher redundancy level, depending on the risk and reliability requirements.
8. Prepare a budgetary estimate for the project.
9. Have a competent consulting engineer prepare specifications and bid packages for equipment purchases and construction contracts. Use only vendors that are familiar with and experienced in the mission critical industry.
10. After bids are opened, select and interview vendors. Take time to carefully choose the right vendor. Make sure you see their work; ask many of questions; verify references; and be sure that everybody is on the same page.

3.8 MISSION CRITICAL FACILITY START-UP

Before the facility goes on-line, it is crucial to resolve all the problems. This will be the construction team's sole opportunity to integrate and commission all the systems, due to the facility's 24/7 mission critical status. At this point in the project, all systems installed were tested at the factory and witnessed by a competent independent test engineer familiar with the equipment.

Once the equipment is delivered, set in place, and wired, it is time for the second phase of certified testing and integration. The importance of this phase is to verify and certify that all components work together and to fine-tune, calibrate, and integrate the systems and gather the documentation so future personnel can be trained. There is a tremendous amount of preparation in this phase. The facilities engineer must work with the factory, field engineers, and independent test consultants to coordinate testing and calibration. Critical circuit breakers must be tested and calibrated prior to placing any critical electrical load on them. When all the tests are completed, the facilities engineer must compile the certified test reports, which will establish a benchmark for all future testing. The last phase is to train the staff on each major piece of equipment. This phase is an ongoing process and actually begins during construction.

3.9 THE EVOLUTION OF MISSION CRITICAL FACILITY DESIGN

To avoid downtime, facilities managers also must understand the trends that affect the utilization of data centers, given the rapid evolution of technology and its requirements on power distribution systems. Unlike industry and common office buildings, the degree of power reliability in a data center will impact the design of the facility infrastructure, the technology plant, system architecture, and end-user connectivity.

Today, data centers are pushed to the limit. Servers are crammed into racks, and the higher-performing processors in these devices all add up to outrageous power consumption. Over the past 4 years, data center power consumption has gone up, on average, about 25% according to Hewlett Packard. At the same time, processor performance has gone up 500%. As equipment footprints shrink, free floor area is populated with more hardware. However, because the smaller equipment rejects the same amount of heat, cooling densities grow dramatically and floor space for cooling equipment increases. Traditional design using watts per square foot has grown enormously and can also be calculated as transactions per watt. All this increased processing generates heat, but if the data center gets too hot, all applications come to a halt.

As companies push to wring more data-crunching ability from the same real estate, the linchpin technology of future datacenters will not necessarily involve greater processing power or more servers, but improved heat dissipation.

To combat high temperatures and maintain the current trend toward more powerful processors, engineers are reintroducing old technology: Liquid cooling was used to cool mainframe computers a decade ago. To successfully reintroduce liquid into computer rooms, standards will need to be developed; another arena where standardization can promote solutions that are more reliable that mitigate risk for industry.

The large footprint now required for reliable power without planned downtime (e.g., switchgear, generators, UPS modules, and batteries) also affects the planning and maintenance of data center facilities. Over the past two decades, the cost of the facility relative to the computer hardware it houses has not grown proportionately. Budget priorities that favor computer hardware over facilities improvement can lead to insufficient performance. The best way to ensure a balanced allocation of capital is to prepare a business analysis that shows the costs associated with the risk of downtime.

Many data center designers (and their clients) would like to build for a 20-year life cycle, yet the reality is that most cannot realistically look beyond 2 to 5 years.

Chapter 4

Mission Critical Electrical Systems Maintenance

4.1 INTRODUCTION

Before a corporation or institution embarks on a design to upgrade an existing building, or erect a new one, key personnel must work together to evaluate how much protection is required. Although the up-front costs for protection can be somewhat startling, the cost and impact of providing inadequate protection for mission critical corporate financial and information systems can lead to millions in lost revenue, risks to life safety, and perhaps a tarnished corporate reputation. The challenge is finding the right level of protection for your building.

However, investing a significant amount of time and money into systems design and equipment to safeguard a building from failure is just the beginning.

An effective maintenance and testing program for your mission critical electrical loads is key to protecting the investment. Maintenance procedures and schedules must be developed, staff properly trained, spare parts provisioned, and mission critical electrical equipment performance tested and evaluated regularly. Predictive maintenance, preventive maintenance, and reliability-centered maintenance (RCM) programs play a critical role in the reliability of the electrical distribution systems.

How often should electrical maintenance be performed? Every 6 months or every 6 years? Part of the answer lies in what level of reliability your company can live with. More accurately, what are your company's expectations in terms of risk tolerance or goals with regard to uptime? As previously discussed, if your company can live with 99% reliability, or 87.6 hours of downtime per year, then the answer would be to run a maintenance program every 3 to 5 years. However, if 99.999% reliability, or 5.25 minutes of downtime per year, is mandatory, then you

Maintaining Mission Critical Systems in a 24/7 Environment By Peter M. Curtis

need to perform an aggressive preventive maintenance program every 6 months. The cost of this hard-line maintenance program could range between $300 and $400 annually per kilowatt (kW), not including the staff to manage the program. The Human Resource cost will vary, depending on the location and complexity of your facility.

There are several excellent resources available for developing the basis of an electrical testing and maintenance program. One is the InterNational Electric Testing Association (NETA), which publishes the *Maintenance Testing Specifications* that recommends frequencies of maintenance tests. The recommendations are based on equipment condition and reliability requirements. Another is the National Fire Protection Association's (NFPA) *70B Recommended Practice for Electrical Equipment Maintenance*. These publications along with manufacturer's recommendation give guidance for testing and for maintenance tasks and schedules to incorporate in your maintenance program.

Over the last 10 years, there have been significant changes in the design and application of infrastructure support systems for mission critical facilities. This has been driven by the desire for higher reliability levels and has been fueled by technological innovations such as automatic static transfer switches (ASTS), information technology (IT) equipment with redundant power supplies and new system architectures. The resulting system configurations result in greater numbers of equipment, but they are more fault-tolerant. Herein lies the opportunity. The response to greater equipment numbers has been to increase preventive maintenance tasks and schedules. However, since these systems are more fault-tolerant, the effect(s) of a failure may be insignificant. Consequently, different approaches to maintenance can be taken. In addition, significant numbers of mission critical failures are directly attributable to human error. Therefore, reducing human intervention during various maintenance tasks can have a substantial impact in improving availability.

Reliability-centered maintenance (RCM) was developed in the 1960s by the aircraft industry when it recognized that the cost of aircraft maintenance was becoming prohibitively expensive, while the reliability results were well below acceptable levels. Since then, RCM strategies have been developed and adapted by many industries.

Traditionally, the goal of a maintenance program has been to reduce and avoid equipment failures. Facilities go to great lengths to prevent the failure of a device or system, regardless of the failure's consequence on the facility's mission. RCM shifts the focus from failure avoidance to understanding and mitigating the failure effect upon the process it protects. This is a major shift from a calendar-based program.

For many facilities, the net effect will be a change from the present labor-intensive process driven by the operating group, as well as by the equipment suppliers, to a more selective method where the understanding of the system responses to component failures plays a critical role.

Although the benefits are proven, the process needs to be supported by thorough analysis. Reliability-centered maintenance analyzes how each component and/or system can functionally fail. The effects of each failure are analyzed and ranked according to their impact on safety, environment, business mission, and cost. Failures that are deemed to have a significant impact are further analyzed to determine the root causes. Finally, preventative or predictive maintenance is assigned based on findings of the analysis with emphasis on condition-based procedures to help ensure the optimal performance level.

4.2 THE HISTORY OF THE MAINTENANCE SUPERVISOR AND THE EVOLUTION OF THE MISSION CRITICAL FACILITIES ENGINEER

Managing a facilities engineering department in a corporation or institution has changed significantly in the past three decades. Thirty years ago, the most important qualification for a facilities manager was exposure to different trades and hands-on experience gained from working up through the ranks of maintenance, beginning at the lowest level and progressing through a craft by way of on-the-job training or various apprentice programs. Upon mastering all the trades and gaining knowledge of the facility from on-the-job experience, upper management would promote this craftsperson to maintenance supervisor.

The maintenance supervisor was an important player in getting equipment and facility up and running quickly, in spite of day-to-day problems, procedures, and policies. This was the key to advancement. That supervisor was regarded as someone who would do whatever it takes to get the equipment operating quickly, regardless of the situation. Operations managers would expect the maintenance supervisor to take $200 out of his own pocket in order to buy a critical part. Coming into the facility at all hours of the night, weekends, and holidays further projected the image that this maintenance supervisor was giving his all to the company. After many years of performing at this level of responsibility, the maintenance supervisor advanced to a facilities management position with responsibility for maintenance supervisors, craftspeople, facilities engineering, and so on.

This promotion was pivotal in his career. Because he had worked hard and made many sacrifices, he reached the highest levels in the facilities engineering industry. The facilities manager continued to get involved with the details of how equipment is repaired and facilities are maintained. He also continued to ensure that he could be called at any hour of the day or night to personally address problems in the facilities.

The secret to his success was reacting to problems and resolving them immediately.

Many facilities managers have tried to adapt from reacting to problems to anticipating and planning for them, but fall short in achieving a true 24/7 mission critical operations department. The most advanced technologies and programs are researched and deployed, and personnel are assigned to utilize them. Yet, with all of these proactive approaches, the majority of facilities organizations fall short of adequate levels of functionality, in some cases due to limited human or monetary resources. The dilemma is that these tools overwhelm many facilities managers—day-to-day management of the facilities becomes a Sisyphusian task.

Today, the facilities manager continues with the tenacious effort and determination to resolve problems, conflicts, and issues. Additionally, the facilities manager must know, understand, and implement the programs, policies, and procedures dictated by state, local, and federal agencies. He must also support corporate governances of his employer. Today's facility manager must formulate capital and operational budgets and report departmental activities to senior management. He must be aware of safety and environmental issues when all the while his role continues to be very critical. However, to be successful when operating mission critical 24/7 operations, a proactive approach (not reactive) is imperative. The days of brute force and reacting to problems are gone. The real key to success is planning, anticipation, and consistency. Today's facility manager is a manager in every sense of the word.

Most facilities organizations appear to run smoothly on the surface: Critical equipment is up and running; craftspeople have more projects than ever; facilities managers and supervisors are juggling the responsibilities of administrative, engineering, technical, and operational duties. Yet, when we look beyond the façade, we can see that the fundamentals of maintenance engineering are changing. In fact, this change is occurring along with developments in technology.

In order to keep up and build a successful mission critical facilities operations department, the facilities manager must obtain, train, and retain the appropriate staff, as well as develop and sustain the following:

- A clear mission, vision, or strategy for the facilities organization
- Narrative descriptions of the design intent and basis of design
- Written control sequences of operation that clearly explain to anyone how the facility is supposed to operate
- Standard and emergency operating procedures
- Detailed, up-to-date drawings
- Effective document management systems
- Maintenance planning and scheduling programs
- Maintenance inventory control
- Competent consultants in all engineering disciplines
- The most advanced communications, hardware, and software
- Adequate capital and operating budgets
- Senior management support

The drive to move to state-of-the-art technologies, practices, and programs often blinds facilities management's ability to ensure that the building blocks of good solid maintenance are applied. A mission critical facilities engineering department is built upon several fundamental factors. The supervisor communicates a clear vision in well-thought-out and clearly delineated maintenance procedures, which illustrate how mission critical facilities engineering is expected to operate. He controls the maintenance spare parts inventory, and he carefully plans and schedules maintenance. These fundamentals cannot be forgotten when new technologies, programs, and practices are implemented.

What often happens is that the energy from facilities management is directed to the research, selection, and implementation of the new initiatives. At the same time, operations management assumes that the fundamentals are well in place and will continue. The reality, however, is that regular maintenance programs most likely were never fully implemented, or the workload on facilities management has let them degrade.

With the increasing complexity of mission critical facilities and the technologies involved, the emphasis has shifted away from the resourceful craftsman who made repairs on the fly. The focus has shifted to the mission critical facilities engineer charged with ensuring 24/7 operation while repairs and preventive maintenance take place through thoughtful design and operation of the facility. Indeed, the fundamental of maintenance engineering and the role of the mission critical facilities engineer continue to evolve and grow in importance with each new technology applied.

4.3 INTERNAL BUILDING DEFICIENCIES AND ANALYSIS

In addition to having a robust UPS system, building design must eliminate the probability of single points of failure occurring simultaneously in several locations throughout the electrical distribution system. Mission critical facilities engineers have exclusive accountability for identifying and eliminating problems that can lead to points of failure in the power distribution infrastructure. Failures can occur anywhere in an electrical distribution system: utility service entrance, main circuit breakers, transformers, disconnect switches, standby generators, automatic transfer switches, branch circuits, and receptacles, to name a few. A reliability block diagram must be constructed and analyzed to quantify these single points of failure.

In a mission critical power distribution system, there are two basic architectures for power distribution to critical loads:

1. Electrical power from the UPS system is connected directly through transformers or redundantly from two sources through automatic static transfer switches to remotely mounted circuit breaker panels with branch circuits running to the critical loads.
2. Power distribution units (PDUs) house transformers and panel boards, with branch circuits running to the critical loads.

Typically in data or operations centers with raised floors, a liquid-tight flexible conduit runs the branch circuits from the PDUs to the critical loads. This allows for easy removal, replacement, and relocation of the circuit when computer equipment is upgraded or changed. In an office application, the typical setup involves a wall-mounted panel board hardwired to a dedicated receptacle that is feeding computer equipment. Where the UPS system output voltage is 480 V, transformers in the PDUs are required to step down voltage to 120/208 V.

The latest computer equipment—new-generation LAN servers, midranges, mainframes, computers, routers, and random access disk (RAD) storage systems—is manufactured with dual power cord power supplies and connectivity. This equipment will accept two sources of power, but in most cases uses only one to operate. In the event that one power sources fails, the load draws on the second source, and the application remains on-line. This is a tremendous step toward computer equipment that is tolerant of power quality problems or internal electrical interruptions. It does, however, challenge the facility to provide two different sources of power supplied to the two cords on each piece of equipment. This can mean the complete duplication of the electrical distribution system from the service entrance to the load, along with the elimination of scheduled downtime for maintenance. Two utility switchboards, along with redundant utility substations, power dual UPS systems that feed dual PDUs. This system topography constitutes what is known as a 2N system.

4.4 EVALUATING YOUR SYSTEM

Read the operation and maintenance manuals for electrical distribution equipment and you will likely find a recommendation for preventive maintenance.

However, can you justify preventive maintenance service for your mission critical equipment? Is comprehensive annual maintenance cost effective for the user? Should all of a mission critical facility's power distribution equipment be serviced on the same schedule? If not, what factors should be considered in establishing a realistic and effective preventive maintenance program for a given facility?

To answer these questions, first evaluate the impact to your business if there were an electrical equipment failure. Then examine the environment in which the electrical equipment operates. Electrical distribution equipment is most reliable when kept dust-free, dry, cool, clean, tight, and, where appropriate, lubricated. Electrical switchgear, for example, if kept in an air-conditioned room, regardless of maintenance practices, will be considerably more reliable than identical equipment installed in the middle of a hot, dirty, and dusty operating floor. In such uncontrolled environmental locations, it would be difficult to pick an appropriate maintenance interval.

The amount of load on the electrical equipment is also critical to its life. Loading governs how much internal heat is produced, which will expedite the aging process of nearly all insulating materials and dry out lubricants in devices such as circuit breakers. High current levels can lead to overheating at cable connectors, circuit breaker lugs, interconnections between bus and draw-out devices, contact surfaces, and so on. An important exception is equipment located outdoors or in a humid environment. In those instances (especially prevalent in 15-kV class equipment), if the electrical load is too light, the heat created by the flow of current will be insufficient to prevent condensation on insulating surfaces. If the proper number and size of equipment heaters are installed and operating in the 15-kV class equipment, condensation is less of a concern. Operating voltage levels should influence the maintenance interval. Tracking across contaminated insulating surfaces can proceed much more rapidly on 15-kV class equipment than at 480 V. Also, higher-voltage equipment is often more critical for the successful operation of a larger portion of the building or plant; therefore, more frequent preventive maintenance can be justified.

Consider the age of electrical equipment. It is important to perform preventive maintenance on older equipment more frequently. Test new equipment before placing it in service, then implement a preventive maintenance schedule based on manufacturer's and experienced consultant recommendations. Certified testing will identify defects that can cause catastrophic failures or other problems. Additionally, it will document problems and repair/replacement, as well as the need for additional testing while still under warranty.

Another consideration is the type and quality of equipment initially installed. Was lowest first cost a primary consideration? If the engineer's design specified high-quality, conservatively rated equipment and it was installed correctly, you are at a good preventive maintenance starting point. It will be less difficult and less expensive to achieve continued reliable operation from a properly designed project. The benefits of quality design and well-thought-out specifications will aid preventive maintenance, because the higher-quality equipment typically has the features that enable service. On the other hand, don't allow simple minor service checks to become the building standard. It is easy to test only the protective relays (because it can be accomplished while in normal operation) and let important shutdown services slide. If minimal service becomes the norm, most of the benefits of effective preventive maintenance will not be realized. Therefore, demand regular, major preventive maintenance on your mission critical equipment.

4.5 CHOOSING A MAINTENANCE APPROACH

NFPA 70B says, "Electrical equipment deterioration is normal, but equipment failure is not inevitable. As soon as new equipment is installed, a process of normal deterioration begins. Unchecked, the deterioration process can cause malfunction or an electrical failure. Effective maintenance programs lower costs and improve reliability. A maintenance program will not eliminate failures but can reduce failures to acceptable levels." There is little doubt that establishing an electrical maintenance program is of paramount importance. There are various approaches for establishing a maintenance program. In most cases, a program will include a blend of the strategies listed below:

- Preventive maintenance (PM) is the completion of tasks performed on defined schedule. The purpose of PM is to extend the life of equipment and detect wear as an indicator of pending failure. Tasks describing the maintenance procedures are fundamental to a PM program. They instruct the technician on what to do, what tools and equipment to use, what to look for, how to do it, and when to do it. Tasks can be created for routine maintenance items or for breakdown repairs.

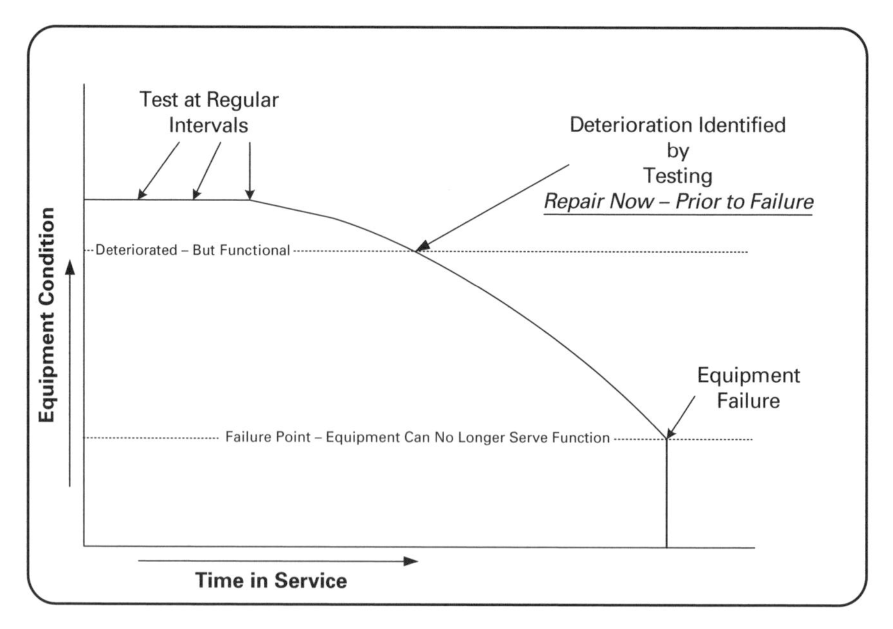

Figure 4.1 The concept of predictive maintenance. (Courtesy of California Data Center Design Group.)

- Predictive maintenance uses instrumentation to detect the condition of equipment and identify pending failures. A PM program uses these equipment condition indices for the purpose of scheduling maintenance tasks.

Figure 4.1 is an illustration of the concept of predictive maintenance.

- Reliability-centered maintenance (RCM) is the analytical approach to optimize reliability and maintenance tasks with respect to the operational requirements of the business. Reliability, as it relates to business goals, is the fundamental objective of the RCM process. RCM is not equipment-centric, but business-centric. Reliability-centered maintenance analyzes each system and how it can functionally fail. The effects of each failure are analyzed and ranked according to their impact on safety, mission, and cost. Those failures, which are deemed to have a significant impact, are further explored to determine the root causes. Finally, maintenance is assigned based on effectiveness, with a focus on condition-based tasks. The objective is to reach the optimum point at which the benefits of reliability are maximized while the cost of maintenance is minimized.

4.6 STANDARDS AND REGULATIONS AFFECTING HOW SAFE ELECTRICAL MAINTENANCE IS PERFORMED

Electrical construction hazards are plentiful. They have several factors including electrical shock, short circuits, arc flash, trauma, and so on. The associated codes/standards include National Electric Code (NFPA 70), Standard for Electrical Safety in the Workplace (NFPA 70E), Occupational Safety and Health Association (OSHA), and IEEE Guide for Performing Arc-Flash Hazard Calculations (IEEE 1584).

1. The National Electric Code, which typically applies to new or renovation work, does not have specific requirements regarding when maintenance may be completed. The NEC requires that a label be applied to switchboards, panelboards, industrial control panels, meter socket enclosures, and motor control centers that are in other than dwelling occupancies and are likely to require examination, adjustment, servicing, or maintenance while energized and shall be field marked to warn qualified persons of potential electric arc flash hazards. Keep in mind that this is a warning label only and no application or site specific information is required.
2. OSHA Regulation 29 CFR Subpart S.1910.333 indicates the following: "Live parts to which an employee may be exposed shall be de-energized before the employee works on or near them. . . ." It goes on to state that working on or near energized electrical equipment is permissible if ". . . the employer can demonstrate that de-energizing introduces additional or increased hazards or is infeasible due to equipment design or operational limitations." The determination of whether work may occur on energized equipment is clearly based on one's definition of "infeasible." This is a discussion that must be documented amongst the individual decision makers regarding maintenance in your facility.
3. Once it is determined that work will occur on or near live parts to which an employee may be exposed, the investigation to determine whether a hazard is present, and what to do, is the responsibility of the employer of the electrical personnel or the employer's delegate. Compliance with the OSHA regulations lead the investigator in part to NFPA 70E-2004 Standard for Electrical Safety Requirements for Employee Workplaces and IEEE 1584–2002 Guide for Performing Arc Flash Hazard Calculations for the Hazard Calculation.
4. NFPA 70E contains many requirements for maintaining safe electrical working practices including protection from shocks and traumatic injury to a worker. It contains tables that may be used for selection of personal protective equipment (PPE) for protection from arc flash burns.
5. Two terms that must be defined regarding electrical faults include bolted faults and arc faults. Traditionally, electrical equipment have been designed and tested to standards identifying its short-circuit rating whether it be the equipment's ability in the condition of an overcurrent protective device to open and clear a fault or for equipment to simply withstand or survive that amount of bolted fault current. Bolted faults are conditions that occur in a test lab and occur in the field on very rare occasions. A bolted fault, as the name implies, includes the solid connection of any combination of energized electrical phases (ungrounded) or an electrical phase and ground.

Equipment currently does not carry ratings related to the equipments ability to withstand or survive arc faults. Arc faults include the inadvertent connection of energized electrical phases (ungrounded) or an electrical phase and ground with the air acting as an electrical path.

a. A bolted fault will typically result in the operation of an upstream overcurrent protective device, resulting in very low equipment damage. An arcing fault will cause lower fault current to travel as compared to a bolted fault and will result in longer clearing times of an upstream overcurrent protective device. This is due to the air, which would typically act as an insulator between phases or phase and ground, acting as a conductor. Arc flash faults will result in high-temperature radiation, molten metal material, shrapnel, sound waves, pressure waves, metallic vapor, and rapid expansion of ionized air.

6. When it is determined that work will be performed on energized equipment, an energized work permit is required. A sample of this document can be found in NFPA 70E. It includes information documenting why and how the energized work will be performed. NFPA 70E presents a table [130.7(C)(9)(a)] for simplified personal protective equipment selection, but they are contingent on compliance with certain criteria at the end of the table related to bolted fault current values and protective device clearing times. Any deviation from the criteria makes the simplified table unusable. A detail analysis must be performed. Formula for detailed analysis can be found in NFPA 70E and IEEE 1584.

4.7 MAINTENANCE OF TYPICAL ELECTRICAL DISTRIBUTION EQUIPMENT

In order to provide more-specific recommendations; let's examine a select portion of a high-rise office building with mission critical loads. The system comprises 26-kV vacuum circuit breakers with protective relays, solid dielectric-shielded feeder cable, and a fused-load break switch. On the 480-V system, there is switchgear with molded-case circuit breakers on the main and feeders. The mission critical loads are then connected to motor control centers and panel boards with molded-case circuit breakers.

4.7.1 Infrared Scanning

First, a high-rise building of this type can easily justify the cost of an infrared scan once per year. Is there a more cost-effective maintenance procedure that can positively identify potential problems quickly and easily and without a shutdown? Its principal limitation is that there are certain areas of the system and equipment where the necessary covers cannot be removed while in service. Without line-of-sight access, the scanning equipment is very limited in its capabilities.

Where electrical panel covers cannot be removed during normal working hours, hinged panel covers can be installed during scheduled building shutdowns; subsequently, all mission critical panel board covers can be changed to hinged panel board covers. This will eliminate a branch circuit breaker from tripping or opening when removing panel board covers. As we know, all it takes is the right outage at the right time to create chaos in a data center. This is a small price to pay for insurance.

For many key businesses—financial, media, petrochemical, oil and gas, telecommunications, large-scale manufacturing, and so on—consequences of sudden unexpected power failures can result in severe financial and/or safety costs. The current accepted method of noncontact thermal inspection has been through thermograph via thermal imaging cameras; however, this only gives a snapshot on the day—problems could arise the next day undetected. Another solution is a thermal monitoring system that is permanently installed within the electrical distribution system. These systems are small, low-cost, accurate, noncontact, infrared sensors that are permanently fitted where they can directly monitor key components like motors, pumps, drive bearings, gearboxes, generators, and even inside enclosures monitoring high- and low-voltage switchgear. The sensors feed back signals to a PC, where automatic data logging provides instant on-screen trend graphs. Two alarm levels per sensor automatically activate in the event that preset levels are exceeded, thus identifying problem components and allowing planned maintenance to be carried out. These sensors, which require no external power, measure the target temperature in relation to ambient with the signal indicating the °C rise on ambient, thereby avoiding inaccuracies due to weather conditions, environment, and so on.

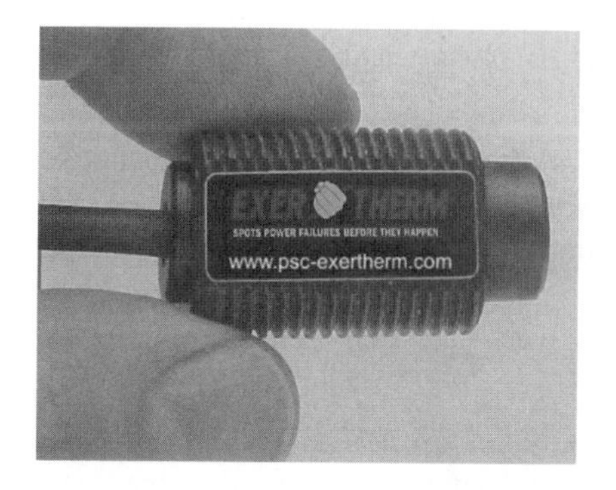

(a)

(b)

Figure 4.2 (a) Thermal imaging sensor for monitoring within an electrical distribution system (b) Typical thermal imaging sensor installation. (Courtesy of Power Service Concepts, LLC.)

4.7.2 15-Kilovolt Class Equipment

Some procedures that should be a part of a preventive maintenance program include the following:

- Insulators should be inspected for damage and/or contamination.
- Cables, buses, and their connections should be visually inspected for dirt, tracking, water damage, overheating, oxidation, or overheating.
- Liquid-filled transformers should be checked for temperature, liquid level, and nitrogen pressure.
- Dry transformers should be checked for temperature and have a physical inspection performed.
- Switchgear encompasses enclosures, switches, interrupting devices, regulating devices, metering, controls, conductors, terminations, and other equipment that should be inspected and have recommended de-energized and energized tests performed.
- Complete testing of the relays should be performed.
- Perform a vacuum integrity test on the breaker interrupters.
- Insulation resistance measurements on the breaker and switch should be taken.
- Do an accurate determination of the contact resistance over the complete current paths of each device. Check for oxidation and carbon buildup.
- Perform operational checks of the mechanical mechanisms, interlocks, auxiliary devices, and so on.
- Inspect and clean contacts and insulators, along with lubrication of certain parts.
- Analysis of all results should be performed by someone with the proper training and experience who is on-site and participating in the work.

Remember that observations and analysis of experienced engineers cannot be replaced by a software program that "trends" the data obtained, or by a later comparison of the results to existing or manufacturers standards.

4.7.3 480-Volt Switchgear

The 480-V switchgear and breakers serve the same function as the higher-voltage switchgear. It is easier to justify less-frequent-than-annual service for this equipment because insulation failure can be a slower process at the lower voltage, and if there is a problem, it will affect a smaller portion of the system. Two years is a common interval, and many buildings choose 3 years between thorough maintenance checkouts on this type of equipment. A comprehensive maintenance program at this level is very similar to the medium-voltage program, above. One change results from the fact that lower-voltage breakers have completely self-contained protection systems rather than separately mounted protective relays, current transformers, and tripping batteries or capacitors.

The best protection confirmation test is primary current testing. A large test device injects high currents to simulate overloads and uses very low voltage to create a fault. The breaker is tested and operated as it must during a fault or overload situation. Use the breaker manufacturer's secondary test device for a less extensive test on breakers with solid-state protective devices. This can range from a test equal to injecting high current levels, to little more than powering up the trip unit and activating an internal self-test feature. The latter test method may be appropriate at alternate service periods, but cannot be considered a comprehensive evaluation.

Low-voltage breaker contacts should be cleaned and inspected after the arc chutes are removed. Precisely measure the resistance of the current path on each pole as well as insulation resistance between poles and ground. Verify proper operation and lubrication of the open/close mechanisms and various interlocks. Again, the inspection and analysis of the results by a knowledgeable and trusted service technician familiar with the equipment and previous history should conclude the test.

Refer to manufacturer recommendations and professional society standards for specific recommendations on the equipment installed at your site.

4.7.4 Motor Control Centers and Panel Boards

Located further downstream are the motor control centers and panel boards. These components should be thoroughly examined with an infrared scanning device once a year. Testing and service beyond that level can vary considerably. To have a testing company come to scan your electrical distribution equipment can cost between $800 and $1600 during normal working hours. Molded case breakers are not designed for servicing; consequently, many buildings limit them to scanning only. An exception is where molded or insulated case breakers are used in place of the air circuit breakers at the substation. There they serve a more critical role, frame sizes can be quite large, and testing is more easily justified. If the breakers are non-draw-out construction, as most circuit breakers are, primary current injection testing is more difficult, especially on the large frame sizes. Secondary testing of solid-state circuit breakers, along with an accurate determination of the contact resistances, might be justified more often than for air circuit breakers. At minimum, exercise molded case breakers by manual opening and closing on a consistent schedule. If so equipped, push the trip button; this will exercise the same mechanism that the internal protective devices do. If periodic testing by primary current injection is available, bring the test equipment to the breaker mounting locations and test them in place, wherever practical

4.7.5 Automatic Transfer Switches

The automatic transfer switch (ATS) is a critical component of the emergency power system. Correct preventive maintenance of an ATS depends on the type of switch and where it is located within the building. There are four basic types of ATSs that will be discussed in Chapter 7.

The following is a list of basic preventive maintenance items for automatic transfer switches.

- Infrared scan the ATS with load to identify hot spots and high resistance connections.
- De-energize the switchgear and isolate it electrically. Make sure the standby power generator is locked out, and tagged out, in a remote location.
- Remove arc chutes and pole covers and conduct visual inspection of the main and arcing contacts.
- Test and recalibrate all trip-sensing and time-delay functions of the ATS. This step will vary, depending on the manufacturer.
- Vacuum the dust from the switchgear and accessory panels. Never use air to blow out dirt; you may force debris into the switch mechanism.
- Inspect for signs of moisture, previous wetness, or dripping.
- Clean grime with solvent approved by manufacturer.
- Inspect all insulating parts for cracks and discoloration caused by excessive heat.
- Lubricate all mechanical slide equipment: rollers, cams, and so on.
- Exercise the switch mechanically and electrically.
- After maintaining the switch, test it for full functionality.

4.7.6 Automatic Static Transfer Switches (ASTS)

- Infrared scan the ASTS with load to identify hot spots and high-resistance connections.
- Visually inspect the cabinet and components of the unit.
- Take voltage reading (phase to phase) of inputs 1 and 2 and the output. Record the voltages and compare the measured readings to the monitor. Calibrate metering as required.
- Retrieve information detected from the event log.
- Clear the event log.
- TEST: Manually switch from preferred source to the alternate source a prescribed number of times.
- Replace any covers or dead fronts and close unit doors.

4.7.7 Power Distribution Units

- Visually inspect the unit: transformer, circuit breakers, and cable connections.
- Check all LED/LCD displays.
- Measure voltages and current.
- Check wires and cables for shorts and tightness.
- Infrared scanning on each unit and visually inspect transformer.
- Check for missing screws and bolts.
- Check all monitors and control voltage.
- Check and calibrate monitors.
- Check high/low thresholds.
- Check for worn or defective parts.
- Check for general cleanliness of units, especially vents.

4.7.8 277/480-Volt Transformers

As long as transformers operate within their designed temperature range and specifications, they will operate reliably. However, if the transformer is not ventilated adequately and is operating hotter than manufacturer's specifications, problems will occur. The most important maintenance function for transformers is to verify proper ventilation and heat removal. Even though there are no moving parts in a transformer, except for fans in larger units, perform the following preventive maintenance:

- Perform infrared scanning for loose connections and hot spots, then re-torque per manufacturer's specifications.
- Remove power.
- Inspect wire for discoloration.
- Vacuum inside transformer and remove dust.
- Verify that the system is properly grounded.

4.7.9 Uninterruptible Power Systems

There are many configurations of uninterruptible power supply installations, such as single module systems, parallel redundant systems, isolated redundant systems, and so on. But there are common tests and maintenance procedures for the major components of these systems. It is important to perform the manufacturer's recommended maintenance in addition to independent testing.

(a) Visual and Mechanical Inspection

- Inspect physical, electrical, and mechanical condition.
- Check for correct anchorage, required area clearances, and correct alignment.

- Verify that fuse sizes and types correspond to drawings.
- Test all electrical and mechanical interlock systems for correct operation and sequencing.
- Inspect all bolted electrical connections for high resistance.
- Verify tightness of accessible bolted electrical connections by calibrated torque-wrench method in accordance with manufacturer's published data.
- Perform thermo-graphic survey at full load.
- Thoroughly clean unit prior to tests, unless as-found and as-left tests are required.
- Check operation of forced ventilation.
- Verify that filters are in place and/or vents are clear.

(b) Electrical Tests

1. Perform resistance measurements through all bolted connections with a low-resistance ohmmeter, if applicable.
2. Test static transfer from inverter to bypass and back at 25%, 50%, 75%, and 100% load.
3. Set free-running frequency of oscillator.
4. Test dc undervoltage trip level on inverter input breaker. Set according to manufacturer's published data.
5. Test alarm circuits.
6. Verify sync indicators for static switch and bypass switches.
7. Perform electrical tests for UPS system breakers.
8. Perform electrical tests for UPS system automatic transfer switches.
9. Perform electrical tests for UPS system batteries for 5 minutes at full load.

(c) Test Values

1. Compare bolted connection resistances to values of similar connections.
2. Bolt-torque levels shall be in accordance with manufacturer's specifications.
3. Micro-ohm or millivolt drop values shall not exceed the high levels of the normal range as indicated in the manufacturer's published data. If manufacturer's data are not available, investigate any values that deviate from similar connections by more than 25% of the lowest value.

4.7.10 A Final Point on Servicing Equipment

Many decisions regarding how and when to service a facility's mission critical electrical power distribution equipment are going to be subjective. It is easy to choose the objective: a high level of safety and reliability from the equipment, components, and systems. But discovering the most cost-effective and practical methods required to get there can be a challenge. Network with colleagues and knowledgeable sources, and review industry and professional standards before choosing the approach best suited to your maintenance goals. Also, keep in mind that the individuals performing the testing and service should have the best education, skills, training, and experience available. You depend on their conscientiousness and decision-making to avoid future problems with perhaps the most crucial equipment in your building. Most importantly, learn from your experiences and from those of others. Maintenace programs should be continuously improving. If a task has historically not identified a problem at the scheduled interval, consider adjusting the schedule respectively. Examine your maintenance programs on a regular basis and make appropriate adjustments.

4.8 BEING PROACTIVE IN EVALUATING THE TEST REPORTS

Acceptance and maintenance testing are pointless unless the test results are evaluated and compared to standards and to previous test reports that have established benchmarks. It is imperative to recognize failing equipment and to take appropriate action as soon as possible. However, it is common practice for maintenance personnel to perform maintenance without reviewing prior maintenance records. This approach defeats the value of benchmarking and trending and must be avoided. The mission critical facilities engineer can then keep objectives in perspective and depend upon his options when faced with a real emergency.

The importance of taking every opportunity to perform preventive maintenance thoroughly and completely, especially in mission critical facilities, cannot be stressed enough. If not, the next opportunity will come at a much higher price: downtime, lost business, and lost potential clients, not to mention the safety issues that come up when technicians rush to fix a maintenance problem. So do it right ahead of time, and don't take shortcuts.

4.9 DATA CENTER RELIABILITY

Data center reliability depends on strategic evaluation of the dangers and available enhancement measures. Avoiding problems and failures also requires a comprehension of the trends that affect the utilization of data centers, given the rapid evolution of technology and its reliance on power distribution systems. And unlike industry and common office buildings, the degree of power reliability in a data center will influence the design of the facility infrastructure, the technology plant, system architecture, and end-user connectivity.

Reliability modeling is an essential tool for designing and evaluating mission critical facilities. The programming/conceptual phase of the design should include full PRA (Probabilistic Risk Assessment) methodology. Each design option needs to quantify performance (reliability and availability) against cost in order to facilitate the main design decisions in the initial phase of the project.

Since the costs associated with each reliability enhancement or redundancy increase are significant, sound decisions can only be made by quantifying the performance benefits and by analyzing the options against the respective cost estimates.

Reliability analysis with PRA software provides a number of significant improvements over earlier, conventional reliability methods.

Reliability predictions are only as good as the ability to model the actual system. In past reliability studies, major insight was gained for various topologies of the electrical distribution system using *IEEE Standard 493-1997, Recommended Practice for the Design of Reliable Industrial and Commercial Power Systems*. It is also the major source of failure and repair rate data for electrical equipment. There are, however, aspects of the electrical distribution system for a critical facility that differ from other industrial and commercial facilities, and internal data accumulated from the Engineer's practical experience is needed to complement the Gold Book information.

Chapter 5

Standby Generators: Technology, Applications, and Maintenance

5.1 INTRODUCTION

Failure of a standby diesel generator to start—which can be due to batteries that are not sufficiently charged, faulty fuel pumps, control or main circuit breaker being left "off," or low coolant level—is inexcusable. In a facility where the diesel generator is being used for life safety or to serve mission critical loads, the cost of generator failure can be immeasurable. Standby generator reliability is contingent on appropriate equipment selection, proper system design and installation, and proper maintenance. To confirm that maintenance is being done correctly, it is imperative to develop a comprehensive documentation and performance plan. Since standby generators run occasionally the schedule for performing specific maintenance tasks is usually stated in terms of daily, weekly, monthly, semiannual, and annual time frames, rather than in hours of operation.

Diesel is one of the most reliable sources of emergency power. Every day, thousands of truckers successfully start a diesel engine without a second thought. An emergency diesel engine should be no exception. However, because truckers operate their engines daily, they tend to be more familiar with the operation and maintenance needs of those engines. With the engines being operated daily, the engines are getting the use and attention they require. Most emergency engines are neglected when they are not being used and therefore not are being maintained. More importantly, the engines are effectively tested under real load conditions every time they are being used. Running an emergency generator on a periodic basis does not provide a significant amount of operational experience. Facilities engineers who

Maintaining Mission Critical Systems in a 24/7 Environment By Peter M. Curtis

have been running one-hour weekly plant exercises or load tests over 10 years have less than 600 hours of operational history. It is not until a disaster occurs, and a long run of 100 consecutive hours or more is needed, that one sees the issues associated with power plant operations and their associated maintenance problems. In this chapter, the reader will become familiar with developing an effective standby generator operations and maintenance program as well as associated procedures.

5.2 THE NECESSITY FOR STANDBY POWER

The necessity for standby power is an essential concern for facility managers today. Traditionally, standby power has been synonymous with emergency generators. With the advent of very sensitive electronic devices, an emergency generator can fall short of meeting requirements for standby power. This does not mean that the role of the emergency generator is extinct. Quite the contrary, in the current environment of utility deregulation, it may serve a greater role than ever. There are a number of ways a standby power system can be provided for a load—namely, battery banks, motor-generator sets, uninterruptible power systems (UPS), and emergency generators. The last two are used most often for providing standby power.

There may be a variety of causes here; the lack of reliable service for the area, utility demand problems, such as the August 2003 Northeast Blackout on the east coast, or weather-related problems. A weather-related problem is normally caused by a loss of transmission service due to high winds, tornadoes, severe snowstorms, or the formation of ice on transmission lines.

Electrical loads in power systems can be roughly divided into two groups. The first group consists of the traditional linear electrical loads, such as lighting, motors, and so on. The second group consists of more sensitive nonlinear electronic devices such as computers. The first group can resume normal operation after short power interruptions, provided that it (a) is not repeated continuously and also has a relatively high tolerance for over- and undervoltages, voltage sags, and voltage swells and (b) operates successfully under moderate levels of noise and harmonics. In other words, the normal variation and power quality problems of the utility service does not affect the loads significantly and as soon as the load is energized with auxiliary power, the equipment will resume its function immediately. For these loads, an emergency generator is sufficient to serve as standby power: After loss of utility service, an emergency generator usually begins to energize the load in about 10 seconds.

On the other hand, the second group of loads—namely, electronic devices—will be affected by power quality problems. Moreover, they cannot sustain a momentary outage without possible hardware failure or software lockup. Therefore, an emergency generator cannot meet the standby power requirements of the equipment. As mentioned previously, a UPS will provide a reliable source of conditioned power where the load remains energized without any interruptions after the loss of utility service. Again, the UPS serves as an on-line power-conditioning device to protect against unwanted power line disturbances.

With the proliferation of real-time computing in LAN, mid-range, and main frame environments, a standby power source is a necessity in many commercial and institutional facilities. A number of power problems can harm the operations of electrical equipment. They include blackouts, brownouts, voltage swells and sags, under- and overvoltage conditions, and power surges and spikes. Regular power received from the utility is usually stable and reliable, so its presence is almost taken for granted. There is a tendency to presume the source of power problems is external, but in reality more than 50% of power troubles stem from in-house equipment.

Although major power blackouts are uncommon, a major electrical outage in July 1996 impacted more than 2 million in the Northwest United States. In August 1996, another power outage affected more than 4 million in five states. Similarly, several areas in New England and Canada were without electric power for in the winter of 1998. Localized blackouts due to weather occur more frequently. For instance, thunderstorms, tornadoes, ice storms, and strong heat waves, as we experienced in the summer of 1999, resulted in power outages. To protect the critical equipment and reduce or eliminate power interruption, organizations rely on backup generators. Table 5.1 indicates power outages over the last few years as well as the causes and impact.

5.3 EMERGENCY, LEGALLY REQUIRED, AND OPTIONAL SYSTEMS

One of the most misunderstood electrical terms applied to the design of electrical distribution is the word "emergency." Although all emergency systems are standby in nature, there are numerous standby systems that are not emergency systems within the meaning of the National Electrical Code. These other systems, whether legally required or optional, are subject to considerably different prerequisites. It is not admissible to select the rules of the NEC to apply the design of an alternate power source to supply the circuits of either an emergency system or a standby system. Once an alternate power source is correctly classified as an emergency power source, conditions that apply to standby systems must not be used to decrease the burden of more stringent wiring. A standby power source is either required by law or required by a legally enforceable administrative agency. Emergency systems and legally required standby systems share a common thread in that both are legally required. If a standby system is not legally required, then it is not an emergency or legally required standby system. Once it is established that the law requires the standby system, then it must be classified as either an emergency system or simply a required standby system. In some cases, the applicable regulation will specify the precise Code article. In other cases, this determination can be difficult to make, since there is some overlap in the two articles. The relative length of time that power can be interrupted without undue hazard is the most useful criterion.

Table 5.1

Location	Recent Power Outages	
	Cause	Effect
Los Angeles	• Massive power outage: utility worker wiring error (9-12-05)	• Traffic and public transportation problems and fears of a terrorist attack
Gulf Coast (Florida/New Orleans)	• 2004/05 hurricanes: Ivan, Charlie, Frances, Katrina, etc.	• Millions of customers without power, water, food and shelter, government records lost due to flooding
China	• 20-million kilowatt power shortage: Equivalent to the typical demand in the entire state of New York (Summer 2005)	• Multiple sporadic brownouts • Government shutdown least energy efficient consumers
Greece	• Temperatures near 104°F • Mismanagement of electric grid (7-12-04)	• Over half of the country left without power
O'Hare Airport	• Electrical explosion (7-12-04)	• Lost power to two terminals • Flight delays over course of a day
Logan Airport	• Electrical substation malfunction (7-5-04)	• Flight delays and security screening shutdown for 4 hours
Italy	• Power line failures • Bad weather (9-29-03)	• Nationwide power outage 57 million people effected
London	• National grid failure (8-29-03)	• Over 250,000 commuters stranded
Northeast, Midwest and Canada	• Human decisions by various organizations, corporate and industry policy deficiencies, inadequate management (8-14-03)	• 50 million people effected due to the 61,800 MW of capacity not being available

In a true emergency system, there are three important considerations in identifying a load as one requiring an emergency source of power, all of which state the basic question of how long an outage would be permissible. The nature of the occupancy is the first consideration, with specific reference to the numbers of people that would be congregated at any one time. A large assembly of people in any single location is ripe for hysteria in the occurrence of a fire, particularly when the area turns dark. Panic must be avoided, since it can develop with intense speed and contribute to more casualties than the fire or other problems that caused it. Therefore, buildings with high occupancy levels, such as high-rise buildings and large auditoriums, usually require emergency systems.

Criticality of the loads is the second consideration. Egress lighting and exit directional signs must be available at all times. Additionally lighting, signaling and communication systems, especially those that are imperative to public safety, must also be available with minimal interruption. Other loads, such as fire pumps or ventilation systems important to life safety, will be incapable to perform their intended function if they are disconnected from the normal power source for any length of time. Danger to staff during an outage is the third consideration. Some industrial processes, although not involving high levels of occupancy, are dangerous to workers if power were to unexpectedly altered.

To the highest degree possible, the applicable guideline will designate the areas or loads that must be served by the emergency system. One document referred to often for this purpose is NFPA 101. It states that all industrial occupancies should have emergency lighting for designated corridors, stairs, and so on, that lead to an exit. There are exceptions for uninhabited operations and those that allow adequate daylight for all egress routes during production. A note advises authorities having jurisdiction (AHJ) to review large locker rooms and laboratories using hazardous chemicals, to be sure that major egress aisles have adequate emergency illumination.

5.4 STANDBY SYSTEMS THAT ARE LEGALLY REQUIRED

Provisions of the NEC apply to the installation, operation, and maintenance of legally required systems other than those classified as emergency systems. The key to separating legally required standby systems from emergency systems is the length of time an outage can be permitted. They are not as critical in terms of time for recovery, although they may be very critical to other items other than personnel safety. They are also directed at the performance of selected electrical loads, instead of the safe exit of personnel. For instance, there are several rules requiring standby power for large sewage treatment facilities. In this case, the facility must remain in operation in order to prevent environmental problems. Another example is that the system pressure of the local water utility must always be maintained for fire protection as well as for public health and safety. Although critical, this is a different type of concern. Allowing a longer time delay between loss and recovery than is permitted for lighting that is crucial to emergency evacuation.

5.5 OPTIONAL STANDBY SYSTEMS

Optional standby systems are unrelated to life safety. Optional standby systems are intended to supply on-site generated power to selected loads such as mission critical electrical and mechanical infrastructure loads either automatically or manually. The big difference between true emergency circuits and those that are served by standby or optional sources is that care must be exercised to select the correct code articles that apply to the system in question.

5.6 UNDERSTANDING YOUR POWER REQUIREMENTS

Managers cannot evaluate the type of backup power requirements for an operation without a thorough understanding of the organization's power requirements. A manager needs to evaluate whether they want to shut down the facility if a disturbance occurs, or whether they need to ride through it. Since the needs of organizations vary greatly, even within the same industry, it is complicated to develop a corrective solution that can be valid for varying situations. Facilities Managers can use answers to the following questions to aid in making the correct decision:

1. What would be the impact of power outages on mission-critical equipment in the organization?
2. Is the impact a nuisance, or does it have major operational and monetary consequences?
3. What is the reliability of the normal power source?
4. What are the common causes of power failures in the organization? Are they overloaded power lines or weather-related, and so on?
5. What is the common duration of power failures? In most cases there are always far more monetary "blips" and other aberrations than true outages (brief interruptions or brownouts/blackouts that can last minutes, hours, or days).
6. Are there common power quality problems, such as voltage swells, voltage sags, ripples, over- or undervoltages, or harmonics?
7. Do the power interruptions stem from secondary power quality problems generated by other in-house equipment?

5.7 MANAGEMENT COMMITMENT AND TRAINING

The items discussed in this chapter or book cannot occur without the support of company senior management and administration. This assistance is critical for the reliable operation of the critical infrastructure. It is possibly the most demanding task for the user to make management aware and responsive of the costs involved with maintaining standby systems. It is the responsibility and mission of the user to work with management to institute reasonable operating budgets and expectations based on the level of risk of your organization. Both budgets and goals are required to be balanced against the ultimate cost of a power loss. Management and the user need to make sure that each is providing the other with realistic expectations and budgets. It is in this area that maintenance records are of vital importance. They form the basis of accurately projecting cost estimates for the future based upon current actual costs. Items that are consuming too much of the maintenance budget can be singled out for replacement. Records kept of power quality problems and the costs avoided because of system reliability are certain to be invaluable to this process. This needs to be a specific plan to monitor and maintain the system, a system of checks to verify that the maintenance is done properly and tests to be sure that the equipment has not deteriorated. There also has to be a commitment to continuous training to keep the knowledge fresh in the heads of critical people who would be in charge of dealing with power failures. One major issue in mission critical facilities is for managers to allow the proper testing of the system, such as an annual plug-the-plug test. In order to accomplish this, you actually need to expose the system to power failure to know that the system and all associated functions, including people, perform accordingly. That does insert a bit of risk into the power reliability, but it is far better to find a problem under controlled conditions than to find it during an emergency.

5.7.1 Lockout/Tagout

Repairs made to equipment and systems using electricity require that the energy be controlled so that workers are not electrocuted. The means of accomplishing this is called lockout/tagout. At points where there are receptacles, circuit breakers, fuse boxes, and switches, facilities must use a lock that prevents the energy from being applied and post a label warning personnel that the receptacle has been de-energized and that the power will not be applied until repairs are made. The lock also prevents the switch from being accidentally being thrown in the closed or "on" position. The locks should be key operated, and the keys should be stored in a safe, secure area. It is also a good idea to have a trustworthy person handle the keys and keep them on a key ring, with each key properly labeled. This person should be in charge of repairs made to power sources.

Before beginning maintenance, be aware that a generator can start without warning and has rotating parts that can cause serious injury to the unwary. Some other safety concerns are:

1. Never wear loose clothing around a generator.
2. Always stay clear of rotating and hot components.
3. Hearing protection and safety glasses must be worn any time the generator set is running.
4. Be aware of high voltages.
5. Only qualified and trained personnel should work on standby generators.

5.7.2 Training

Facilities Managers should be aware of and take advantage of available resources to help companies meet the challenges of safety issues and keep informed as the regulations change from year to year. The issue is that since power failures are rare, it is nearly impossible for people to remember what to do when they occur. So training is needed to keep actions fresh and to provide practice so they are properly performed. One such remedy is enhanced training. Whether the training is done in house or performed by an outside source, a complete and thorough safety training session should be implemented. Some safety topics require more time to finish, but that should not be a factor in the amount spent learning about ways to keep workers safe. After all, workers are an organization's most precious assets. Always have regular safety training sessions, usually every year, and practice sessions quarterly. This schedule will instill basic safety requirements, while helping the worker remember procedures in wearing and removing personal protective equipment. Practice sessions can be done in-house with ease. There should be a formal program documenting that each critical person did the training, was tested on proper operations, and performed properly. Safety products made with durable material suited for rough applications can make the difference between worker productivity and worker death. It also is wise to buy any upgrades to the safety equipment because technology changes rapidly. Up-to-date information will help ensure complete worker safety. Managers must remember that safety pays. Cutting back on worker safety is not only detrimental to productivity, it is also expensive. There are several factors to include: How much is the company willing to pay for medical expenses and lost time versus keeping in the clear with OSHA?

In the 1970s, worker deaths were at an all-time high. Now, increased awareness and safety training has reduced these figures. But that does not mean that OSHA doesn't still target companies, especially repeat offenders. Remember that the cost of being safe is insignificant in comparison to the value of a human life.

5.8 STANDBY GENERATOR SYSTEMS MAINTENANCE PROCEDURES

When a standby generator fails to start, it is usually due to oversight of the people accountable for its maintenance. Generator sets are reliable devices, but failures often occur because of accessory failures. Managers must take a system approach to power to get more reliability. As stated before, people play a big part of this. They need to be trained and monitored, and their performance must be documented. Standby generators are too dependable and too easily maintained for a failure to occur. If facilities engineers do their part to establish maintenance and testing programs to prove its reliability, the standby generator can be counted on to perform in an emergency situation. As dependable as standby generators are, that reliability factor is only as good as its ongoing maintenance and testing programs.

5.8.1 Maintenance Record Keeping and Data Trending

One of the most important elements in any successful quality assurance program is the generation and storage of relevant information. Relevant information can include items such as procurement contracts, certificates of conformance, analysis results, system modification, operational events, operational logs, drawings, emergency procedures, contact information, etc. It is important to realize that by generating and maintaining the appropriate documentation, it is possible to review important trends and significant events or maintenance activities for their impact on system operation.

Suggested maintenance items for stand by generators are:

Engine

1. Verify and record oil pressure and water temperature.
2. Inspect air intake system including air filter condition, crankcase breather, and turbocharger.
3. Inspect muffler system and drain condensation trap (if applicable) and verify rain cap operation.
4. Inspect engine-starting system, and verify cable integrity and connections.
5. Inspect exhaust flex coupling and piping for leaks and proper connection.
6. Check for abnormal vibration or noise.

Coolant System

1. Inspect clamps, verify condition of all hoses, and identify any visual leaks.
2. Check temperature gauges for proper operation of engine jacket water heater.
3. Test coolant's freezing point and verify coolant level.
4. Test coolant additive package for proper corrosion inhibitors.
5. Inspect belt condition and tension, and correct as required.
6. Inspect radiator core for visual blockage or obstructions. Keep it clean and look for junk laying around that might blow into it and block it.
7. Inspect for proper operation of intake louvers, motorized or gravity (if applicable).
8. Verify proper operation of remote radiator motor and belt condition (if applicable).

Control System

1. Verify and record output voltage and adjust voltage regulator, if necessary. However the only people who should be adjusting voltage regulators, governing and protecting equipment are those who are specifically trained in how to make the adjustments.
2. Calibrate control meters.
3. Verify and record output frequency and adjust governor if necessary.
4. Verify operation of all lamps on control panel.
5. Inspect for any loose connections, or terminals, or check for discoloration. Thermographic inspection would work well here.

Generator (Figure 5.1)

1. Inspect and lubricate generator and ball bearing. (*Note*: This is usually only necessary for older generators; most new models come lubricated for life. Also, when possible, this should be done by a qualified technician.)

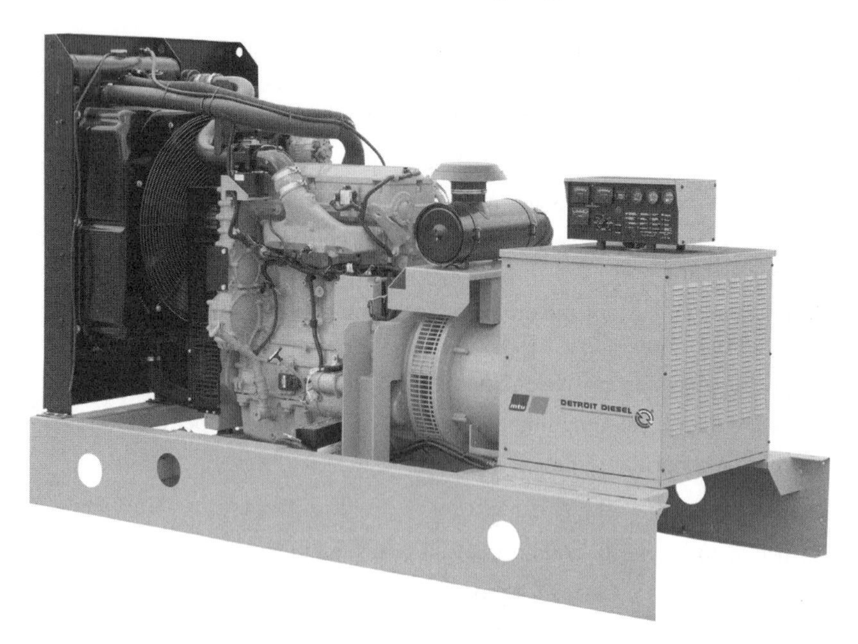

Figure 5.1 Generator. (Courtesy of ADDA.)

2. Look for blocked cooling air passages around the alternator and general condition of generator.
3. Inspect for abnormal vibration.
4. Verify connections and insulation condition.
5. Verify that the ground is properly attached.
6. Verify proper operation of shunt trip on mainline circuit breaker (if applicable).

Automatic and Manual Switchgear

7. Verify proper operation of exercise clock (adjust if necessary).
8. Visual inspection of all contacts and connection points.
9. Perform building load test (if practical).
10. Verify operation of all lamps on control.

5.8.2 Load Bank Testing

Load bank testing (Figure 5.2) exercises a single piece of equipment. Tests verify proper operation of generators and associated equipment such as transfer switches. Exercise is part of a maintenance process to verify load carrying capability and sometimes to remove deposits from injection systems that can occur due to light

Figure 5.2 Load bank testing. (Courtesy of ADDA)

load operation. Perform a test for approximately 4 hours that includes a detailed report on the performance of your system. The duration of the test can be shorter in warm climates, and it may need to be longer in colder ones. The important factor is to run the test long enough that all temperatures (engine and generator) stabilize. (In addition to ensuring a proper test, this will help drive off condensation.) During the test, all vital areas are monitored and recorded, including electrical power output, cooling system performance, fuel delivery system, and instrumentation. All of this is done with no interruption of normal power! Load bank testing is the only sure way to test your system. Manufacturers, such as MTU Detroit Diesel, recommend that testing always be done at full load.

5.9 DOCUMENTATION PLAN

5.9.1 Proper Documentation and Forms

When the generator is operating, whether in an "exercise" mode or the load "test" mode, it must be checked for proper performance conditions. Many facilities have chosen an automated weekly "exercise" timer. One of the unpleasant aspects that is found in this type of system is that the generator is rarely checked when its weekly "exercising" is automated. Anytime the generator is being run, it must be checked and all operating parameters verified and thereafter documented. It is through the exercising and testing of the generator that deficiencies become evident. Yet many facilities with automated weekly exercise timer on the automatic transfer switches don't know that the generator has run, except for the reading of the incremental run hour clock. If your facility selects this automated feature, be sure that the building engineer or maintenance technician is also available, so that the generator never runs without being checked and having the proper documentation completed. A weekly exercise is recommended but not always practical. However, manufacturers recommend that generators be exercised at least once per month.

5.9.2 Record Keeping

Keeping good records of inspection can give better insights for potential problems that might not be obvious. In other words, certain types of deterioration occur so gradually that they will be hard to detect in a single inspection. However, combined with prior inspection data, a particular trend of potential failure might become evident. For instance, measuring the insulation value of a cable without any past data might appear to be satisfactory, but when compared to a prior test, the insulation value has dropped a certain percentage during every test. It will be easy to estimate an approximate failure time for the cable in the future. Therefore, the cable needs to be replaced before then. The success of an effective PM program is based on good planning. When deciding when to schedule a shutdown, pick a time that will constitute minimal impact for the operation. The continual proliferation of electronic equipment has created higher dependence on electrical power systems. As the hardware for electronic devices becomes more robust, the reliability of the electrical distribution may become the block in deciding the overall system capability. Many elaborate reliability considerations that were typically considered for mission critical applications exclusively would be more commonplace for examining facilities.

Facilities managers play a more critical role than ever in the operational and financial success of a company. Now, when facilities managers speak about the electrical distribution system, facility executives are much more likely to listen. The Internet is also becoming a key tool in shaping the way businesses communicate with customers. A growing number of maintenance operations are setting up home pages designed, among other things, to provide information to and foster communications with customers. The most active in this regard have been colleges and universities. Additionally, a web-based application can be set up so everybody can keep track of scheduled maintenance as well as to provide pertinent documents and spreadsheets.

5.10 EMERGENCY PROCEDURES

Even with the best of testing and maintenance procedures, it is wise to prepare for a potential generator-set failure so that its affects are minimized. The generator-set emergency "stops" and "shut-offs" should be easily identifiable and so labeled. A posted basic trouble-shooting guide should be readily available right beside the unit with emergency call numbers of affected areas and repair contractors. A clean copy of the generators operation and maintenance (O&M) manuals, complete with schematics, should be kept in the generator room at all times. Never leave the original (O&M) manuals in the generator room because it might get soiled and unusable. Keep the (O&M) originals in a safe place for making additional copies or for an emergency if any of the copies get misplaced or otherwise become unusable. The emergency "stop" and any contingency plan also need to be tested to prove their proper operation and effectiveness. This regular testing and review is key because responsible people demonstrate their understanding of the system design and their ability to respond to the common failure modes in the system.

It is recommended that the gen-set logbook contain the previously noted emergency telephone numbers, basic trouble-shooting guidelines, and contingency plans. While this may be redundant with the posted procedures, it provides an extra safeguard that is strongly recommended. With the emergency procedures detailed in the gen-set logbook, more precise information and instructions can be provided to the operating technician in case of an emergency situation.

5.11 COLD START AND LOAD ACCEPTANCE

An important concern of the standby generator design engineer is how much time it takes for the standby or emergency generator system to sense and react to a loss of power or other power quality problems. A "cold start" does not mean a completely cold engine, but rather an engine that is not at operating temperature. There is a stipulation in certain codes and standards that state that emergency generators must be able to pick up the emergency loads within 10 seconds in the United States and 15 seconds in Canada following a power failure. Once these loads are on-line, then other critical loads can be connected. In most cases, there are two different standby generator systems. The first is for emergency loads, such as the life safety systems. The second is for critical systems that support the data centers critical, electrical

Figure 5.3 Generator control cabinet. (Courtesy of ADDA.)

and mechanical loads. The design criteria from a 10-second cold start is to equip the generator with coolant heaters or block heaters. The coolant heaters are necessary to start the machine and enable it to pick up a full rated load in one step.

5.12 NONLINEAR LOAD PROBLEMS

If the facility's electrical loads, such as computer power supplies, variable speed drives, electronic lighting ballasts, or other similar nonlinear electrical equipment, is furnished with switch-mode power supplies, it is imperative that you advise the generator supplier of this situation so that proper steps can be taken to avoid equipment overheating or other problems due to harmonics. Some generator manufacturers recommend low-impedance generators and have developed winding design techniques to reduce the effects of the harmonic currents generated. The key issue is maintaining the ability to produce a stable voltage waveform and a stable frequency. This is mostly an issue of voltage regulation system design, but can also be impacted by the governing (fuel control) system design. The winding arrangement is not as critical as the impedance of the machine relative to the utility service. The closer it is to what the utility provides, the more like the utility it will operate and the lower probability of problems. In some instances, the generator may have to be de-rated and the neutral size increased to safely supply power to nonlinear loads.

When a standby generator and UPS are integrated in a system together, problems occur that typically do not exist with a UPS system or generator when they are operating alone in a system. Problems arise only when the UPS and standby generator are required to function together. Neither the UPS nor the standby generator manufacturer is at fault, and both manufacturers would probably need to work together to solve the problem. The following are common problems and solutions when applying a design that incorporates a standby generator and UPS.

1. Line Notches and Harmonic Current. The UPS manufacturer using a properly designed passive filter can address the problem of both line notches and harmonic currents. Most generator manufacturers have de-rating information to solve harmonic heating problems. However, an input filter on the UPS that reduces the harmonics to less than 10% at full load eliminates the need for de-rating the generator.

2. Step Loading. When a generator turns on and the ATS switch connecting it to the UPS closes, the instantaneous application of the load to the generator will cause sudden swings in both voltage and frequency. This condition can generally be evaded by verifying that the UPS has a walk-in feature. This requires that the UPS rectifier have some means of controlling power flow so that the power draw of the UPS can slowly be applied to the generator between a 10- and 20-second timeframe.

3. Voltage Rise. This is an application problem that transpires when a generator is closely designed and sized to the UPS and there is little or no other electrical load on the generator. When the UPS is first connected to the generator by the ATS, its charger has turned off so that it may begin its power "walk-in" routine. If the input filter is the only load on the generator, it may provide increased excitation energy for the generator. The issue is the capability of the alternator to absorb the reactive power generated by the filters. The amount that can be absorbed varies considerably between different machines from the same manufacturer. System designers should evaluate the capability in the initial design of the system to avoid problems. The outcome is that the voltage roams up without control to approximately 120% by some fundamental generator design constraint, typically magnetic saturation of the generator iron. If the value does hit 120% of the nominal, it will be damaging or disruptive to the system operation. However, a UPS that disconnects its filter when its charger is off avoids this predicament altogether.

4. Frequency Fluctuation. Generators possess inherent limitations on how closely they can manage frequency regarding their response to changing electrical loads. The function is complicated and not only involves generator features, such as rotational inertia and governor speed response, but also involves the electrical load's reaction to frequency changes. The UPS charger, conversely, also has inherent limitations on how closely it can control its power needs from a source with fluctuations in voltage and frequency. Since both the generator controls and the UPS charger controls are affected by and respond to the frequency, an otherwise small frequency fluctuation may be exasperated. The most noticeable effect of this fluctuation is a recurring alarm that is found on the bypass of the UPS, announcing that the generator frequency is changing faster than a UPS inverter can follow. In order to minimize or eliminate frequency fluctuation problems, good control design from both the engine generator and UPS manufacturer are required. The engine must have a responsive governor, appropriately sized and adjusted for the system. The UPS manufacturer should have a control responsive to fast frequency fluctuations.

5. Synchronizing to Bypass. Some applications require the UPS to synchronize to bypass so that the critical load may be transferred to the generator. This generally places tighter demands on the generator for frequency and voltage stability. When this is the case, the system integration problem may be intensified. As described above, good control design can usually reverse this problem.

6. Automatic Transfer Switch. Most generator/UPS projects incorporate automatic transfer switches that switch the UPS back to utility power once it becomes available again. The speed of transfer can be an obstacle and may result in a failed transfer. This in turn will lead to nuisance tripping of circuit breakers or damage to loads. If the ATS switch also has motor loads, such as HVAC systems, the UPS input filter will supply excitation energy during the transfer. This excitation source turns these motors into generators using their inertia as an energy source and also their alternator field strength. If the transfer occurs too fast, causing an unexpected phase change in the voltage, the consequences can be devastating for both the motors and the UPS. One of the best solutions is to simply slow the transfer switch operation speed so that the damaging condition does not exist. Rather than switching from source to source in one-tenth of a second, slowing to one-half of a second will resolve the problem. UPS manufacturers can resolve this problem by providing a fast means of detecting the transfer and disconnecting the filter.

5.13 CONCLUSIONS

This chapter has discussed selected ways to enhance the time factor in the terms generally used to describe reliability. This is not, and should not be, regarded as the final statement on reliable design and operation of standby generators. There are many areas of this subject matter that warrant further assessment. There is also a considerable lack of hard data on this subject, making practical recommendations difficult to substantiate. For example, there is virtually no data on the trade-offs between maintenance and forced outages. It is self-evident that some maintenance is required, but what is the optimum level? This question is easily answered if your organization's level of risk has been evaluated. However, there are certain common elements in both reliable and unreliable systems that I think were relevant, and I listed them accordingly in this chapter. These suggestions, when seriously considered, will help to increase the level of reliability of the standby system.

Chapter 6

Fuel Systems and Design and Maintenance for Fuel Oil

Howard L. Chesneau, Edward English III, and Ron Ritorto

6.1 FUEL SYSTEMS AND FUEL OIL

Since this information is primarily for mission critical systems, the fuel systems discussed will be diesel fuel. In many applications, diesel fuel is chosen because it offers easy on-site storage, has reduced fire hazards, and diesel engines are capable of more operating hours between overhauls. Also, modern diesel engines offer many advantages over other types of prime movers. A disadvantage of diesel fuel is its low volatility at low ambient temperatures. For generators to be reliable starters, they need to be warm; one step in ensuring this would be to install engine block heaters to maintain the water jacket temperature at 90 °C or higher.

Diesel engines are designed to operate on diesel fuel. A diesel engine doesn't have igniters or spark plugs to ensure proper fuel ignition as compared to gasoline engines or jet turbines. Instead the diesel depends totally on the compression stroke of the piston to create high temperatures to ignite the air–fuel mixture. This unique characteristic of a diesel requires the ignition to occur very close to the top of the piston stroke such that the expanding burning fuel–air mixture produces the power stroke. Consequently, fuel quality plays a critical role in engine performance. Diesel fuel is less flammable than gasoline and yet much more flammable than water. Consequently, an inadvertent delivery of gasoline (even high octane), water, or other contaminate-laden fuel significantly diminishes the diesel's ability to produce power. Normally our fuel is tested after delivery, but during an extended run the results of the test are not received before the fuel is utilized. A reliable long-term fuel supplier is the best answer. For the best reliability and performance, use ASTM D975 fuel in the engines, not boiler fuel.

Maintaining Mission Critical Systems in a 24/7 Environment By Peter M. Curtis

Unlike other fuels like gasoline and natural gas, diesel can actually support microorganisms that consume fuel and flourish within the fuel tank, especially in humid areas where water condensation can occur. If the filters become clogged with biological waste or fuel is contaminated enough, the engine will sputter and surge. This can produce voltage or frequency transients as the engine struggles to sustain the critical load. There are biocide additives accessible that avert this type of contamination and are useful when your fuel reserve has gradual turn over. A fuel polishing system that circulates stored fuel through a polisher or centrifuge and returns it to the stored tank is a means of preventing contamination. This system can run on a periodic basis or be kept on-line continuously. Regular engine testing does not provide for significant fuel reserve turnover. Therefore, most transfers from receiving tank to the supply or day tank are performed by manual operation of the transfer pumps. What if during an extended emergency run, it was found that when the day tanks became depleted, the automatic feature of our transfer pumps did not always operate. This also stresses the need to have a properly designed facility. The facility should always be tested regularly after commissioning. If the system is properly designed and tested during commissioning, you will not have these problems. Events like this should make you more aware of the fuel supply system, which should then prompt a thorough investigation of the fuel delivery system. You would be surprised what is found that can ultimately interrupt service.

6.1.1 Fuel Supply Maintenance Items

a. Inspect all fuel lines and hoses for visual leaks and general condition.
b. Inspect day tank and level control assembly for proper operation and leaks.
c. Inspect fuel pump for proper operation and leaks.
d. Verify proper operation of governor and governor linkage (where applicable).
e. Check level of fuel in primary supply tank (when practical).
f. Drain water and sediment from fuel filter and fuel supply tanks (when practical).

6.1.2 Fuel Supply Typical Design Criteria

The diesel fuel system is considered the life blood of the emergency power system. Without a clean readily available source of quality diesel fuel, the emergency power system is doomed to failure. Yet there is so little understanding about designing and maintaining this system that it goes completely neglected from conceptual design to implementation of a fuel quality and maintenance program. All aspects of the fuel storage and distribution system including selection of the main bulk storage tanks, piping distribution system, controls system, fuel maintenance system, and an ongoing program of fuel testing and maintenance is an absolute must in order to keep the emergency generators running efficiently.

6.2 BULK STORAGE TANK SELECTION

Selecting the main bulk storage tanks is an important first step in system design. There are many factors that must be evaluated in order to establish the best configuration: aboveground or underground, single wall or double wall, steel or fiberglass. Since 1998, federal and state codes that govern the installation of underground tanks made it more attractive to look at aboveground bulk storage tanks. From an environmental standpoint, underground tanks have become a difficult management issue. In 1982 the president signed into effect the reauthorization of the Resource Conservation and Recovery Act (RCRA). Federal and state codes now require very sophisticated storage and monitoring systems to be placed into operation in order to protect the nation's ground water supply. In some instances, property owners prohibit the installation of petroleum underground storage tanks on their properties. On the other hand, aboveground storage tanks occupy large amounts of real estate and may not be esthetically pleasing. Here are some factors to consider when deciding on aboveground or underground bulk storage tanks.

Aboveground tanks are environmentally safer than underground tanks. The tanks and piping systems are exposed aboveground, more stable, and easier to monitor for leaks than underground systems. Since the tanks and piping are not subject to ground movement caused by settlement and frost, there is less likelihood of damage to the system and consequently less of a likelihood of an undetected leak. However, if large tanks are required, the aboveground storage system can be awkward and will occupy large quantities of real estate. Code officials, especially fire departments, may have special requirements or often completely forbid the installation of large quantities of petroleum products in aboveground tanks. Day tanks and fuel headers may be located at the same elevation or occasionally lower than the bulk storage tank, making gravity drains from these devices impossible to configure unless the generators are on or above the second floor of the building. Filling the tanks may require special equipment, and in some cases codes have even required secondary containment systems for the delivery truck because it is felt that the "pumped fill" is more vulnerable to catastrophic failures that can empty a compartment of the delivery truck. Additionally, constant thermal changes to the fuel may stress the fuel, causing it to become unstable. Diesel fuel that is exposed to subfreezing temperatures requires special conditioning in order to make it suitable for low-temperature operability. If the fuel is left unconditioned, coagulation may occur, causing the fuel to plug filters and fuel lines.

Aboveground tanks are almost always fabricated from mild carbon steel. The outer surface of the tank is normally painted white in order to prevent the fuel from becoming heated do to exposure to the sun. The tank should always include a manway and ladder in order to allow access to the tank for cleaning, inspection, and repair. Epoxy internal coatings compatible with diesel fuel provide extra protection to the fuel from the eventual contamination from internal rust. It also provides extra corrosion protection to the internal tank surface. All aboveground tanks should be fabricated to UL142 or UL2085 standards and should bear the appropriate label. The UL142 tank is an unprotected steel tank in either single- or double-wall configuration. The UL2085 tank is a special tank that is encased in concrete. The concrete provides a listed fire rating.

Modern underground tanks and piping systems have become more reliable, easier to install, and in general easier to operate over the past several years. Since the tanks are usually buried under parking lots or driveways, they occupy little or no real estate that is usable for other purposes. Drain and return lines from fuel headers and day tanks can drain by gravity back to their respective bulk storage tank, making for more efficient piping systems. On the down side, federal, state, and local codes usually impose construction standards that may be difficult to cope with. The underground storage tank must be designed and installed in accordance with the code of federal regulations, 40CFR280. In addition, the appropriate state and local codes must also be followed. The state and local government must follow, at a minimum, the federal code; however, state and local governments may implement local codes that are more stringent than the federal code. Obtaining permits can be a long and tedious process, with many agencies involved in the approval process. Most important, since the tank and piping system is out of site, leaks can go undetected for long periods of time if monitoring systems and inventory control are not meticulously maintained.

Underground tanks are most commonly fabricated from mild carbon steel or fiberglass-reinforced epoxy. Steel tanks must be specially coated with either fiberglass-reinforced epoxy or urethane coatings to protect them from corrosion. In addition to the coating, some steel tanks are additionally provided with cathodic protection. Fiberglass tanks are usually not coated. The most common manufacturing standard for steel tanks is UL58 and for fiberglass tanks, UL1316. The tank should always include a manway and ladder in order to allow access to the tank for cleaning, inspection, and repair. For steel tanks, epoxy internal coatings compatible with diesel fuel provide extra protection to the fuel from the eventual contamination from internal rust. It also provides extra corrosion protection to the internal tank surface.

6.3 CODES AND STANDARDS

Codes and standards should always be reviewed prior to any discussion on type and location of the bulk storage tank. A manager should understand that there is a huge risk in an improperly designed fuel system, not just with regard to reliability, but also with respect to environmental concerns. An improperly installed or maintained fuel tank can expose a company to literally millions of dollars liability for cleanup of spilled fuel. It is a good idea to have a preliminary meeting with the governing code agencies in order to discuss the conceptual design and requirements of the system. Don't assume that merely because the state building code allows the use of aboveground storage tanks, there won't be some opposition from one of the other agencies that may have jurisdiction. Fire marshals in particular may be completely opposed to the installation of large aboveground diesel storage tanks, especially when they are located in metropolitan areas. The usual sequence to follow when establishing the priority of compliance is to start with the state building code. The state building code will normally reference either NFPA or the uniform fire code for the installation of flammable and combustible storage systems. NFPA 30 Flammable and Combustible Liquids Code addresses the installation of storage tanks and piping systems. NFPA 37, Standard for the Installation and Use of Stationary Combustion Engines and Gas Turbines, addresses the piping systems, day tanks, flow control systems, and, in some instances, references NFPA 30. The uniform Fire Code Article 79, Flammable and Combustible Liquids, addresses the installation of flammable liquid storage systems. In addition, the federal EPA has codes in place to address performance standards for both aboveground and underground storage systems. Underground storage tank standards are covered in 40 CFR 280. This code sets minimum standards that must be followed by all states. The individual states have the option of requiring more stringent standards. Aboveground storage tanks are covered in 40 CFR 112. This code also covers certain topics relative to threshold storage limits for both aboveground and underground that require the owner to maintain a document known as a Spill Prevention Control and Countermeasure Plan.

6.4 RECOMMENDED PRACTICES FOR ALL TANKS

There are some **recommended practices** that should be adhered to regardless of what the codes require. Bulk storage tanks should always be provided with some form of secondary containment, regardless of the requirement by the local authority. Anyone who has ever been involved in a remediation project connected to a leaking underground storage tank understands the astronomical costs that can be incurred. In addition, data centers are usually high-profile facilities. The bad press that may be associated with a petroleum spill contaminating underground water supplies is horrific. The most popular system and the easiest to maintain is the double-wall tank. Open containment dikes for aboveground tanks are a maintenance headache, especially in cold climates where removing snow and ice is very difficult. Similarly, for underground tanks cutoff walls and flexible liners are not effective or efficient ways to provide secondary containment. Instead the double-wall tank is efficient and easiest to maintain.

Piping systems for underground tanks can also be furnished in a multitude of different configurations. Secondary containment should always be provided for underground piping. Similar to the underground tank, the piping is out of site, and can cause grave environmental damage if a leak goes undetected for long periods of time. Some of the choices for piping are double wall carbon steel, double wall stainless steel and double wall fiberglass to mention a few. Probably the most popular system in use today is the double wall flexible piping system. It is uncomplicated to install and UL listed for use with underground petroleum storage systems. Sinces underground piping runs each have only two connections, one at the tank, the other where the piping leaves the ground and attaches either to a device or an aboveground pipe, there is little opportunity for leaks at joints. There are complete systems available including high density polyethylene termination sumps that provide secondary containment where the piping attaches to the tank and also allows access from grade without excavation to service the termination fittings, as well as submersible pumps, and any other equipment such as tank monitoring equipment.

Figure 6.1 describes basic installation practices for all tanks, whether aboveground or underground. The tank should be installed with a slight pitch to encourage collection of accumulated water at a low point in the tank. A water pump out connection should be provided as close as practical to the low end of the tank. A spill containment fill box should be located at the opposite end of the tank with a drop tube terminating close to the bottom of the tank. Introducing new fuel at this end of the tank will help to move any water accumulated at the bottom of the tank to the opposite end where the water pump out connection is located. An access ladder and manhole allows access to the tank for cleaning and maintenance. The fuel maintenance system should draw fuel from the lowest point possible in the tank and return it to the opposite end of the tank.

Piping systems for aboveground tanks provide less choices. Carbon steel and stainless steel are the primary choices. The most common is carbon steel. When the piping runs are in open areas such as generator rooms and machine runs, it is generally installed as a single-wall pipe. If the piping runs are in more critical areas or where codes require secondary containment, it is usually installed as a double-wall piping system. However, there are some problems associated with the double-wall system. If a leak develops in the primary pipe, it can be difficult to locate the source of the problem since the leak tends to fill the annular space between the primary and secondary pipe. Another concern is the inability to contain

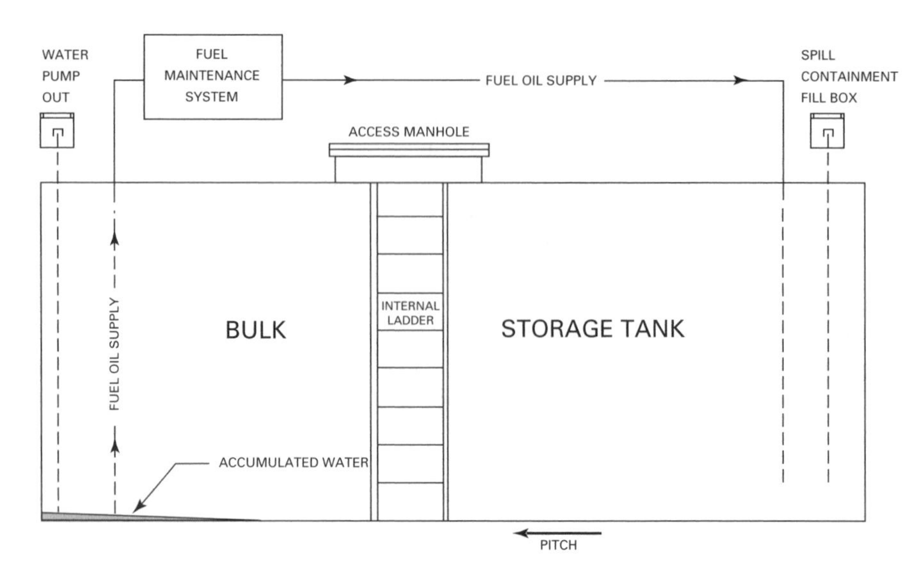

Figure 6.1 Basic installation practices for all tanks, whether aboveground or underground. (Courtesy of Mission Critical Fuel Systems.)

valves, flexible connectors, and other vulnerable fittings. Typically, the double-wall pipe is terminated with a bulkhead at valves and flexible connectors, leaving these devices unprotected. Since the piping systems are under rather low pressure and the product, diesel fuel, is not very corrosive, leaks are usually minor. For this reason, wherever possible, install single-wall carbon steel pipe and either place it into trenches or provide curbs around the area where it is installed. The area is then monitored with leak detection equipment. If a leak develops, an alarm warns the facility operations personnel of the problem. Since the piping is exposed, the source of the leak can be readily determined and corrected. The pipe trench or curb will also contain the leak, thereby preventing widespread damage.

Fuel distribution system piping materials are of major concern. Many of the performance problems that I have seen with diesel fuel systems are directly related to the materials of construction. There are two categories of materials that should be avoided. The first group is described in NFPA 30. These materials include low-melting-point materials that may soften and fail under exposure to fire. They include aluminum, copper, brass, cast iron, and plastics. There is, however, one exception to this rule. NFPA 30 allows the use of nonmetallic piping systems, including piping systems that incorporate secondary containment. These systems may be used outdoors underground if built to recognized standards and installed and used within the scope of Underwriters Laboratory Inc.'s S*tandard for Nonmetalic Underground Piping for liquids*, UL 971. The next group is materials that contain copper or zinc. Although these materials may be commonly found in many fuel distribution systems, their contact with diesel fuel should be avoided. These materials form gummy substances that can cause filter clogging.

The most popular and practical piping material is carbon steel. The piping should comply with the applicable sections of ANSI B31, *American National Standard Code for Pressure Piping*. Pipe joints, wherever possible, should be either butt-welded or socket-welded. Where mechanical joints are required, flanges provide the best seal. Flange gaskets should be fiber-impregnated with viton. Threaded connections should be avoided wherever possible. Flexible connectors should be installed wherever movement or vibration exists between piping and machinery. The flexible connectors should be listed for the appropriate service.

Careful consideration should be taken to the type and placement of isolation valves. Carbon steel or stainless steel ball valves with teflon trim and viton seals provide excellent service. Once again the avoidance of threaded end connections should be considered. Socket-welded or flanged end connections provide excellent service. If the end connections are welded, consider the use of three-piece ball valves in order to facilitate valve service and repair without disturbing the welded joint. Fire-safe valve designs should be considered, especially where codes may mandate their use.

Fuel maintenance systems or fuel polishing systems are frequently included in new installations where diesel fuel is stored for prolonged periods of time—that is, one year or longer. These systems were not common several decades ago.

However, with the recent decline in diesel fuel quality and stability, they are becoming a must for anyone who stores diesel fuel for emergency standby power. Fuel quality will be discussed in greater detail later in this chapter. The system should be installed as shown in Figure 6.1. The tank is pitched to a low area to encourage settlement of water. The suction stub for the fuel maintenance system is terminated as close to the bottom of the tank as possible at this end of the tank. When the system operates, fuel is pumped through a series of particulate filters and water separators, then returned to the tank at the opposite end. The operation of the system is timed in order to circulate approximately 20% to 25% of the tank volume once per week.

The tank fill connection should be in a spill containment fill box, the capacity of which is dictated by the authority having jurisdiction. The fill box should be placed at the high end of the tank encouraging the movement of any sediment and water to the opposite end of the tank, where it will be removed by the fuel maintenance system.

A connection similar to the fill connection and spill containment fill box should be located as close as possible to the opposite or low end of the tank. From this connection, any accumulated water may be removed by manually pumping it from the tank.

An access manhole and internal ladder provide convenient access to the tank for cleaning and maintenance. For underground tanks the manhole should be accessible through an access chamber that terminates at final grade.

The bulk storage tanks should be provided with an automatic gauging and monitoring system. Systems range from a simple direct reading gauge to very sophisticated electronic monitoring systems that will provide automatic inventory control, temperature compensated delivery reports, system leak monitoring, and underground tank precision testing. Many systems include gateways that allow them to interface with many of the popular building automation systems. These systems are especially beneficial since any unusual conditions are easily identified at the BAS console. An evaluation should be performed to determine the appropriate system for the application. Prior to deciding on a tank monitoring system, a compliance check should be performed to establish if the local authority has special requirements for tank monitoring.

6.5 FUEL DISTRIBUTION SYSTEM CONFIGURATION

The next step is assembling the pieces into a fuel storage and distribution system. It is strongly advised to divide the bulk fuel storage into at least two tanks. This is a practical approach to providing a means to quarantine new fuel deliveries and allow a laboratory analysis to be performed on it prior to introducing it into the active fuel system. Fuel testing and maintenance will be covered later in the chapter. Additionally, if it is necessary to empty one of the tanks for maintenance purposes, or if the fuel becomes contaminated in one of the tanks, or for any other reason one of the tanks becomes unusable, switching to the alternate tank is a simple task.

However, without the alternate tank, it may be necessary to install a temporary tank and piping system, which can be a major undertaking.

The simplest of fuel distribution systems is a single tank associated with a single generator. Common configurations are underground tank, aboveground tank, and generator sub-base tank. Suction and return piping must be sized within the operating limits of the generator's fuel pump. There is a common error often encountered when calculating the size of the piping. The fuel flow rate, used for calculating the pipe size, is approximately three times the fuel consumption rate. The additional fuel is used to cool the diesel engine injectors and then returned through the return line to the storage tank.

A more common system is a combination of bulk storage tanks and smaller day tanks located closer to the generators. The day tanks can be furnished in several different configurations. Generally, the configuration is one day tank per generator. This makes the generator system autonomous for better reliability. However, day tanks can provide fuel to several generators. The size of the day tank is dependent on several factors, including a predetermined period of fuel supply usually calculated at generator full load. Other determining factors are maximum fuel quantity threshold limits as specified by the governing code. Figure 6.2 shows a typical fuel storage and distribution system flow diagram utilizing two bulk storage tanks and three day tanks.

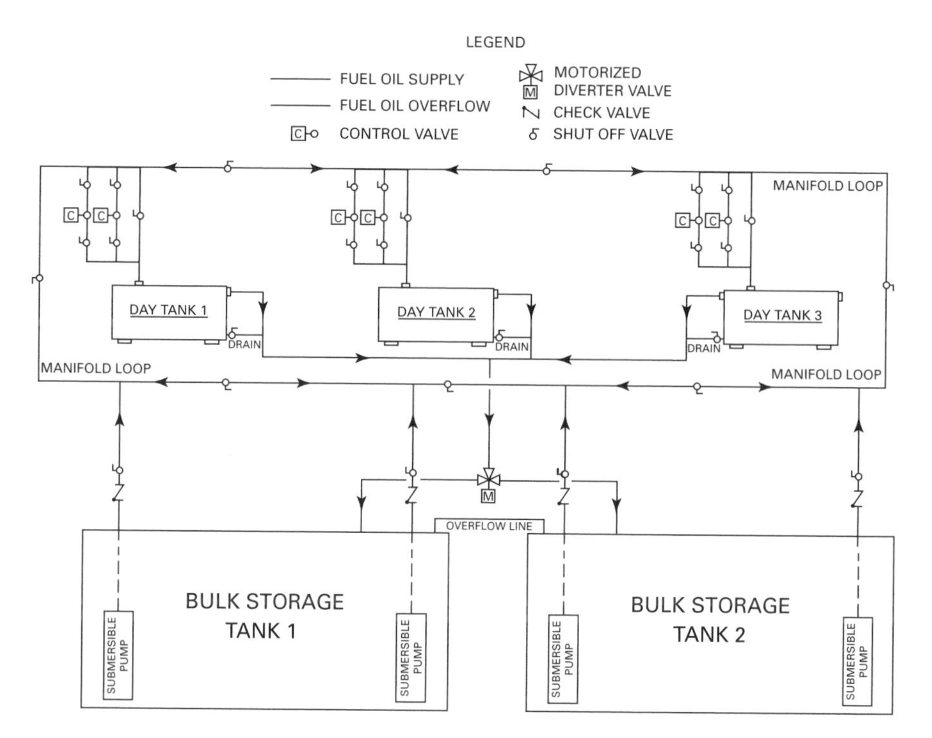

Figure 6.2 Typical fuel storage and distribution system flow diagram.

The fuel transfer pumping system consists of two submersible petroleum transfer pumps in each of the bulk storage tanks. Systems utilizing submersible petroleum pumps are efficient and dependable, and the pumps are more easily selected. The alternative is the positive displacement fuel transfer pump, which typically is located inside the facility, usually in an equipment room. If a positive displacement pump is selected, it is vital that the pressure loss in the fuel suction line from the bulk storage tank to the pump inlet be carefully calculated within the pump performance limits. Each of the submersible fuel transfer pumps should contain a dedicated supply pipe from the pump to the manifold loop that supplies the day tanks. In particular, if the bulk storage tanks are located underground, the supply piping is also underground and vulnerable to damage from ground movement do to settling and frost. It may also be damaged by excavating. This is a good place to mention that underground piping, especially nonmetallic piping systems, is very fragile. Therefore it is good practice to protect the piping with concrete top slabs and early warning tape.

The fuel oil supply manifold shown in Figure 6.2 is piped in a loop configuration. This allows fuel flow to the day tanks from two separate directions. By placing isolation valves at strategic locations in the manifold, sections of the manifold can be isolated for maintenance or repair without disabling the entire system.

Final fuel distribution is accomplished by supplying fuel via two control valves in parallel piping circuits to each day tank. Since a failed control valve will disable the day tank that it services, providing a redundant valve in a parallel circuit eliminates the single point of failure created if a single control valve is used. Control valves should always fail closed during a power failure in order to prevent a day tank overflow. As a backup to the control valves, there should be a manual bypass valve that will allow the tank to be filled manually in case a catastrophic control failure disables both control valves or the day tank control panel. Finally manual isolation valves should always be provided around the control valves. This allows isolation of the control valve for servicing without disabling the remainder of the system.

A day tank overflow line should be provided as a final spill prevention device. In the event of a control system or control valve failure, the overflow line should be adequately sized to return fuel to the bulk storage tank and prevent the eventual release of diesel fuel into the facility or to the environment. The applicable code should be closely followed in order to comply. If a diverter valve, shown in Figure 6.2, is required to divert fuel to the selected supply tank, the ports should be drilled for transflow. This will prevent the valve from stopping flow if the actuator stalls between valve positions. An overflow line can be provided between the bulk storage tanks to prevent an overflow if the diverter valve actuator fails, leaving the valve sequenced to the wrong tank. This will prevent the overflow of one of the bulk storage tanks if fuel is supplied from one tank and inadvertently returned to the other. The overflow line should not be confused with an equalization line. An equalization line enters both bulk storage tanks through the top and terminates near the bottom. The line is then primed with fuel, causing a siphon to begin which allows the fuel level in both tanks to equalize. It also allows contaminated fuel or water from one tank to transfer to the other tank. For this reason the use of an equalization line should be avoided.

6.5.1 Day Tank Control System

There are many commercially available day tank control systems available. Systems vary from simple packaged systems that are best suited to control a single day tank associated with a single bulk storage tank to very elaborate packaged systems that provide control of multiple pumps associated with multiple bulk storage tanks and multiple day tanks. There are several important steps that should be followed when establishing the appropriate system, beginning with the expectations of the degree of reliability. The bulk of discussion will be directed to the highest degree of reliability. From that level, "optional equipment" can be eliminated or tailored to fit the budget and the degree of reliability necessary.

Step 1 is to establish the day tank capacity. Assuming that the generators and day tanks are located inside of the facility, as opposed to the generators being packaged and located in a parking lot, code restrictions will dictate the threshold limit of fuel that may be placed inside the facility. There are a few techniques that may increase that threshold limit. One technique is to place the day tanks in fire-rated rooms. In addition to increasing the threshold limit, this will also provide some additional physical protection for the day tanks. There is also an exception that was added to NFPA 37 in the 1998 addition, stating that *"Fuel tanks of any size shall be permitted within engine rooms or mechanical spaces provided the engine or mechanical room is designed using recognized engineering practices with suitable fire detection, fire suppression, and containment means to prevent the spread of fire beyond the room of origin."* Common day tank capacities range from 20 minutes to as much as 24 hours of fuel supply at full load. Between 2 and 4 hours, fuel supply is a very common range. An important consideration the amount of fuel remaining after a low-level alarm is activated. If the facility is manned 7×24, 1 hour of fuel supply after the low-level alarm is activated is usually adequate. The day tanks should be provided with some form of secondary containment, either double-wall, open-top, or closed-top dike. Open-top dikes are difficult to keep clean. The space between the tank and the dike wall is normally only a few inches, making it difficult to clean. On the other hand, if a leak develops at a device that is installed through the top of the tank, the open-top dike will capture it while closed-top dikes and double-wall tanks will not. The tank should be labeled either UL142 or, in some special cases as required by the permitting authority, UL2085.

Step 2 is to establish a pumping and piping system configuration. As mentioned earlier, submersible petroleum pumps are efficient and dependable, and the pumps are more easily selected than positive displacement pumps. The pumps should be labeled in accordance with UL 79 standard. When establishing the piping system configuration, it is most efficient to place bulk storage tanks and day tanks in a configuration that allows the day tank overflow and day tank drain to flow by gravity back to the bulk storage tanks. By doing this, commissioning and periodic testing of the control system is simplified since fuel must be drained from the day tanks in order to adequately exercise and test the level control system. If the piping system does not allow for a gravity drain a pumped overflow will be necessary. Prior to selecting a pumped overflow, discuss the concept with the permitting agency.

Step 3 is to establish a control system. The control system that affords the best reliability utilizes separate control panels for each of the day tanks and pump controls. An important concept in the control system design is the ability to allow manual operation of the entire system in the event there is a catastrophic failure of the control system. This includes manual operation of pumps, bypass valves that will allow fuel flow around control valves, and manual tank gauging. Keeping in mind that a day tank may contain several hours, supply of fuel for the connected generator, it is possible to start a pump, manually open a bypass valve, and fill the day tank in just a few minutes. With the day tank now full, the facility operator can more easily assess the control problem and resolve it. Figure 6.3 shows a poorly arranged system. This type of arrangement depends on the control panel to control pumps and control valves. A failure of something as simple as the power feed to the control panel will disable the entire system.

Figure 6.4 describes a system using the same components, but arranging them into a configuration that allows the system to operate in a manual mode if the control panel fails. Additionally, the motor starters each have a diverse electrical feed which increases the reliability further. This system, however, still depends on a single control panel for logic for the entire system.

A further improvement would be to provide individual control panels for each of the day tanks with a separate panel to control the pumps. This will allow the day tanks to function individually, preventing a single point of failure at the fuel system control panel. With this system, if the pump controls fail, a pump could be started manually and the day tanks will continue to cycle with their independent control panels. Communication between all panels can either be hard-wired or via a communication bus. See Figure 6.5.

Figure 6.6 further increases reliability by adding redundancy for critical day tank components. The first location is a redundant control valve. Control valves are notoriously prone to failure. Providing two valves in parallel piping circuits increases reliability. Motor-actuated ball valves are less troublesome than solenoid

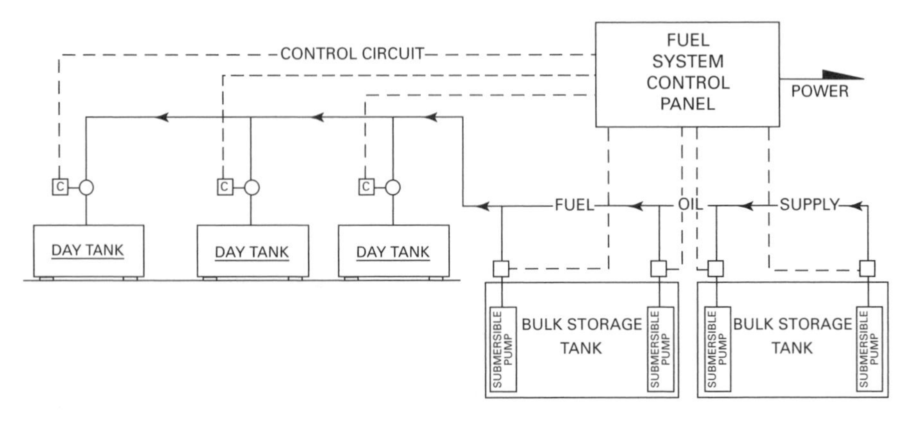

Figure 6.3 Poorly arranged system. (Courtesy of Mission Critical Fuel Systems.)

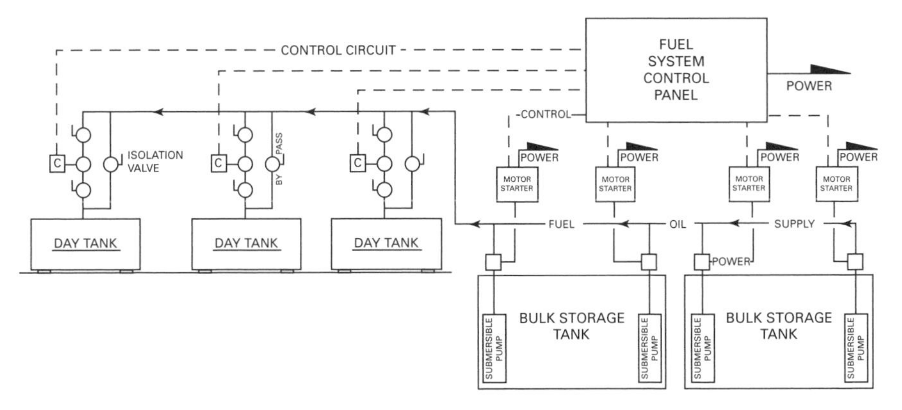

Figure 6.4 System using the same components as in Figure 6.3, but arranging them into a configuration that allows the system to operate in a manual mode if the control panel fails. (Courtesy of Mission Critical Fuel Systems.)

valves, but come at a higher cost. Control valves should fail closed to prevent day tank overflow during power failure. The control valves should also have separate power circuits with separate fuse protection. A common single point of failure is to provide two control valves, both of which are wired to the same circuit in the control panel. A short circuit in one valve actuator will disable both control valves, if they are being fed from the same circuit.

When redundant devices are incorporated into the controls scheme, it is of utmost importance to provide continuous monitoring for each device. Getting back to the control valve scenario, if there are two control valves in parallel fluid circuits, as shown in Figure 6.6, a failure of one of the valves would not be apparent without valve monitoring. The day tank would appear to function normally with diesel fuel flowing through the operating valve. The first indication of a problem appears when

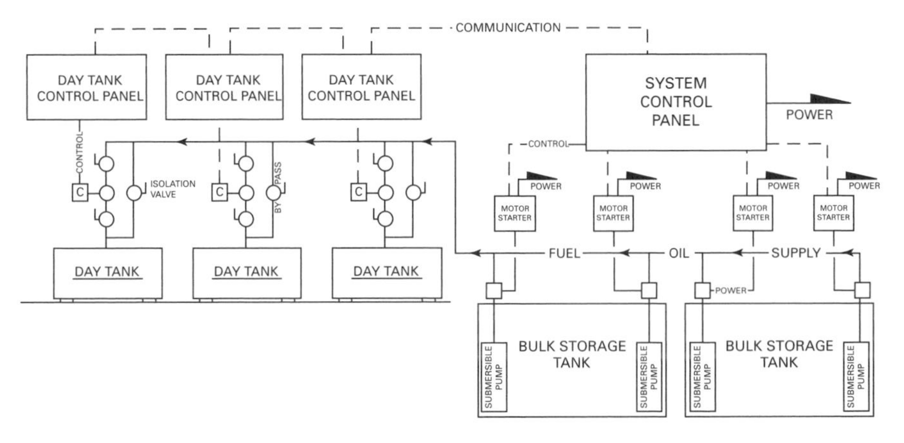

Figure 6.5 System using individual components to prevent a single point of failure. (Courtesy of Mission Critical Fuel Systems.)

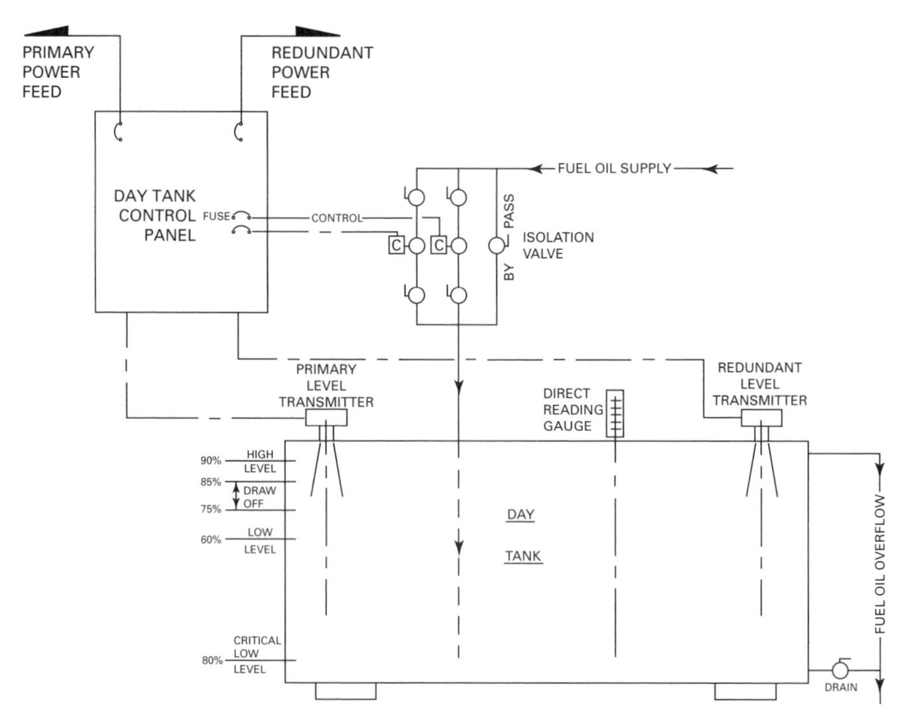

Figure 6.6 System that further increases reliability by adding redundancy for critical day tank components. (Courtesy of Mission Critical Fuel Systems.)

the redundant valve also fails. Monitoring devices such as actuator end switches or flow switches should be used to monitor the devices for failure.

The last item is selection of day-tank-level control devices. Float-type controls are the most common device used for day tank control. If this type of device is used, the floats should be either stainless steel or another high-quality material compatible with diesel fuel. Avoid the use of copper or copper-containing alloys for reasons stated earlier in this chapter. There is an increased popularity with non-contact-level measuring devices such as ultrasonic transmitters. Due to the decline in diesel fuel quality over the past few decades, gummy substances may begin to form on float controllers, rendering them inoperative. The noncontact devices are less likely to be affected by this condition.

Fuel Quality Assurance Program

The basis for every successful fuel quality assurance program is a clear understanding of the program objective(s) coupled with a reasonable approach to achieve them. In the case of standby diesel generators the program objective is fairly clear: Ensure 100% operability and reliability of the fuel when called upon for service.

In most diesel engine applications, the fuel that is specified for use is a Grade No. 2-D diesel fuel; in rare circumstances, local environmental regulations may require the use of a Grade Low Sulfur No. 2-D instead. Typically, diesel fuel is manufactured, released to market, and consumed within a relatively short period of time (e.g., < 3 months). During the period of time between manufacture and consumption, the diesel fuel typically maintains its fuel properties and thus is available for use. However, for standby diesel generator systems, this is not the case. Because a standby diesel generator is operated much less frequently and even fewer hours, the fuel tends to remain in inventory for a much longer period of time, thus becoming increasingly susceptible to instability and microbial contamination and ultimately causing system failure.

There are several issues that have the potential to impact the quality and longevity of the diesel fuel. As seen in Table 6.1, something as simple as receiving the wrong product can create serious problems with the existing fuel inventory. Secondly, not all Grade No. 2-D diesel fuels are the same. A Grade No. 2-D fuel produced at two different refineries may not be compatible with each other in the same fuel storage system, although both comply with ASTM D975, Standard Specification of Fuel Oils. Lastly, long-term storage conditions, outside contaminants, inherent fuel deficiencies, and system design flaws can contribute to the degradation of fuel quality over time.

Fortunately, a good fuel quality assurance program has the quality checks at each stage of the process that helps to ensure the long-term availability and use of the fuel. The basic quality check points include: fuel procurement program, pre-receipt analysis, post receipt analysis, routine surveillance, remediation, regulatory compliance, and record keeping.

Fuel Procurement Specifications

In order to ensure that a shipment of diesel fuel complies with the fuel quality requirements at the time of delivery, the procurement process must ensure that the fuel is purchased in accordance with industry standard fuel specifications, ensure

Table 6.1 The Elements of a Fuel Quality Test Program

Contamination Product	Off-Spec Property	Test Method
Gasoline	Flash point	ASTM D93
Jet fuel (Jet A) or kerosene	Flash Point	ASTM D93
Jet fuel (JP-4)	Viscosity	ASTM D445
No. 1 fuel	Carbon residue	ASTM D4176
Fuel oil (residual)	Water and sediment	ASTM D4176
Biodiesel	Flash point, free and total glycerin	ASTM D6751
Other products	Water and sediment	ASTM D4176

that the fuel vendor provide a certificate of conformance for the delivered product, allow the purchaser to validate the fuel shipment prior to off-loading product into the main storage tank, and allow the buyer the right to reject the fuel shipment without penalty in the event the product shipment fails to comply with any of the above mentioned quality checks.

The fuel procurement process is the first step in the fuel quality assurance program. Fuel specifications for five grades of diesel fuel are detailed in ASTM D975, Standard Specifications for Fuel Oils, including both low sulfur and regular No. 2 diesel. A Certificate of Conformance should accompany the shipment and be given to the owner/operator or designee at the time of delivery. The Certificate of Conformance should provide sufficient information that the fuel delivery conforms with the purchase specifications. In addition, the procurement specification must also allow for quality checks of the product compartments of the transport vehicle and pre-receipt analysis of the fuel from the transport truck in order to qualify the product for delivery. If possible, the fuel should contain one of the additive combinations described by the U.S. military specification MIL-S 53021, consisting of an approved multifunctional stability additive and an approved biocide. These additives have been approved for use by the United States Environmental Protection Agency (USEPA). If these additives are not added by the supplier, then they should be added by the owner/operator after the pre-receipt analysis when the fuel is unloaded into the storage tank. The elements of a proposed fuel quality assurance program are outlined in Table 6.2.

Fuel Delivery

The process of receiving a fuel shipment is where the product can be evaluated and accepted or rejected before it has the opportunity to contaminate the existing bulk fuel inventory. It is at the fuel delivery step where the owner/operator of the standby diesel generator has the opportunity to review the Certificate of Conformance to ensure compliance with the purchase specifications, visually inspect the fuel compartments and fuel while still in the transport truck, and obtain a sample of the diesel fuel for pre-receipt analysis prior to the transfer of the fuel shipment into the main fuel storage system. In addition, during the inspection step the owner/operator can also obtain a sample for a fuel quality analysis by ASTM D975.

This step is necessary to guard against delivery of the wrong product or accepting a diesel fuel shipment that has been cross-contaminated by another fuel type somewhere in the distribution system. In today's market, transport trucks are rarely dedicated to the carriage of a single fuel product; therefore, the probability of receiving the wrong product or contaminated product is possible.

Pre-Receipt Inspection and Analysis

Pre-receipt inspection & analysis of the diesel fuel shipment should be done to evaluate the diesel fuel shipment to confirm that the product conforms to the specifications of the procurement contract and ADTM D975 for grade of diesel fuel

Table 6.2 Fuel Quality Assurance Program

Description	Test/Requirement	Specification
	Procurement Process	
Diesel fuel compliance with fuel grade specified	All tests	ASTM D975
Allow for pre-receipt analyses and inspections	See below	
	Pre-Receipt Analysis and Inspections	
Pre-receipt inspection of transport truck prior to off-loading	Visual inspection of fuel compartments for dirt and debris	ASTM D4176
Obtain samples for pre-receipt analysis and post-receipt analysis	Review all test requirements to determine the volume of fuel samples required for all testing	ASTM D4057
Pre-receipt analysis:		
Relative density	Relative density	ASTM D1298
Kinematic viscosity	Kinematic viscosity	ASTM D445
Flash point	Flash point	ASTM D93
Water and sediment	Water and sediment	ASTM D2709
	Post Receipt Fuel Quality Confirmation Analysis	
Pre-receipt fuel sample	All tests	See ASTM D975
Flash point		
Water and sediment		
Distillation temperature		
Kinematic viscosity		
Ash % mass		
Copper strip corrosion		
Cetane number,		
Cloud point		
Ramsbottom carbon		
	Within 24 Hours of Delivery	
Check storage tank sump for sediment and water by the Clear and Bright test, ASTM D2198. Sampling and flushing should continue until samples pass.		
	Quarterly Surveillance	
Fuel sampling (manual or automatic)		ASTM D4057 or 4177
Fuel grade conformance		ASTM D975
Fuel stability		ASTM D2274
Particulates		ASTM 2276
Microbial contamination		ASTM Manual 47

desired. The tests that should be performed prior to off-loading of the shipment are indicated in Table 6.2. Fuel deliveries not meeting the specifications listed above should be rejected for cause. All observations, conversations, and analysis results should be documented and maintained on file for future reference.

24 Hours After Delivery

Within 24 hours of receipt of the diesel fuel shipment the sump of the storage tank should be checked for sediment and water. The sample taken from the sump should indicate clear and bright as per ASTM D2198. If the sample has positive indications for water and sediment, continue to flush product from the sump until the samples pass by clear and bright test. All observations, conversations, and analysis results should be documented and maintained on file for future reference.

Post Receipt Fuel Quality Confirmation Analysis

Within 30 days of the fuel delivery, complete the analysis of sample obtained during pre-receipt inspection and analysis with ASTM D975. All observations, conversations, and analysis results should be documented and maintained on file for future reference.

Quarterly Analysis

At least quarterly, fuel samples should be obtained from the main storage tank and any downstream day and/or skid tanks, to analyze and monitor the effects of long-term storage to the on-going quality of the existing bulk fuel inventory. The test data will provide the owner/operator the opportunity review data trends for indication that corrective actions or remediation are needed. All observations, conversations, and analysis results should be documented and maintained on file for future reference.

ASTM D975, Standard Specification for Fuel Oils, monitors the quality of the fuel stored in the main storage tank to ensure that conditions in the main storage tank have not changed.

ASTM D2274, Test Method for the Oxidation Stability of Distillate Fuel Oil (Accelerated Method), is a predictive test that evaluates the inherent instability of the fuel under specified oxidizing conditions. This test can be very helpful in predicting the propensity of the fuel toward instability but will not predict when the fuel will become unstable. Problems can arise when the fuel inventory is comprised of a number of fuel shipments from various crude oil sources and refining processes that create instability and compatibility issues.

ASTM D2276, Test Method for Particulate Contamination in Aviation Fuel, is a real-time test that analyzes the fuel for the formation of particulates. Over time, these particulates could achieve a sufficient diameter and number to cause potential filter plugging issues and terminate fuel flow to the diesel engine.

Microbial Analysis, ASTM Manual 47, provides the owner/operator with an introduction to fuel microbiology, sampling methods for detecting microbial contamination in fuel tank and systems, remediation techniques, and standards to guard against fuel degradation, filter plugging, fuel storage system corrosion, and physical and chemical changes to the fuel inventory.

Remediation

Fuel that indicates microbial contamination, shows a trend toward instability, or is unstable needs to be corrected. There are several actions that can be taken by the owner/operator of the system to address these issues. For example, as mentioned under "Fuel Procurement Specifications," it is best if the purchased fuel contains the additive combinations described by the U.S. military specification MIL-S 53021; however, the owner/operator can add these packages to the fuel shipment at the time of each delivery. If used as directed, the fuel has a much better chance of resisting the problems associated with long-term storage conditions. If analytical data are reviewed for adverse trends, then the system can be treated proactively before the system is out of control and restored to service without impacting the availability or operability of the system.

In other cases the problem may accelerate rapidly and without warning and result in a catastrophic situation. In these situations the problem(s) may involve a tank cleaning and conditioning of the fuel by a reputable company. Afterward, the fuel inventory is treated with the stabilizer additive and microbicide as discussed above.

Regulatory Compliance

When considering the use of fuel stabilizers and microbicides, the user must take into consideration the regulatory status of the products. The first consideration of the owner/operator should be the regulatory approval status of the intended biocide, registered uses, and use sites. If the owner/operator is uncertain of the regulatory status of the biocide, they are encouraged to contact their local or state environmental agencies for assistance or visit the USEPA website at http://www.epa.gov/pesticides to access the consumer hotline, contact the National Pesticide Information Center (NPIC) at 1-800-858-7378 to obtain information regarding product registration and site use(s). To make this even easier, you should contract a reputable supplier and they will take care of this for you.

Chapter 7

Automatic Transfer Switch Technology, Application, and Maintenance

7.1 INTRODUCTION

An automatic transfer switch (ATS) typically controls the engine start and transfers a load automatically to an alternate source of power, in case of normal power failure. The ATS is one of the basic building blocks of a backup power system, and performs a key role in power reliability. Although a standby generator typically includes internal engine starting and operating controls, the actual control of timing of engine starting, load transfer, and shut down is typically external to the generator and is located in one or more ATSs or in separate generator control switchgear.

In any operation where continuous electrical power is required, an interruption in the flow of power could result in significant losses unless backup power comes online quickly. A prolonged power outage could disastrously effect numerous aspects of the operation, from data center critical loads to emergency systems, including life safety. The ATS quickly and automatically transfers power to a generator or other power source, eliminating the need to manually switch from utility service to backup. When power is restored, the ATS automatically transfers back to the utility service or normal power source, after an adjustable time delay to allow for utility stabilization. For automatic transfer switches to work automatically, quickly, and dependably, they must be properly selected, sized and installed and, of course, properly maintained.

The ATS is typically applied where two sources of power are available to support critical infrastructure. However, different types of transfer switches are available with various options. Selection of a new or replacement ATS may seem

Maintaining Mission Critical Systems in a 24/7 Environment By Peter M. Curtis

to be a minor purchase, but it requires a thorough understanding of the facility where the switch will be installed and the technical attributes of ATSs. An evaluation of the facility's system helps determine the ATS specifications that satisfy the technical requirements and cost considerations for a particular location and application.

Providing safe maintenance and repair of an ATS requires either shutdown of both the normal and emergency power sources, or preferably, installation of a maintenance bypass isolation switch. The bypass switch isolates the ATS while maintaining power to critical infrastructure. Some ATS manufacturers provide an integrated ATS bypass switch. Switchgear configurations can also provide an ATS maintenance bypass isolation function, which often has advantages over an integral bypass switch. Some bypass switches allow for load transfer while on bypass; others do not.

An ATS is a mechanical load switching device. Therefore it cannot switch IT (computer) loads fast enough, even if switching between two live sources, to avoid a power interruption affecting the IT loads. For high-speed source switching applications, such as switching between two UPS systems, an electronic static transfer switch (STS) is required. Whereas a mechanical ATS switches within several cycles, an STS detects and switches in a half cycle or less.

The critical nature of emergency and standby power systems applications dictates the importance of rigorously maintaining the equipment involved. A good preventive maintenance program that includes operator training and maintenance and testing of ATSs and the larger integrated system will maximize system reliability.

7.2 OVERVIEW

Blackouts, such as the one that plunged the northeast United States and parts of Canada into darkness on August 14, 2003, drove home one very important aspect of emergency power systems: Emergency systems must be understood, tested, inspected, maintained, and documented. Facilities that followed regular maintenance programs had few, if any, problems when the power went out, while many of those who did not follow routine maintenance practices, had inaccurate drawings, insufficiently documented procedures, or had insufficiently trained operating staff had major problems, with some complete failures.

The National Electrical Manufacturers Association (NEMA) recognizes the need for preventive maintenance in their ICS 23: "A maintenance program and schedule should be established for each particular installation to assure minimum down time. The program should include periodic testing, tightening of connections, inspection for evidence of over-heating and excessive contact erosion, removal of dust and dirt, and replacement of contacts when required."

Additionally, the National Fire Protection Association (NFPA) 7084 describes electrical preventive maintenance as "the practice of conducting routine inspections tests, and the servicing of electrical equipment so that impending troubles can be detected and reduced, or eliminated." It also states, "The purpose of this recommended practice is to reduce hazard to life and property that can result from failure or malfunction of industrial-type electrical systems and equipment."

NFPA-76A6 requires that an emergency power transfer system be tested for at least 30 minutes at least every 30 days; although, 30 minutes per week is ideal. While many standards stress the importance of a comprehensive maintenance program, few provide sufficient detail. The original equipment manufacturer (OEM) is your best source for detailed recommended practice, formulated to maintain reliability, and protect warranty.

Automatic transfer switches are the nuclei of the emergency system; it is up to the ATS to continuously monitor the normal power supply then transfer to backup during a power failure, and back to normal once power is restored. If the ATS does not function properly, loads could be left high and dry, no matter how sophisticated the rest of the system may be.

A preventive maintenance program should be comprehensive enough to reduce the likelihood of experiencing a major malfunction in the emergency power distribution system.

7.3 TRANSFER SWITCH TECHNOLOGY AND APPLICATIONS

Automatic transfer switch ratings range from 30 to 4000 A, and primarily serve emergency and standby power generation systems rated 600 V or less. When the ATS senses a significant drop in voltage (typically 80% of nominal voltage), the ATS begins the process, often after a few seconds intentional time delay, to signal starting a standby power source. Once the standby source is ready to accept load the ATS transfers the load to the alternate power source. Where a generator set supplies backup power, the ATS control will

1. Signal the generator set to start after an adjustable time delay to prevent unnecessary engine operation for momentary power outages.
2. Monitor generator voltage until it reaches an acceptable voltage and frequency level.
3. When the predetermined voltage requirements are satisfied, the ATS disconnects the load from the normal power source and connects it to the alternate source (open transition ATS) or momentarily connects the normal and alternate sources together to transfer load (closed transition ATS).
4. When the normal power supply is restored to a predetermined voltage level, and after an adjustable time delay, the load is transferred back to the normal power source.
5. Send a stop signal to the generator set after an adjustable time delay to allow the generator set to cool down properly.

This transfer and retransfer of the load are the two most basic functions of a transfer switch. Features specified for the ATS can address particular applications and expand or decrease the capabilities of the switch. The ATS application often drives the basic type of switch required. For example, healthcare applications typically require closed transition operation to reduce power disruptions during frequent testing. Data center applications, on the other hand typically require three pole, open transition switches as the bulk of the loads are balanced three-phase UPS and HVAC equipment that do not require neutral conductors and are designed to handle brief power interruptions during ATS testing. ATSs can also be applied for redundancy purposes, to switch between two utility sources, between two generator sources, or for small loads downstream of two larger ATSs. If an alternate source other than a diesel or turbine generator is to be used (most commonly an uninterruptible power supply), the manufacturer must understand the requirement at the time the ATS is ordered, in case special engineering is required for the application.

7.3.1 Types of Transfer Switches

7.3.1.1 Manual Transfer Switches

Loads can be transferred with a manual transfer switch, but an operator must be present. The manual transfer switch has a quick-make quick-break (if closed transition type) operating handle or manually initiated electrically operated controls. The handle or manual controls must be accessible from the exterior of the switch enclosure. The switch has two positions: normal and emergency. Setting the handle to the normal position causes the normal power source to be connected to the load. Setting the handle to the emergency position causes the emergency power source to be connected to the load. If the emergency source is a generator set, it must be started and an acceptable voltage and frequency must be established before transfer.

Remember: emergency systems as defined in article 517 and article 700 of the *National Electrical Code* and NFPA 99 must automatically supply alternate power to equipment that is vital to the protection of human life. Electrical power must be automatically restored within 10 seconds of power interruption. Manual transfer switches will not meet these code requirements and automatic transfer switches are required.

7.3.1.2 Automatic Transfer Switches

These switches are available in 30 to 4000 amperes for low-voltage (< 600 VAC) applications (see Figure 7.1.) Different types of transfer switches are available with various options, including bypass-isolation switches. ATSs equipped with bypass isolation allow the automatic portion of the ATS to be inspected, tested, and maintained (with some designs the automatic portion can be drawn out and serviced or replaced) without any interruption of power to the load.

Open transition switch An OT switch briefly interrupts load during transition between sources; typically 3 to 5 cycles. Some ATS manufacturer's products

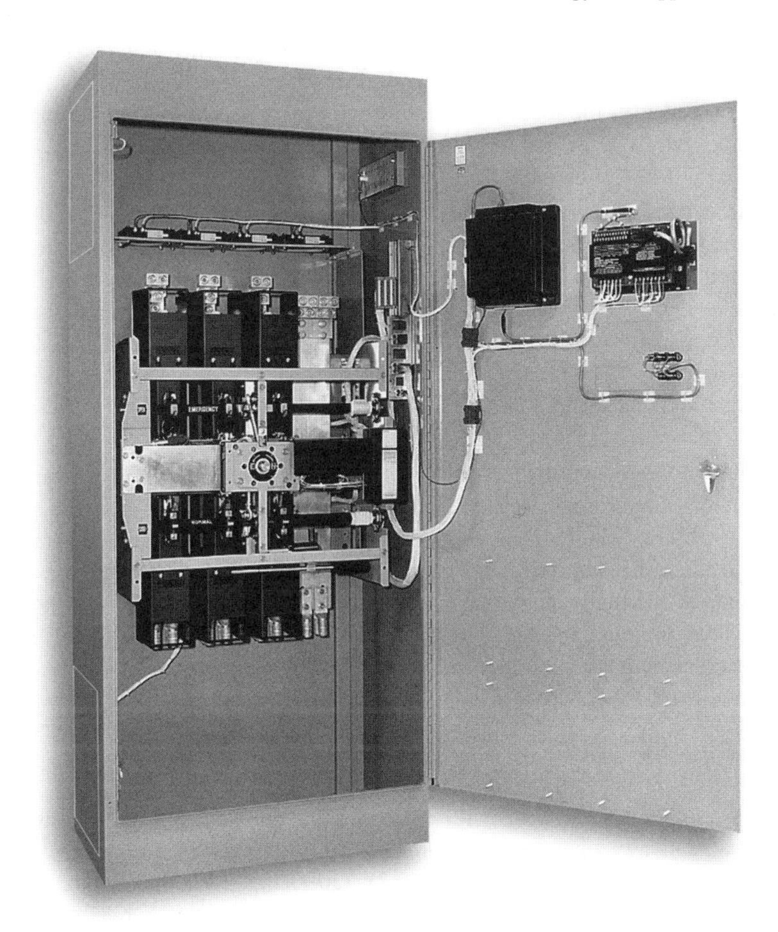

Figure 7.1 Automatic transfer switch.

have safety interlocks to prevent improper operation. Basic switches may retransfer loads quickly between two live sources that are not synchronized. For many applications such as UPS systems and inductive motor loads, synchronizing the sources, or at least using synch check monitors (in-phase monitors) allow for a smooth transfer of sources that are in synch. OT switches are typically less expensive and can be more reliable for certain applications such as data centers.

Closed transition switch Also known as a nonload break design, the CT switch transitions between sources without load interruption. This switch utilizes a make-before-break operation that permits transferring without load interruption. This switch is ideal for loads that cannot tolerate any loss of power, such as healthcare applications not supported by UPS systems. CT switches are typically more expensive as they need to safely connect the normal utility and standby generator sources together. In the event of problems, the CT switch carries a higher risk of upstream circuit breaker nuisance tripping or otherwise disconnecting all sources of power to the load. Power utility company requirements will need to be understood and followed for the safety of utility service personnel.

Delayed transition switch Delayed transition switches are designed to provide open transition transfer of loads between power sources with an intentional disconnection of the load for an adjustable period of time to allow any stored energy in loads to dissipate. The switch has two operators with a programmable delay or open time between the transfer from "normal" to "emergency." Applications include large inductive motor loads, variable frequency drives, UPS systems and specialized medical equipment.

Soft load closed transition switch Soft-load closed transition switches provide a make-before-break transfer of a building load from a utility source to an alternate in-house standby generator. This system brings the standby generator into sync with the utility source, and then gradually shifts the load over to the standby generator with virtually no voltage or frequency fluctuations. This is especially useful where the engine generator is of marginal size or uses soft fuel, such as natural gas or methane.

7.3.2 Bypass-Isolation Transfer Switches

A bypass-isolation switch, also referred to as a maintenance bypass switch, is actually a manually operated (nonautomatic) transfer switch in parallel with an electrically actuated ATS (see Figure 7.2). Operating the manual bypass switch isolates the ATS for periodic maintenance, repair, or testing while maintaining power to the load from either power source. A bypass-isolation switch allows the automatic portion of the ATS to be temporarily removed for repair without physically disconnecting power conductors.

Bypass-isolation switches are vital in critical circuits of a backup power system, found in data centers, hospitals, communication facilities, and airports. These facilities have used bypass-isolation switches or their equivalent for years to maintain continuity of power during corrective maintenance, and to permit routine preventive maintenance. More recent applications often use standard, nonbypass-isolation ATSs with an external MTS or external switchgear for maintenance isolation. Other applications include 100%, dual path redundancy of the switchgear, such that one ATS provides the maintenance bypass for the other ATS.

Because the bypass switch must carry the same load as the ATS, it must be rated for the same load and have the same withstand and interrupting rating. If this switch is not sized correctly for its particular power distribution system, facilities with a 24/7 operation could experience unnecessary outages during corrective maintenance or may have to forego recommended preventive maintenance if power cannot be interrupted.

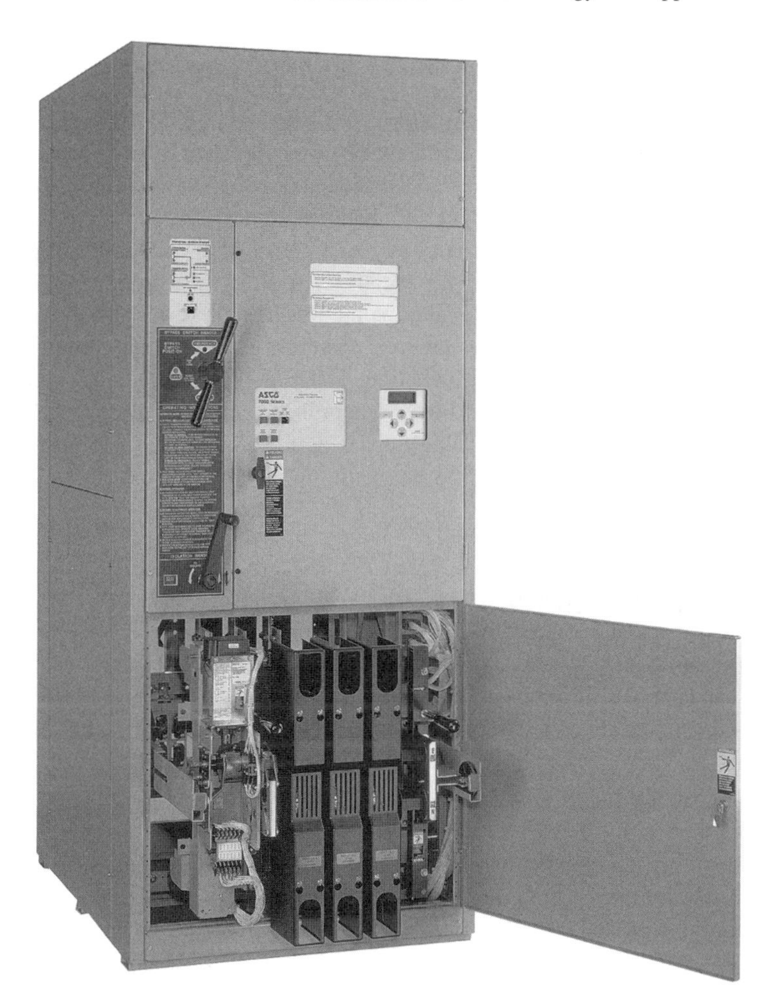

Figure 7.2 Automatic transfer and bypass-isolation switch.

Because the goal is to bypass safely without interruption to the load, the bypass portion must be wired in parallel to the automatic portion. Only when manual bypass is activated can the ATS be placed in either the "test" or "isolate" position. In the test position, qualified personnel may electrically test this portion without affecting the load, as the main poles are disconnected. In the isolate position, the automatic portion may be totally removed for repair or replacement. While the automatic portion is in the test or isolate position, the manual bypass mechanism can transfer the load (manually actuated) to either utility power or backup. This procedure would be identical on all other types of bypass-isolation switches.

Bypass-isolation switches come as complete assemblies with either open transition (OT) or closed transition (CT) switches.

7.3.3 Breaker Pair ATSs

Most standard, cataloged ATS designs are essentially large contactors. The contactor ATS design carries the advantages for most applications of lower cost and physical size as well as simplicity and the resulting higher reliability. However, there are some standard ATS designs and many customized ATS designs that use a pair of automatic or nonautomatic circuit breakers instead of contactors. The circuit breakers are motor operated and are typically controlled by a programmable logic controller (PLC). Advantages of breaker pair ATSs include better adjustment capability of timing and overlapping sequences, easier to get high fault current ratings, and more reliable for large load switching (4000 A). Disadvantages of breaker pair ATSs include complexity and lower reliability of motor operators, PLCs, programming issues, etc. Except for specialized applications, it is recommended to avoid customized breaker pair ATSs as operation and service can be an ongoing challenge.

7.4 CONTROL DEVICES

Most major ATS manufacturers have various options for their products. The options selected depend on cost, different conditions and load requirements.

7.4.1 Time Delays

Time delays are crucial in many ATS applications, and can range from a few seconds to 30 minutes. Often included as standard on ATSs, time delays allow for power stabilization and coordination of equipment.

7.4.1.1 Start Time Delay

Start time delays can override a monitored deviation in the power supply. They prevent the starting of an engine or transfer of load during momentary voltage dips. Timing starts at the instant of normal power interruption or dip. If the duration of the interruption exceeds the time delay, the ATS signals the generator set to start. Start time delay should not be set greater than a few seconds for emergency power because the NEC requires automatic restoration of electrical power within 10 seconds of power interruption, and you need to allow several seconds for the engine to start and come up to speed in addition to the starting time delay. Delays should be long enough to prevent a response to voltage sags caused by nearby faults or a momentary loss of voltage from the operation of reclosers in the electrical distribution system. However, in specialized applications such as UPS systems with limited energy storage backup time, especially in the case of batteryless flywheel systems, engine start delays need to be minimal, as low as 1 second or less.

7.4.1.2 Transfer Time Delay

Transfer time delays allow adequate time for stabilizing of the generator set at the rated voltage and frequency before the load is applied. The range for adjustment is 0 to 5 minutes. *Retransfer time delay* allows time to stabilize the normal power source before returning the load. This also allows the generator set to operate under load for a minimum time and prevents multiple transfers back and forth during temporary power recoveries. The range for adjustment is 0 to 30 minutes. Some facility operators prefer to turn the ATS to a manual or "test" position once the facility is on generator following a power outage. The philosophy here is that if utility power fails once, it may fail again soon after returning. Or simply recognizing that there is a small risk with the actual retransfer of power back to normal. Then they will run on generator until the storm has passed, for the rest of the day until the bulk of the work shift is over, etc., then place the ATS back into "auto" and observe time out and retransfer of the standby system back to normal.

7.4.1.3 Engine Warm-Up Time Delay

Engine warm-up time delays allow time for automatic warm-up of engines prior to loading during testing. This feature is not available from all ATS vendors. Frequent loading of cold engines (including those with jacket water heaters) may increase wear and thermal stresses. Because standby engines normally accumulate a relatively small number of operating hours, the added wear and stress may not be of concern, especially for small and medium-sized engines.

7.4.1.4 Stop-Time Delay

Stop-time delays allow the generator set to cool down at no-load conditions before stopping, and should be specified on all ATSs associated with diesel generators. This feature minimizes the possibility of a trip and in some cases an engine lockout due to an increase of the cooling-water temperature after shutdown of a hot engine. The delay logic does permit an immediate transfer of load back to the generator set if normal power source is lost during the cool down period. ATSs with stop time delay also conform to vendor recommendations for operation of larger engines. The range for adjustment is between 0 to 10 minutes.

7.4.2 In-Phase Monitor

This device monitors the relative voltage and phase angle between the normal and alternate power sources. It permits a transfer from normal to alternate and from alternate to normal only when values of voltage and phase angle difference are acceptable. The advantage of the in-phase transfer is that connected motors continue to operate with little disturbance to the electrical system or to the process.

An in-phase monitor prevents residual voltage of the motors from causing an out-of-phase condition, which can cause serious damage to the motors or driven loads. Also, transferring power to a source that is out of phase can result in an excessive current transient, which may trip the overcurrent protective device or blow fuses. In-phase transfer is principally suited for low-slip motors driving high inertia constant loads on secondary distribution systems. In-phase monitor transfer is also suitable for most UPS applications. In-phase monitors are typically inexpensive compared to closed-transition ATSs or delayed-transition ATSs.

Some vendors provide different in-phase monitor modules, depending on whether the alternate power source (i.e., diesel generator) has a governor, which operates in isochronous or droop modes. Consider this option for all ATS applications that will supply motor loads and will not tolerate momentary interruptions. The cost for this option is relatively low and, in some switch designs, the cost of a later retrofit may be quite high.

In-phase monitors are passive devices and become permissive only when the two sources coincidently align in the allowable bandwidth and phase displacement. It is important that the generator be adjusted to provide an offset in frequency from the mains and thereby create slip between the two sources. Typically, this frequency offset occurs without effort, however, if the generator is operating at exactly the same frequency as the normal source (but not synchronized) it can become "stuck" out of phase for an undesirably long period of time during the retransfer operation. Alternatively, a synchronizer may be employed to "drive" the emergency generator into phase if so desired. This has to be engineered by the manufacturer at the time of order.

7.4.3 Programmed (Delayed) Transition

Programmed (delayed) transition extends the operation time of the transfer switch mechanism. A field-adjustable setting can add an appropriate time delay that will briefly stop the switch in a neutral position before transferring to the alternate energized source. In forward transfer from the failed normal source to the standby source, enough delay is typically inherently present such that additional programmed delay is not needed. The programmed delay occurs primarily during the retransfer operation, when the ATS must transfer between two live sources (Figure 7.3). This allows motors to coast down and the transformer fields to collapse before connection to the oncoming source. This also reduces the possibility of circuit breakers tripping and fuses blowing when transferring large inductive loads. Programmed delayed transition is typically more expensive than fast transition with an in-phase monitor, however in some ways delayed transition is more reliable. This device is another design option that can be used with UPS applications as described in the preceding section. The delayed transition feature avoids the need for control cables and devices to be tripped (which are required by a motor load disconnect control device). However, appropriate control circuitry or sequencing equipment will be required for motors to restart. Delayed transition or any other UPS control scheme that results in additional transfers of the UPS to battery operation, or prolonged battery operation (even for a few more seconds) can be seen as decreasing reliability. Battery operation remains as one of the highest risk of failure for UPS systems.

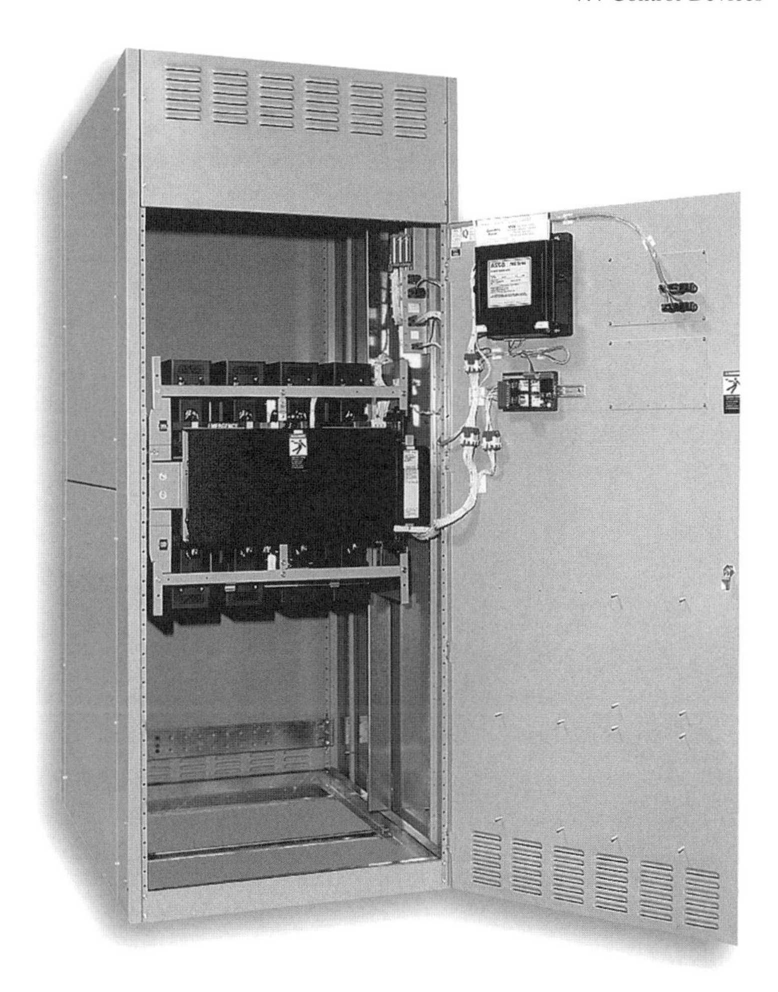

Figure 7.3 Delayed transition ATS.

7.4.4 Closed Transition Transfer (Parallel Transfer)

Closed transition transfer instantaneously parallels the two power sources during transfer from either direction. It closes to one source before opening to another. In this application, additional devices must protect the power sources during transfer, when both sources are briefly connected together in parallel. An in-phase monitor or an active synchronizer synchronizes voltage, frequency, and phase angle. Without proper synchronization, closed transition transfer can damage both power supplies, and may pose a safety concern for personnel. Power utility company requirements need to be understood and followed for the safety of utility service personnel. Applications for closed transition transfers include transferring the system in anticipation of a potential power outage, frequent testing (especially for health care facilities) load switching for electronic loads, motors, and load-curtailment systems. More advanced closed transition ATS applications include distributed generation, where the ATS remains closed on both sources for a period of time to allow selling standby power back to the utility during peak needs. Additional load sharing controls and power utility company interface (and approval) is needed for distributed generation applications. (see Figures 7.4 and 7.5.)

7.4.5 Test Switches

Regular testing is necessary to both maintain the backup power system and to verify its reliability. Regular testing can uncover problems that could cause a malfunction during an actual power failure. A manually initiated test switch simulates failure of the normal power source and will start the transfer of the load to the alternate

Figure 7.4 Closed transition ATS.

Figure 7.5 Soft-load closed transition ATS.

power source, and then retransfer either when the test switch is returned to normal or after a selected time delay. The test switch does not require shutting off the normal source. The test switch transfers the load between two live sources, avoiding the several seconds of power loss during time out and engine start. The advantage of a test switch is that it does not cause a total loss of power to connected loads as if the main disconnect is opened ("pull-the-plug test") to simulate a failure. Therefore, the test switch reduces risk of complications, including a normal source breaker failure to reclose. It also allows the system to be proven under controlled conditions. It is common for the connected load characteristics to change over time, either by planned expansion of the system or by inadvertent connection of new loads. A periodic controlled function test insures the performance of the system. (If the ATS is not closed transition, the load will be briefly affected when the ATS transfers to emergency and returns to normal.)

7.4.6 Exercise Clock

An exercise clock is a programmable time switch. It automatically initiates starting and stopping of the generator set for scheduled exercise periods. The transfer of load from the normal to alternate source is not necessary, but is highly recommended in most cases. As mentioned previously, the standby generator system should be exercised as often as 30 minutes per week, or at least monthly. Operating diesel engines repeatedly or for prolonged periods of time without load can cause wet stacking of unburned diesel fuel oil in the exhaust system. Wet stacking can lead to other diesel engine problems including explosions in the exhaust or muffler system. Many operating engineers prefer to transfer building load onto the engine once every four tests to reduce wet stacking. Many operating engineers either prefer to be present during each and every automatic test of the standby power system or they disable the automatic exerciser so that they can manually initiate the test.

7.4.7 Voltage and Frequency Sensing Controls

Voltage and frequency sensors, controls, and local and remote metering add additional useful features to an ATS. The voltage and frequency sensing controls usually have adjustable settings for "under" and "over" conditions. The meters offer precise readings of voltage and frequency, and the sensing controls provide flexibility in various applications. Enhanced ATS metering functions include load metering. This adds cost as current transformers (CTs) need to be added to the power conductors. However it is beneficial for the operating engineer to be able to see and monitor the load on the ATS at any given time, including when operating on the normal source. If load metering is included in the ATS then it might not be necessary in downstream switchboards. Local metering can be captured remotely through a wide range of options, the most common using modbus. Remote metering (especially load metering) is important for building automation systems and load shed schemes.

7.5 OPTIONAL ACCESSORIES AND FEATURES

To make inspection easy and safe, additional features are desirable.

ATS switches rated 100 A or larger are typically mechanically held in both the normal and emergency positions. In such switches, the main contact pressure relies on the mechanical locking mechanism of the main operator. Automatic operation of these switches in either direction requires electrical control power. Therefore, the main poles of the transfer switch ought to be the type that are held closed mechanically and energized even when the control circuit is de-energized. In this type of design, the load remains energized while the sensing and control circuitry is de-energized for inspection.

Sensing and control circuitry should be located on a panel detached from the main transfer switch panel. Compact and lightweight sensing and control panels permit mounting on an enclosed door that swings out and away from the energized main poles of the ATS. A harness containing a means of disconnect allows the sensing and control panels to be completely de-energized. Then, the control relay contacts, time delays, and pick-up and dropout values of the sensing circuits can be inspected. With the harness disconnected, the ATS is incapable of automatically responding to a power outage. Control circuitry will typically include integral a small battery backup. The battery must be regularly serviced (replaced) for reliable ATS operation. Therefore, emergency procedures must be written and personnel must be ready to respond if an emergency should occur. Being prepared, personnel can save valuable time in manually transferring the loads to the suitable emergency source.

7.6 ATS REQUIRED CAPABILITIES

The ATS is typically located at the main or secondary distribution bus that feeds the branch circuits. The specification of the switch should include several important design considerations.

7.6.1 Close Against High In-Rush Currents

When a transfer switch closes on an alternate source, the contacts may be required to handle a substantial in-rush of current. The amount of current depends on the loads. In the case of a motor load, inrush current can be six to eight times the normal current. If the two power sources are not synchronized, or the residual voltage of the motor is (worst case) 180° out of phase with the source, the motor may draw as much as 15 times the normal running current.

A minimum contact bounce should occur during switching. The contact material must have adequate thermal capacity so that melting will not occur as a result of high in-rush current. UL requires that a transfer switch be capable of withstanding in-rush current 20 times full load rating.

7.6.2 Withstand and Closing Rating (WCR)

When a short circuit occurs in a power system, a surge of current will flow through the system to the fault until the fault clears by a protective device opening. If the device supplying the transfer switch opens, the transfer switch will sense a loss of power and initiate transfer to an alternate source. On closing to the alternate source, the transfer switch closes a source into the fault. The entire available fault current from that source may flow through the switch. Devices designed to maintain the fault current have a withstand and closing rating, or WCR. Transfer switches should be able to not only safely carry current under fault conditions, but also close into the fault and remain operational. UL sets a minimum acceptable WCR for transfer switches. The correct ATS WCR must be specified depending on the available fault current at the point of ATS application. Standard ATS WCR ratings available from manufacturers typically increase as the continuous ampere ratings increase. However, available fault currents are generally rising over time and ATSs are increasingly applied in smaller sizes in large electrical systems (typically with higher available fault current). Therefore it can be challenging to apply an ATS with a low continuous ampere rating and a high fault current rating. Breaker pair type ATSs, series rated breakers ahead of the ATS, currently limiting fuses, impedance increasing devices (transformers, reactors) and other design options are available.

7.6.3 Carry Full Rated Current Continuously

A transfer switch must continuously carry current to critical loads for a minimum life of 20 years. Perform complete temperature tests on the ATS to ensure that overheating will not occur. Periodic infra-red thermal scanning of ATS load-carrying components, when loaded from both the normal and emergency sources is recommended. The maximum continuous load current should be estimated in selecting a transfer switch. The selected transfer switch must be able to carry continuous load current equal to or greater than the calculated value and must be rated for the class of the connected load.

7.6.4 Interrupt Current

An arc is drawn when the contacts switch from one source to another under load. The duration of the arc is longer when either:

1. The voltage is higher and the power factor is lower or
2. The amperage is higher.

This arc must be properly extinguished prior to the contacts connecting to the other source. Arc interrupting devices such as arc splitters are often used. Wide contact gaps can also extinguish an arc. Unless no-load transfers are used exclusively, the ATS must be able to interrupt load current.

7.7 ADDITIONAL CHARACTERISTICS AND RATINGS OF ATSs

7.7.1 NEMA Classification

Transfer switches are classified as NEMA Type A or NEMA Type B. A Type A transfer switch will not provide the required overcurrent protection, so an overcurrent protection device should be installed ahead of the ATS to clear fault current. A Type B transfer switch is capable of providing the required overcurrent protection within the transfer switch itself.

7.7.2 System Voltage Ratings

The system voltage rating of transfer switches is normally 120, 208, 240, 480, or 600 VAC, single phase or polyphase. Standard frequencies are 50–60 hertz.

7.7.3 ATS Sizing

For most electrical and mechanical equipment, it is generally more efficient and cost effective to use fewer, larger sizes of equipment. However for ATS applications, the opposite is generally true. ATS cost, physical size and reliability concerns increase dramatically as the ampere size increases. Rather than using one large 2000 or 3000 A ATS, breaking the loads into smaller increments and using ATSs rated at 225–1200 A is often much better. Contactor type ATSs are arguably less feasible above 3000 A and breaker pair type designs or customized switchgear should be used if large ATS applications of 4000 A or larger are required.

7.7.4 Seismic Requirement

Some DOE sites have specific seismic requirements for certain facilities. Normally, ATS manufacturers do not have their ATS seismic qualified. However, many vendors specialize in seismic qualification, and will test an ATS at their facility.

7.8 INSTALLATION, MAINTENANCE, AND SAFETY

7.8.1 Installation Procedures

When the ATS arrives at the job site, preventive maintenance begins with the installer or contractor. It is his responsibility to ensure that the shipping carton or crate is delivered undamaged; notify the shipper immediately if it is. Look for handling instructions on the outside of the crate, so the switch can be moved and unpacked without damage to the contents. For instance, the carton may be top-heavy and will be labeled as such.

The equipment should be placed in a safe, dry area and partially unpacked for inspection and removal of the manufacturer's handling and installation instructions. If the equipment is to be stored for an extended period, protect it from dirt, water, and physical damage. Heat should be used to prevent condensation inside the equipment.

If a forklift is used after the equipment is unpacked, transport the unit in the upright position and take care not to damage any handles, knobs, or other protrusions. If the equipment is to be lifted, lifting plates should be fastened to the frame with hardware as recommended in the mounting instructions. Lifting cables or chains should be long enough so that the distance between the top of the equipment and the hoist is greater than one-half the longest horizontal dimension of the equipment.

The installer or contractor should closely follow the manufacturer's instructions. If there is any question on any part of the unpacking or installation procedure, contact the manufacturer, or manufacturer's representative. It is extremely important to protect the controls and transfer switch mechanism from metal shavings, drillings, and other debris during installation. Failure to do so may result in equipment failure and personal injury.

After the ATS is installed, it should be thoroughly cleaned and vacuumed, taking special care to remove any loose metal particles that may have collected during installation. Before energizing the circuits, the manually operate the switch in accordance with the manufacturer's recommendations. Have the manufacturer's representative perform the recommended start up procedure. This includes wiring, mechanical, and visual inspections before power is applied; adjustment of parameters; and finally testing after initial energization. The next important item, the phase rotation between the normal source and the emergency source should be checked and noted on the ATS. Additional load performance testing and integrated system commissioning may be required for high reliability ATS applications, in addition to and following field startup.

7.8.2 Maintenance Safety

Maintenance is most effective when it is preventative or predictive. However, when maintenance is reactive the cause for a condition must be determined and corrected. If equipment conditions show signs of deterioration, the equipment should be repaired or replaced with manufacturer-recommended renewal parts. Use the diagnostic information in a manufacturer's instruction manual, and follow instructions to the letter.

Observe safety precautions when performing maintenance on the ATS. Due to their design, many ATSs have exposed energized buses or operating mechanisms inside the cabinet door.

Hazardous voltages will be present in the ATS during inspection and testing. Wear proper Personal Protective Equipment (PPE) at all times when handling equipment for arc-flash safety. Review the NFPA 70E-2004 Section 130.7 for more information.

When transferring an energized ATS, workers should never stand in front of the ATS cabinet door. Instead, they should stand to the hinged side of the ATS cabinet door. AN electrical arc or blast can occur during switching if there is component failure. A strong fault can create a fireball and the pressure can swing the door open. This blast can potentially cause death or severe burns.

Maintenance personnel should also never stand in front of an open ATS cabinet door during switch operation. Most new ATS designs have control components mounted on the door, away from power-carrying conductors. These control devices can then be easily disconnected for troubleshooting. The use of recording devices may occasionally be necessary to perform measurements during troubleshooting of control circuits in older switches where control components are hard-wired and are adjacent to high power conductors and moving components.

Remember, an ATS is connected to two power sources. When performing work requiring a de-energized ATS, make sure that both power sources are disconnected, locked out, and tagged out. Always verify absence of voltage before starting work.

7.8.3 Maintenance

Exercising and performance testing should be performed per the manufacturer's recommended procedures and by a qualified testing and maintenance organization. Basic procedures include cleaning, meter calibration, contact inspection, load transfer testing, timer testing and infra-red thermal scanning under load.

7.8.4 Drawings and Manuals

Before attempting to test, inspect, or maintain an emergency power transfer system, at least one set of the manufacturer's drawings and manuals should be on hand. Keep these in a clean and readily accessible location. For convenience, additional copies should be readily accessible; they can usually be easily obtained from the manufacturer.

Accurate, as-built drawings and manuals are an indispensable aid in

- Understanding the sequence of operation of the equipment
- Establishing test and inspection procedures
- Determining the type and frequency of tests
- Troubleshooting
- Stocking spare parts
- Conforming to code requirements for periodic preventive maintenance

7.8.5 Testing and Training

Regular testing of an emergency power system and training of operators will help to uncover potential problems that could cause a malfunction or misoperation during an actual power failure. Furthermore, regular testing gives maintenance personnel an opportunity to become familiar with the equipment and observe the sequence of operation. By being familiar with the equipment, personnel on duty during an actual power failure can respond more quickly and correctly if a malfunction occurs.

An automatic emergency power system should be tested under conditions simulating a normal utility source failure—field surveys have proven this. Experience shows that the rate of discovery of potential problems when a system is tested automatically with the emergency generating equipment under load for at least one-half' hour, is almost twice the rate as when systems are tested by manually starting the generating equipment and letting it run unloaded.

A test switch on the transfer switch will accurately simulate a normal power failure, and some manufacturers supply this as standard equipment. The advantage of a test switch is that it does not cause a loss of power to connected loads, as is caused if the main disconnect is opened to simulate a failure. A test switch opens one of the sensing circuits and the controls react as if there were a normal power failure. In the meantime, the loads stay connected to the normal source. When replicating an actual failure, power is interrupted to the loads only when the load is being transferred from the normal to the emergency source and vice versa. This interruption is only for a few cycles.

However, if the main disconnect is opened to "simulate" loss of normal power (sometimes referred to as a "pull-the-plug" test), then power to the load could be interrupted for up to 10 seconds. The interruption could be much longer if a malfunction occurs during start-up of the emergency generator, or if there is a problem re-closing the normal feeder switch or circuit breaker. When a test switch is used, fewer maintenance personnel are needed because the potential for an extended power outage during the test is significantly reduced. However, using the test switch is not as thorough of a performance test.

Though it is not always recommended, the control circuits of an emergency power system can be designed so that a clock will automatically initiate a test of the system at a designated hour on a specific day for a predetermined test period. This type of test can be restricted to running the engine generator set unloaded or can include an actual transfer of loads to the generator output. Using a clock to initiate testing under load or no-load conditions is *not* recommended for the following reasons.

- With no load interruption, the engine generator set is tested unloaded, which can adversely affect engine performance. In addition, the transfer switch is not tested, and the interconnecting control wires are not proven. The transfer switch is a mechanical device that has been lubricated for life. However, to keep the lubrication effective, the switch should be periodically operated.
- There may be installations where automatic tests do not easily integrate into the operation of a building, and it may be difficult to stop the clock.
- A clock malfunction or improper setting could result in a test at a most inopportune time.
- Clocks are usually set to initiate a test on weekends or during early morning hours. If a malfunction occurs, skeleton crews are usually not equipped to handle it.

When a test switch is used to test an emergency power transfer system in the automatic mode, all relays, timing, and sensing circuits operate in accordance with the engineered sequence of operation. If the test procedure is written in accordance with this predictable sequence of operation, then precious steps and person-hours can be saved. It is possible for one person to initiate a test and then move to another location to initiate another action or monitor and record certain test results. The automatic sequence of operation should not vary greatly over time, so maintenance personnel movements can be correlated to that regular sequence.

Further, test procedures can include emergency procedures to cover any malfunction. This is added insurance. Key people should be trained to react almost instinctively when certain malfunctions occur. For example, in a multiple-engine standby generator system with automatic load-shed capabilities, an overspeed condition on one set may cause certain loads to be shed—retransferring them from the emergency source of power back to the normal. The test procedure should simulate such a condition and include steps to

1. Confirm that the loads were retransferred.
2. Reset the failed generator and put it back on the line.
3. Confirm that the load is again transferred to the emergency source of power.
4. Be aware of vulnerabilities of ATS load equipment including chillers and other HVAC equipment (especially those powered by variable frequency drives) and UPS equipment. UPS systems are always vulnerable during ATS transfer and re-transfer operations. Although UPS batteries may be designed and rated to carry the load for power interruptions of 15 minutes or more, numerous failure modes exist for the UPS to fail immediately or within seconds of power loss. Many facility engineers believe that live ATS load transfer testing provides a good test of UPS batteries, since the batteries carry the load for several seconds. This is not a reliable test of UPS batteries and connections. Although the UPS batteries may pass a test which requires operation for five or ten seconds, the batteries may not survive a test for fifteen seconds. Many UPS systems successfully survive the forward transfer to standby power during testing or an actual power outage, only to drop the load during retransfer to normal because a marginal component failed or the UPS batteries were on the edge of failure during the forward transfer.

Note that some systems frequently use this type of load-dump relay (LDR) on preselected transfer switches to dump some of the load back on the normal source when one or more generator fails on a multiple generator system. This arrangement permits the dumped loads to be fed by the normal source, when available and perhaps avoid major shutdowns of dumped loads during test periods.

A minimum checklist for an emergency power system test should note the following items, as a minimum:

- Loose control wires and cables
- The time from when the test switch is activated until the engine starts to crank

- The time required for the engine to achieve nominal voltage and frequency after the test switch is activated
- The time the transfer switch remains on the emergency generator after the test switch is released
- The time the generator system runs unloaded after the load circuits have been retransferred back to the normal source of power
- Proper shutdown of the engine generator set and engine auxiliary devices

7.9 GENERAL RECOMMENDATIONS

It is important for users to establish their practical needs for a given system. The selection process is easiest when the system requirements are pre-determined. Should a facility choose not to use an automatic transfer switch or to use only a manual transfer switch for its backup system, processes may be stopped due to the power interruption. Some processes cannot be restarted automatically due to their design, and may require extensive rework to restart. The costs of restoring a facility to normal conditions can exceed the cost of an automatic transfer switch.

On the other hand, some processes are continually staffed and can recover from short-term outages easily. In these cases, a transfer switch may not be required, or a less expensive manual transfer switch may be adequate.

When selecting an ATS and hardware, the following criteria are recommended:

1. Unsupervised transfer switches can transfer the load within 6 cycles. This might cause damage to or miss-operation of motors, variable speed drives or UPS operations if additional synchronized devices or controls are not used. Incorrect ATS application can result in unnecessary tripping of drives or UPSs. When the transfer is out of phase, the motors can experience severe mechanical and electrical stress, which can cause motor winding insulation to fail. Excessive torque may also cause damage to motor coupling or loads. To prevent this damage, protective devices may be actuated, but may cause loss of load.
2. Cost will increase as more hardware and options are added. The cost of wear and maintenance to other components such as diesel engines may justify the use of timing modules, which facilitate routine testing.
3. Maintain continuous operation of a UPS system by selecting a closed transition ATS or an open transition ATS with an in-phase monitor, motor load disconnect control devices or ATS with programmed (delayed) transition.
4. Take future expansion and load growth into account (and revisit the issue periodically).
5. Equipment design must consider maintenance requirements. Bypass-isolation switches are relatively high-cost options, but may be necessary to perform proper maintenance when loads cannot be de-energized. The control

circuit design should permit safe and thorough troubleshooting without de-energizing loads or requiring electricians to work in close proximity to high-power, high-voltage conductors.

6. For applications where downtime cannot be tolerated and electrical power to the load is critical, an external or integral bypass-isolation switch (nonload break design) used with an ATS is the best choice.

Chapter 8

The Static Transfer Switch

8.1 INTRODUCTION

The UPS and the static transfer switch (STS) are the alpha and omega of the power system in mission critical sites. The UPS and its accompanying backup power plant/ATS, as covered in previous chapters, is the first line of power quality defense. The UPS removes all power anomalies generated external to the site. The STS is the last line of power quality defense and removes all power problems that occur in the site between the UPS and the STS. The STS does not protect against power component failures downstream of the STS; it merely transfers between sources so that the customer's loads are preserved. In this manner, the static switch is used to protect the information and/or financial business that is transacted on a computer server.

The STS receives power from two or more independent power flow paths, and generally the paths derive power from one or more UPSs. If one power path fails, then the function of the STS is to transfer the load to the other power path. The transfer of the load between the two power paths that lead to the static switch is fast enough to ensure that the load (customers information/business that is transacted over the servers supported by the switch) is always preserved. These independent power paths that feed the static switch are set up so that they do not affect components such as circuit breakers, power panels, distribution panels, transformers, conduit wiring ducts or wiring trays.

The STS performs four major functions:

- It allows maintenance and repair of the upstream site components without affecting the load.
- It prevents upstream component failure from effecting the load.
- It reduces transformer inrush current from transformers connected to the STS output, which reduces circuit-breaker tripping and UPS bypassing.

Maintaining Mission Critical Systems in a 24/7 Environment By Peter M. Curtis

- The STS must provide data to help management meet the key elements of the Sox, Base 12, and NFPA 1600—that is, provide data for production risk management, production upset analysis and for security management, as discussed in Chapter 2.

8.2 OVERVIEW

8.2.1 Major Components

The STS contains only three major components, excluding the sheet metal; these are:

- Protective and isolating circuit breakers or MCSWs
- Logic, typically tri-redundant to ensure the safety of the loads
- Two sets of SCRs, configured as single- or three-phase switches

8.2.1.1 Circuit Breakers

The STS can be supplied with standard circuit breakers, but generally the STS is supplied with circuit breakers containing only instantaneous trips. These circuit breakers, with only instantaneous trips, are referred to as nonautomatic circuit breakers, or molded case switches (MCSW). The circuit breakers, or MCSWs, are selected to have ampere squared·second (I^2t) rating to protect the SCRs from damage when the STS output is shorted.

If the STS is supplied with MCSWs, upstream source circuit breakers or MCSWs must be rated to protect the STS internal power conductors and even the MCSW frames. The continuous current rating of the STS must be equal or less than the continuous rating of the upstream protective device.

To provide Hot Swap capability, all the MCSWs are plug-in or draw-out. Most STS are supplied with two paralleled isolation circuit breakers or MCSWs for two reasons:

- Redundancy
- Spare MCSW; if all MCSWs supplied are identical, one isolation MCSW can be unplugged and used to temporarily replace any of the other MCSWs

8.2.1.2 Logic

Since the logic is the heart of the STS, it is typically specified to be tri-redundant with voting circuits. This redundancy ensures that the switch supports the loads (business operating on the servers located downstream of the switch). Part of the logic function is the power supplies that power the logic elements. These logic power supplies are also typically specified tri-redundant for the same reasons listed above.

The SCR gate drivers are generally specified to be redundant for the S1 SCRs and redundant for the S2 SCRs, thereby requiring four different sets of gate drivers. Again, this is designed so that the switch fails safe and supports the loads.

The logic provides the following functions:

- Monitors both sources and output power and determines if and when the load should be transferred.
- Provides signals to the SCRs to turn them "on" and "off" when transferring or retransferring the load from one source to the other.
- Provides the interface, generally multi-screen and graphical, between the operator and the STS.
- Provides metering, status indicators, and alarm indicators, locally on the operator interface and remotely via commutations links.
- The logic inhibits transfers of short circuit currents, or transfers to a source that is worse (in terms of power and/or power quality) than the connected source.
- Learn the environment and adjust transfer parameters to prevent continuous nuisance transferred due to incoming power "noise."

8.2.1.3 SCRs

The SCRs (silicon controlled rectifiers) are the electronic elements that control the power flow between the source and the load and therefore control which source will provide power to the load. The SCR can be turned "on" or gated "on" by the logic; however, turning "off" the SCR requires no gate drive and no power or the "reversing" of power through the SCR. The "no" power or "reversing" of power through the SCR can be accomplished by natural or forced commutation.

The SCRs must be rated for the available I^2t let through of the MCSWs and for the continuous rated current of the STS.

8.3 TYPICAL STATIC SWITCH ONE LINE

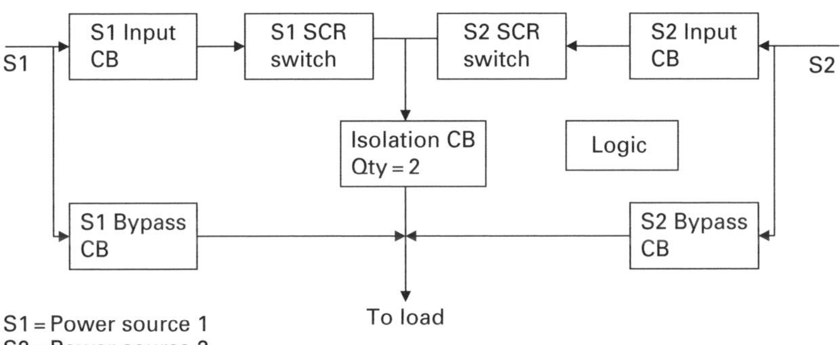

8.3.1 Normal Operation

Refer to the diagram on page 115. Power is supplied to the STS from two sources, S1 and S2. In normal operation, both of the bypass circuit breakers or MCSWs are open and all other circuit breakers or MCSWs are closed.

The logic can control the SCR switches to be "ON" or "OFF."

Normally, one of the SCR switches is "ON" and is supplying power to the load. If the source supplying power to the load has an anomaly, the logic will disconnect the failing source and turn "ON" the SCRs being powered by the "good" source. The transfer of the load by commutated "OFF" one set of SCRs, and turning "ON" the other set of SCRs can be accomplished in four milliseconds or less; thus the transfer is transparent to the load.

From the discussion above, the STS is a dual position switch designed to connect a critical load to one of two separate sources . . . . The source to which the load is normally connected is designated the "Preferred" source and the other, the "Alternate" source. Either source can be designated as the "Preferred" or "Alternate" via operator selection. The load is transferred to the "Preferred" source whenever the source is within specified tolerance unless retransfer is inhibited via an operator selection. When the retransfer is inhibited, the load is only transferred from one source to the other when the connected source deviates outside the specified tolerance.

Bypass Operation

If the SCRs or logic must be serviced or upgraded by a factory authorized technician, the STS can be placed in bypass operation by closing one of the bypass circuit breakers shown in the one line. After the bypass circuit breaker is closed, the two-input and the isolation circuit breakers or MCSWs can be opened, thus isolating the SCRs and logic for servicing without disrupting power to the critical loads. Following the reverse procedure, the load can be transferred from bypass to normal operation without disrupting power to the critical loads. This type of operation should only be performed by a trained technician because the improper sequencing of this procedure may compromise the power feeding the servers.

The circuit breakers or MCSWs are interlocked in the following manner to prevent the interconnecting of the two sources:

- The bypass circuit breakers are interlocked so that only one can be closed at the same time.
- Source 1 input circuit breakers or MCSWs and source 2 bypass circuit breakers or MCSWs are interlocked so that only one can be closed at the same time.
- Source 2 input circuit breakers or MCSWs and source 1 bypass circuit breakers or MCSWs are interlocked so that only one can be closed at the same time.

8.3.2 STS and STS/Transformer Configurations

The STS can be specified as a stand-alone unit, but it is generally specified as part of a STS/transformer system.

There are two STS/transformer configurations:

- Primary system where the transformer is connected to the STS output
- Secondary system where each STS input is fed from a transformer.

There are many iterations of these configurations and different trade-offs made when choosing between stand-alone, primary systems and secondary systems. Some of these trade-offs are outlined in Section 8.4.2.

8.4 STS TECHNOLOGY AND APPLICATION

8.4.1 General Parameters

This STS/transformer system is the last line of defense, and the STS logic provides the transfer algorithm, the operator interface, status data, alarm data, and forensics data. The STS or the STS/transformer system must have or perform the following:

- The STS must be as electrically close to the load as possible.
- The STS must provide data to help management meet the key elements of the policies and regulation identified in Chapter 2 (i.e., production risk management data, production upset data, etc.) and provide production security.
- The STS must communicate locally and remotely: status including waveform capture, possible problems (predictive maintenance) and immediate problems (alarms) and forensics (data for failure analysis).
- The STS and the overall system design must be resilient to human error, for both operator and maintenance personnel.
- The STS must have data logging of any configuration changes and record when and by whom any changes are made.
- The STS and/or the STS/transformer system must have 99.999% availability.
- The STS and/or the STS/transformer system should minimize the mean time to repair (MTTR).
- The STS and/or the STS/transformer system must be internally fault tolerant, must be fail safe, and must not fail due to external faults and abnormalities.
- The STS and/or the STS/transformer system must be safely repairable while maintaining load power.
- The STS and/or the STS/transformer system should have hot swappable modules and components.

8.4.2 STS Location and Type

The STS should be as electrically close to the load as possible. A rack mount STS is used to supply this dual-power path redundancy as close as possible to rack-mounted loads, such as servers. Typically, these systems are single-phase static switches.

Most data centers will have three-phase static switches in their design. The server rack systems are typically supplied single-phase power from a remote power panel (RPP) or a remote distribution unit (RDU), which are in turn fed power from a STS or STS system. The STS or STS system can be stand-alone or part of a primary or secondary static switch power distribution system as previously outlined. These STSs or STS systems will typically feed the RDU, which in turn feed the servers in the racks in a data center.

8.4.3 Advantages and Disadvantages of the Primary and Secondary STS/Transformer Systems

Each configuration has advantages and disadvantages, and the decision is different for every data center. Some of the decision-making criteria are listed below as a sample:

Primary System Advantages

- Less costly since only one transformer is required
- Less floor area required since there is only one transformer

Secondary System Advantages

- High level of maintainability: Since there are two transformers, one can be serviced or replaced while the other transformer is powering the load.
- Higher level of redundancy: The STS is closer to the load, so a transformer failure will not cause a power outage.
- The possibility of transformer inrush is much less. The STS in primary system should have a low inrush transfer algorithm; this algorithm is not required in a secondary system.

There are many opinions and theories about which system is better than the other, but the best fit will be a result of considering all of the design criteria for the each individual data center.

8.4.4 Monitoring and Data Logging and Data Management

The STS should incorporate features that allow a company's management to meet the key elements of Sox, Base 12, NFPA, etc. (i.e., production risk management data, production upset data, etc.) and provide production security.

The STS should communicate (locally and remotely) status, including waveform capture, possible problems (predictive maintenance), immediate problems (alarms), and forensics (data for failure analysis).

Most STS have a dynamic, multiscreen, graphic display rather than a text display. The multiscreen, graphic display can display information as icons, images and text, this combination allows the operator to absorb the information quicker than with a text display. Another advantage of the graphic display is the ability for the operator to view the waveforms of the two sources and of the STS output.

The graphical/touch screen operator interface is complex, and a failure leaves the operator blind. A simple backup interface with LED indicators and manual control switches is an important backup operator interface and ensures that the switch can be properly operated if the monitor were to fail for any reason.

8.4.5 STS Remote Communication

It is becoming increasingly important for power quality products to provide information along with insuring the quality of the power being provided. If the product is being overloaded or performance degradation is occurring, the operator/maintenance personnel must be alerted (Alarms). If there is a failure of any component in the system, data center personnel need to be able to establish what happened (Forensics) in order to prevent future occurrences. The STS monitor should provide a log of events, alarms, voltage and current values, frequency, waveform capture and other information. The information is useful when displayed locally, but most data center/facility/IT managers will want to view the information remotely. The data needs to be communicated in a universal language in order to be viewed over the site management system. There are several standard languages for these communications.

The STS should have, as a standard or optional feature, all of the necessary hardware and software for this monitoring of the STS status and alarm data and the accompanying waveforms. The typical protocols needed are:

- Modbus RTU or ASCII
- Modbus TCP—Modbus over the Ethernet
- SNMP (Simple Network Management Protocol)
- Web enabled communication
- Data available using a Web browser, like Internet Explorer
- Summary data that are e-mailed at specified times

Other features sometimes requested are:

- Cell or land-line modem to alert users of alarms and status changes

8.4.6 Security

Security is important for all equipment in mission critical sites and especially for both the UPS and STS systems due to the nature of the information that they are feeding power to. The quality of the power protection afforded by the UPS and STS is largely determined by operating parameters that are set in each unit. The method of protecting the UPS and STS operating parameters against unauthorized changes is a primary concern when setting up an internal/external firewall for the security system.

Two other levels of security that will need to be addressed for the STS system concern access to the mechanical devices in the switch. The first level starts with the doors of the units, which should be secured via a lock with a specific key. The second level starts with the circuit breakers or MCSWs and limits the ability of an untrained person to toggle or change their status, which may cause power losses. The caveat for the first-level or door security is that the doors must be opened in emergency situations to operate bypass circuit breakers or MCSWs; hence, the keys to equipment doors must be readily available.

There are two issues to be addressed in security management: preventing problems and preventing the recurrence of problems. Prevention of problems requires physical and psychological barriers. A lock on a door is a physical barrier, and alarming a door with a sign indicating that the door is alarmed is a psychological barrier; both are effective. A system involving both a psychological barrier and a requirement of a specific personnel identification device is especially effective.

Preventing the recurrence of problems is simply logging enough data to make a full forensics analysis of a security problem. From the forensics analysis, a solution can be formulated and implemented.

Some of the security features offered by STS manufacturers are:

- The "sign-in" process to change the STS operating parameters should include a password and a PIN so the operator who is "signed-in" can be identified and logged with a time/date stamp. The password is physical security; the PIN is psychological security and data collection.
- The "sign-in" is only good for a predetermined amount of time.
- Use a fingerprint and/or swipe card device instead of a password and PIN number to allow changes to the operation of the STS.
- Provide data to help management meet the security elements mentioned in Chapter 2 and provide data for security management.
- Use door opening alarms if the user does not "sign in" first, entry to be logged time/date stamped. Since the change of status of circuit breaker is logged and time date is stamped, the operator changing circuit breaker status could be identified.

The security features should be an important part of the specification, used to procure an STS.

8.4.7 Human Engineering and Eliminating Human Errors

It has been stated that 40–60% of mission critical site problems are caused by human errors. It has also been written that a human has difficulty in following written instruction in high-stress situations. Operators under stress respond well to visual text instructions, with icons and simultaneous voiced instructions.

STS suppliers address human errors with the following methods:

- Operating instructions written in text on a graphics screen, as well as voiced to avoid mis-operation
- Screen icons to guide operators in a step-by-step fashion through the command steps
- One display for STS/transformer systems, so that the operator has a common place to view data
- STS and STS/transformer systems with door alarms for psychological barriers
- Password and PIN or finger print/swipe card for psychological barriers

Since human engineering can help eliminate 40–60% of mission critical site problems, the human engineering features should be an important part of the specification used to procure an STS.

8.4.8 Reliability and Availability

As mentioned in Chapter 3, availability is a function of reliability or mean time between failures (MTBF) and mean time to repair (MTTR).

$$\text{Availability} = \text{MTBF}/(\text{MTBF} + \text{MTTR})$$

When availability is equal to one, there is no downtime. Availability approaches one if MTBF is infinity or repair time is zero.

MTBF is a function of the number of components and the stress levels on the components. The STS logic is generally the limiting factor of MTBF due to the large number of components mounted on the PCBs. A new failure mode is becoming apparent.

Radiation from space is becoming a problem as the junction areas in the new FPGA and other ICs are becoming smaller and smaller and operating frequencies are becoming higher and higher. The logic circuits and memories become prone to switching unexpectedly when hit by neutrons and/or other space particles. The neutrons are especially dangerous since it may take 10 feet of concrete to shield chips from neutron radiation.

The most effective method of increasing MTBF is via redundancy. Most STS manufacturers can provide the following redundancies:

- Redundant logic power supplies
- Redundant operator interface. The graphical/touch screen operator interface is complex and a failure leaves the operator blind. A simple backup interface with LED indicators and manual controls switch is important.

- Redundant SCR gate drivers. This would mean having two drivers for S1 SCRs and two drivers for S2. Since the drivers are in parallel, a failure of one cannot affect the other.
- Redundant logic

Tri-redundancy with voting is the most recognized redundancy method. Tri-redundancy is accomplished by having three circuits generate signals, and the three signals are routed to a voting circuit. The output of the voting is based on at least two signals being the same.

The implementation of tri-redundancy has problems as illustrated in the following example:

EXAMPLE

Assume that the nonredundant STS logic contains 500 components on each of three PCBs and has a logic and STS mean time to first failure (MTTF) of 9 years, and thus assume that the logic failure would cause a loss of load power.

In an effort to generate tri-redundancy, the designer could use three sets of PCBs (nine total) and add a simple voting circuit. This would probably increase the STS MTTF to 100 years. However, part of the logic would fail in 3 years, since there are three times the components. The logic failure would not cause an STS output failure, but the logic would have to be repaired to return the system to the tri-redundant configuration. Therefore, using this method of tri-redundancy would cause too much downtime for repair. The service costs would also increase.

An example of one way to implement tri-redundancy is to use a new system on a chip (SoC), use ICs to reduce the parts count to achieve a logic mean time to first failure of 8–10 years, and use STS mean time to first failure of 80–100 years.

A quick way to determine logic parts count is by measuring the PCB area, excluding the logic power supplies (the magnetics require a large area).

The logic of a typical nonredundant STS may contain 400–450 square inches of PCB area. The logic of a tri-redundant STS should contain only 5–10% more square inches of PCB area than the nonredundant logic.

8.4.9 Reparability and Maintainability

The STS and/or the STS/transformer system should have hot-swappable modules, PCBs, and components. If the STS has to be placed in bypass mode for maintenance or repair, all protection is lost for the servicing period. If the component is truly hot-swappable, the load will be protected while the component is swapped.

To be hot-swappable, the component must have the follow attributes:

- The STS continues to protect the load within CBEMA and/or ITIC power profiles while hot-swapping.
- The hot-swapping process should take less than 3 minutes.

- Fasteners securing the component must be captive or nonconductive, so if the fastener is dropped, there would be no consequence.
- The removed connector must have guarded pins so that there is no event if the connector pin touches metal or the pins on another connector.
- The exposed section, required to hot-swap the component, should not contain any voltage higher than 42 V (UL allows humans to be exposed to voltages up to 42 V).

Even if the STS is placed in bypass for the service period, the components should have the hot-swappable attributes so that service is quick and safe to personnel and the load.

8.4.10 Fault Tolerance and Abnormal Operation

The STS and/or the STS/transformer system should be internally fault tolerant, fail safe, and externally fault tolerant/fail safe due to fault and abnormalities.

Redundancy does not necessarily make the STS fault tolerant. For example, if a shorted capacitor on one PCB or module could reduce the logic power to other PCBs, tri-redundancy does little good, since all circuits will fail.

Abnormal operation is the operation of the STS when abnormal situations occur, such as:

- Loading short circuits: The STS should not transfer the short to the other source due to low voltage.
- Both sources have reduced voltage, beyond 15%. The load should transfer to the best source.
- Transferring magnetic loads between out-of-phase sources. The inrush should not trip any internal or external circuit breakers or MCSWs.

Fault tolerant and abnormal operation must be specified and tested on the purchased STS.

8.5 TESTING

The following tests should be performed on a sample or production STS:

- Standard performance test, including transfer times, alarm and status indications, and monitor functions
- Redundancy testing
  - Disconnect all but one logic power supply while the STS is operating and verify alarms.
  - Disconnect one logic power supply and then open each power source one at a time and verify alarms and operation.
  - Disconnect enough fans to check redundancy when operating at full load and verify alarms and operation.

  - Disconnect one SCR gate driver on each source at the same time and verify that load transfer is proper and verify alarms and operation.
  - Remove the connector supplying logic power on each PCB and module, verify STS operation, and verify alarms and operation.
  - Verify the backup operator interface.
  - If there are serial signal buses, like CAN bus, break each serial loop and verify operation and alarms.
- Reduce the voltage on one source 20% and then reduce the voltage on the other source 22%. The STS should transfer to the best source.
- Short the STS output, verify that the STS does not transfer to the other side, and by reclosing the circuit breakers or MCSWs, verify the STS is fully operational. Generally, factory test stations only have less than 12 kA available, and all STS have short-circuit rating of 22 kA or higher.
- Verify the low inrush algorithm used in the STS by connecting a transformer on the STS output and manually transferring the transformer load with the sources 0, 15, 120, and 180 degrees out of phase.
- Repairability: Remove and install several PCB or modules with the load hot in bypass or via hot swap.
- Verify internal fault tolerance.
  - Short the logic power supplies on each PCB or module; the load must be supported.
  - Remove or stop the crystal or oscillator on each processor, DSP or FPGA; the load must be supported.

8.6 CONCLUSION

The previous sections outline the operation and some of the other important parameters of the STS that should be addressed when designing and specifying for a data center.

Some of the site problems that can be eliminated with a properly designed site and properly utilized static transfer switch system are as follows:

- A reduction of the potential inrush currents which naturally flow out of a transformer when energized; by reducing this with a switch or a switch system, the risk of tripping breakers and/or causing the UPS to go to bypass can be minimized.
- Further ensure the loads against the risk of an outage due to the improperly made crimped or bolted connections in the power system.
- Further ensure the loads against the risk of an outage due to a circuit breaker failure.
- Further ensure the loads against the risk of an outage due to the opening, shorting, or overheating of site transformers.

- Further ensure the loads against the risk of an outage due to overheating and/or failure of the conductors.
- Further ensure the loads against the risk of an outage due to a failure of the upstream UPS by quickly switching to the other leg of power.

Other important features beside the inherent performance characteristics of the static switch as outlined above are:

- Monitoring and data collection
- Remote communications
- Redundancy
- Maintainability
- Reliability
- Fault tolerance

If all the above features are considered the STS procured will indeed be the last line of defense in the mission critical power system.

Chapter 9

The Fundamentals of Power Quality and their Associated Problems

9.1 INTRODUCTION

You wouldn't run your high-performance car on crude oil, so why run your mission-critical computer system on low-quality power? Your high-performance computer system needs high-octane power to keep it revving at optimum capacity. Just as a high-performance car needs quality fuel to run at peak performance, today's computer systems can falter when subjected to less-than-perfect power.

As a network manager, facilities engineer, or chief engineer, you need to know how to avoid potential power pitfalls that can put undue wear and tear on your computers, data, and mission critical infrastructure. Do not take the power from your wall outlet or local power station for granted. A common misconception is that utility power is consistent and steady. It isn't.

The power we receive from the grid is 99.9% available, which translates into approximately 8 hours of downtime per year. That level was once acceptable, but today's data centers and other 24/7 operations cannot tolerate or afford that level. For those facilities, the requirement is closer to 99.9999% available—less than one minute of downtime a year. The newer-generation loads, with microprocessor-based control and a multitude of electronic devices, are extremely sensitive to power quality variations than the equipment used in the past. No matter where you're located, spikes, surges, brownouts, and other power problems are potential hazards to your computer network, the engine that runs your business. Table 9.1 shows the average cost of downtime for a range of industry sectors.

Maintaining Mission Critical Systems in a 24/7 Environment By Peter M. Curtis

Table 9.1 Cost of Downtime

Industry	Typical Loss
Financial	$6,000,000/event
Semi-conductor manufacturing	$3,800,000/event
Computer	$750,000/event
Telecommunications	$30,000/event
Data processing	$10,000/event
Steel/heavy manufacturing	$300,000/event
Plastics manufacturing	$10,000/event

Source: *The cost of Power Quality*, Copper Development Association, March 2001

Some power problems stem from obvious sources, such as lightning. According to the National Weather Service, most regions in the United States experience at least 40 thunderstorms a year. These storms, sometimes accompanied by high winds and flooding, can cause damage to aboveground and underground wires, thus interfering with the high-quality power your mission critical infrastructure requires.

While the outside world poses numerous threats to your power quality, what happens inside your building can be just as dangerous. In fact, about 70% of all power quality disturbances are generated inside the facility, by the very equipment you purchase to run your business. Microprocessor-based equipment and computer systems oftentimes inject harmonics and other power quality events into the power infrastructure. If your office building is more than 10 years old, chances are that its wiring was not designed to meet the demands of a growing number of PCs, laser printers, scanners, LANs, midranges, and mainframes. These older buildings are wired for loads such as elevators, copiers, and mechanical equipment. As more nonlinear loads, or switching loads, are implemented into the existing building design, overloading of the electrical distribution systems occurs more frequently. Thus, more power quality problems and failures that can harm valuable data and result in equipment loss or malfunction. Furthermore, electronic devices and other electrical equipment could be feeding power anomalies back through your power distribution system. This produces an effect similar to viruses that have the potential to spread and infect sensitive electronics. Many devices interconnect via a network, which means that the failure of any component has far-reaching consequences.

While you may not have control over nature, insufficient wiring, and the unpredictability of electricity, you can shield your mission critical infrastructure and data center from these hazards.

9.2 ELECTRICITY BASICS

Electricity must flow from a positive to a negative terminal, and the amount of electricity produced depends on the difference in potential between these two terminals.

When a path forms, electrons travel to reach the protons and create electricity. The difference of potential depends upon the amount of electrons on the negative terminal: The greater the charge, the higher the difference in potential.

Imagine two tanks of water, equal in size and height, set apart from each other but connected at the bottom by a pipe and a valve, currently closed. The first tank is full of water and creating a maximum pressure at the valve. The second tank is completely empty. When the valve opens, the water flows into the second tank until it reaches equilibrium. The rate at which the water flows depends on the pressure difference between the two points. With electrons, the movement depends on the difference in electrical potential, or voltage, across the circuit.

Voltage is the force or "pressure" that is available to push electrons through the conductor from the negative to the positive side of the circuit. This force, measured in volts, is and represented by the symbol E. While the voltage is pushing electricity, current measures the rate.

Current, measured in amps, is represented by the symbol I. Current is determined by the voltage difference. In a water pipe, if the pressure differential is great, then the flow of water is swift, but as the pressure decreases, the rate of flow slowly dwindles until it reaches equilibrium. Voltage and current are directly proportional when resistance is constant (Ohm's Law).

Resistance restricts the flow of electricity by inner friction. This resistance to flow, represented by R, is measured in ohms.

Ohm's Law: Each of these three quantities is affected by the other two and can be represented in the equation $E = I * R$. This equation can also be understood as $R = E/I$ or $I = E/R$. For example, a circuit has a voltage of 12 V and a resistance of 4 ohms. The current can be found by using the equation $I = E/R$. So 12 volts/4 ohms is equal to 3 amps of current.

Another term used to describe electrical quantity is power expressed as volt-amperes (VA) (or 1 kVA = 1000 VA). It is possible to have a very large voltage; but without current flow, the power realized is insignificant. Conversely, a large current flow with only a moderate voltage can have a catastrophic effect, as in a ground-faulted circuit.

Another manner of expressing true power is in watts (W) or kilowatts (kW). In simplified terms, wattage is equal to voltage multiplied by amperage times a power factor that accounts for the type of load (inductive, capacitive, or resistive). Typical power factors range from 0.8 to 0.95.

9.2.1 Basic Circuit

An electric circuit comprises a power source, wires, the load, and a means of controlling the power, like a switch. Direct current (DC) flows in only one direction. The traveling electrons only go from negative to positive. For current to flow, the circuit must be closed. The switch provides a means for opening and closing the circuit. Remember, current always travels through the path of least resistance.

Manufacturers rate each piece of equipment for a given voltage, current limit, operating duty, horsepower or kW power consumption, kVA, temperature, enclosure, and so on. Equipment voltage ratings are the maximum voltage that can be safely applied continuously. The type and quantity of insulation and the physical spacing between electrically energized parts determine the voltage rating. Actual operating voltages are normally lower.

The maximum allowable operating temperature at which the components can properly operate continuously determines the current rating. Most equipment is protected internally against overheating using fuses as circuit breakers.

9.3 TRANSMISSION OF POWER

Power disturbances are increasing, not only because of the utility, but also because of the stress applied to the nation's electrical grid. Starting from the utility company to the workplace, the journey that power takes can be affected in many different ways. Facilities managers must identify a power quality problem in order to implement a proper corrective action. But first, the cycle of power must be understood. This section gives a brief background on the production and distribution of electricity.

9.3.1 Life Cycle of Electricity

There are four phases of the cycle of electricity: generation, transmission, distribution, and end use. Generation is the actual process of converting different types of energy to electrical energy. This process usually occurs at a power-generating plant. Utility companies generate their electrical power from many different types of energy sources including nuclear, coal, oil, and hydroelectric plants. The next step, transmission, transports the electricity long distances to supply towns and cities. The electrical energy is then distributed to the residential and commercial consumers, or end-users. Electricity is very rarely the end product; it is used for kinetic work (e.g., to turn the blades of a fan), in solid-state devices, or converted to other forms of energy, such as light and heat.

In reference to power quality, transmission and distribution are most important. These two phases of the journey produce the most significant disturbances of power quality.

Once the power is generated, transformers step up the voltage. The purpose of increasing the voltage from 300,000 to 750,000 V is to minimize the current flow. Current is proportional to heat in an electrical circuit. By keeping the voltage high and the current low, there will be fewer disruptions caused by overheating.

Once the electricity reaches the town, a transformer will step down the voltage to 66,000 V (typical) to distribute locally. The voltage then gets stepped down again to about 13,000 V (typical) at smaller substations before reaching homes and buildings, where the electricity is stepped down to a usable 480 or 208 V. Each panelboard or load center then feeds branch circuits, sending 120 V of convenience power using 15-amp branch circuits to the wall receptacles. Figure 9.1 lays out

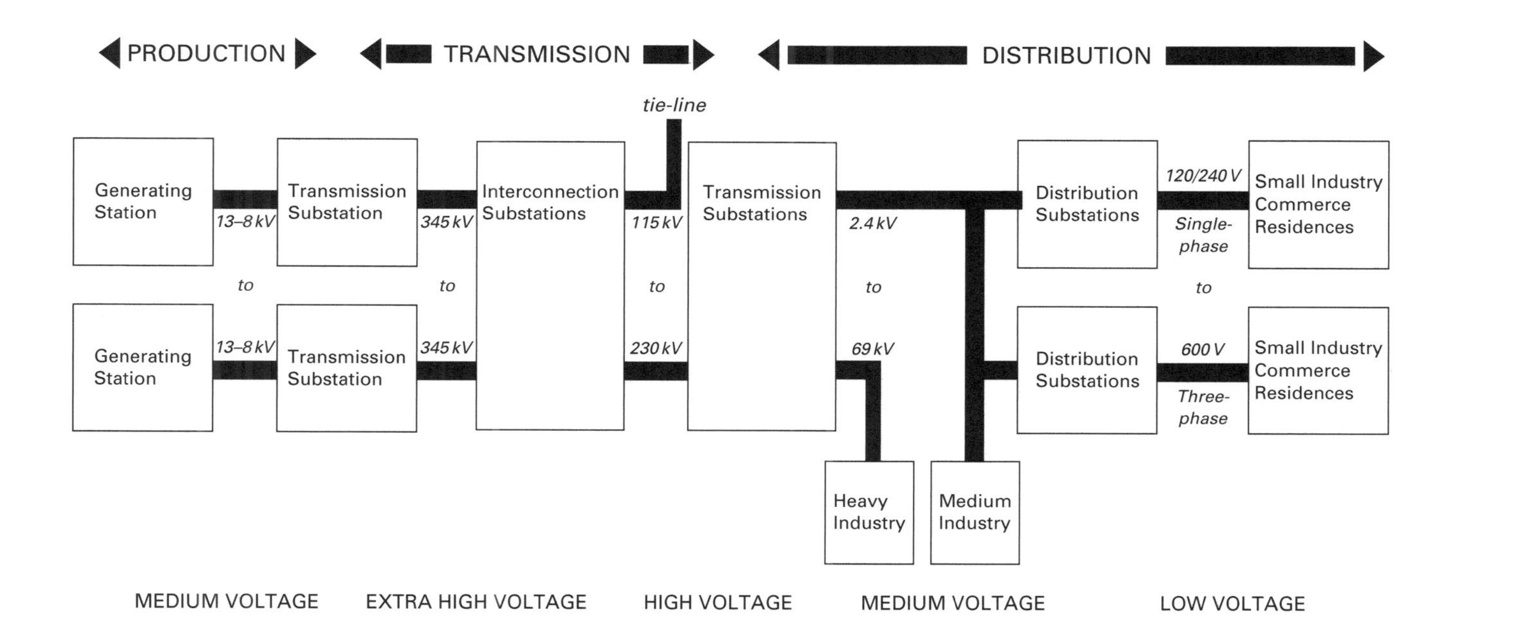

Figure 9.1 Typical electric system. (Courtesy of Dranetz-BMI.)

the transmission and distribution process, demonstrating the flow of electricity from the generating station to the end-user such as industry, commercial, and residences.

Alternating current (AC) is preferred over direct current (DC) because it is easy to generate, and transformers can adjust the voltage for many different power requirements.

Weather can be crucial to reliable transmission of power. When winds are high, aboveground lines sway too close together, creating an arc and turbulent voltage variations. An arc occurs when two objects of different potential create a spark between them—a flow of current through the air.

9.3.2 Single- and Three-Phase Power Basics

You cannot see electricity, but you can record it. Figure 9.2 shows what one cycle of undistorted electricity looks like. In the United States, a cycle occurs 60 times per second. If a 1-second recording shows 60 turns, look at the wave on a scale of milliseconds or even microseconds to see the distortions. Sophisticated recording equipment can capture the waveform graphically, so it can be reviewed for disturbances. This equipment is referred to as disturbance or power monitoring equipment.

The cycle shown in Figure 9.2 is a sine wave with a fundamental frequency of 60 Hz—the waveform model for reliable power.

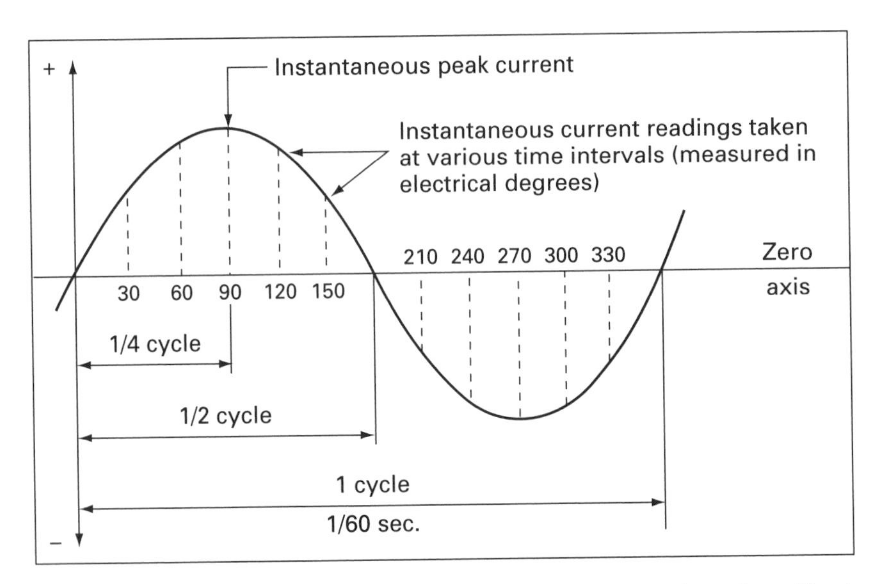

Figure 9.2 This undistorted sine wave is also known as the fundamental waveform. (Courtesy of Dranetz-BMI.)

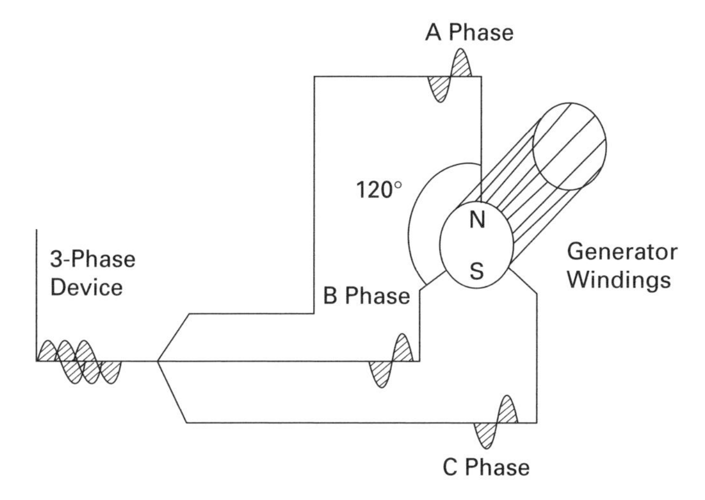

Figure 9.3 Three-phase power is produced from the rotating windings of a generator. (Courtesy of Dranetz-BMI.)

9.3.2.1 Single-Phase Power

Some of the most common service designs using single-phase power are:

- 120-V, single-phase, two-wire used for small residences and isolated small loads up to about 6 kVA
- 120/240-V, single-phase, three-wire used for larger loads up to about 19 kVA

Depending upon the rating of the service transformer system, voltages can be 120/240, 115/230, or 110/220. The 120/240-V, single-phase systems come from a center-tapped single transformer. When the two hot legs and a neutral of a three-phase system are used, it is possible to have 120/208-V, single-phase, three-wire systems, which are commonly found within residential and commercial buildings.

9.3.2.2 Three-Phase Power

Three-phase power takes three single-phase sine waves and places them 120 degrees apart. A three-phase wiring system powers 1.73 times as much load as a single-phase. The result is an almost smooth line instead of an up-and-down (on/off) effect.

Some of the most common service designs using three-phase power are as follows:

- 120/208-V, three-phase, four-wire systems are a common service arrangement from the local electric utility for large buildings.
- 277/480-V, three-phase, four-wire systems use the 277 V level for fluorescent lights, 480 V for heavy machinery, and relatively small dry-type closet-installed transformers to step the voltage down to 120 V for convenience

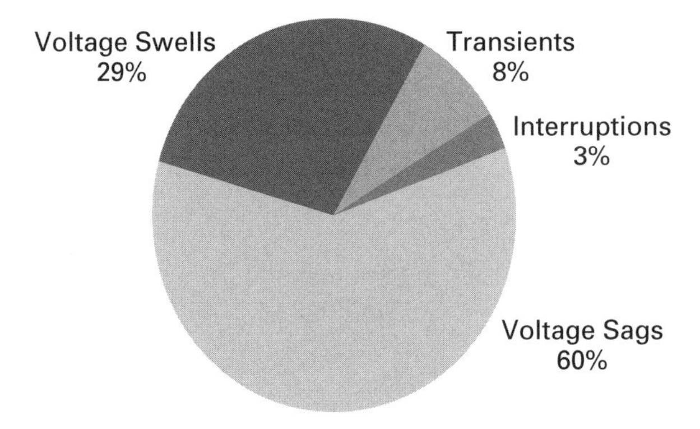

Figure 9.4 Types and relative frequency of power quality disturbances. (Courtesy of Georgia Power.)

receptacles and other loads. This system is ideal for high-rise office buildings and large industrial buildings.

- 2400/4160-V, three-phase, four-wire systems are only used in very large commercial or industrial applications where on-site equipment mandates this incoming system design (e.g., large paper mills and manufacturing plants).

9.3.3 Unreliable Power Versus Reliable Power

Reliable power is clean power that makes equipment function properly with little or no stress. When measuring reliable power, the monitor would record an unchanging or stable sine wave at a rate of 60 cycles per second. A continuous sine wave that never changes and never ends would never cause problems, but is almost impossible to find.

Unreliable power can be described as having any type of problem, such as a sag, swell, flicker, dimming of lights, continuous malfunctions that happen frequently, and even deficient amounts of power, making equipment work harder. Unreliable power problems can affect many day-to-day activities. When measuring this type of power, the monitor may record so many disruptions that the sine wave is unrecognizable. Figure 9.4 identifies the most common forms of power quality disturbances. Note that voltage sags are by far the most common, although spikes (transients) are the ones people protect against most commonly, through the addition of surge protectors.

9.4 UNDERSTANDING POWER PROBLEMS

A wide range of businesses and organizations rely on computer systems, and a multitude of networked equipment to carry out mission critical functions. Often, network links fail because the equipment is not adequately protected from poor power quality. When a facility is subjected to poor power quality, the electrical devices and equipment are stressed beyond their design limits, which leads to premature failures and poor overall performance.

Power supply problems can cause significant consequences to business operations. Damaged equipment, lost or incorrect data, and even risks to life safety can all result from power quality problems. In order to affect appropriate solutions, facilities engineers must clearly understand the causes and effects of specific power problems that may interrupt the flow of data across an enterprise.

IEEE 1159 *IEEE Recommended Practice for the Transfer of Power Quality Data* is the prevailing U.S. standard for categorizing power quality disturbances:

- Transients
- RMS variations (include sags, swell, and interruptions)
  - Short-duration variations
  - Long-duration variations
  - Sustained variations
- Waveform distortions
  - DC offset
  - Harmonics
  - Interharmonics
  - Notching
- Voltage fluctuations
- Power frequency variations

Table 9.2 outlines the causes and effects of common power problems.

9.4.1 Power Quality Transients

Transients, which are common to many electrical distribution systems, can permanently damage equipment. Transients are changes in voltage and/or current, usually initiated by some type of switching event, such as energizing a capacitor bank or electric motors, or interference in the utility company's power lines. Transients are often called "ghosts" in the system, because they come and go intermittently and may or may not impact your equipment. Transients are much less than a full cycle, oftentimes a millisecond or less.

Characterization can help the skilled user determine what happened and specifically how to mitigate. Broadly speaking, we classify transients as either impulsive or oscillatory. An impulsive transient is a sudden frequency change in the steady-state condition of voltage or current, or both, that is either positive (adds energy) or negative (takes away energy). Oscillatory transients include both positive and negative polarity.

Transients on AC lines, also known as surges, glitches, sags, spikes, and impulses, occur very quickly. Events like lightning and the switching of a reactive load commonly cause these transients. While lightning generates the most impressive transients, switching large motors and transformers or capacitors create transients big enough to affect today's delicate microprocessor-based equipment.

Table 9.2 Causes and Effects of Power Problems

Power Problem	Possible Causes	Effects
High-voltage transients (spikes and surges)	Lightning Utility grid switching Heavy industrial equipment Arcing faults	Equipment failure System lockup Data loss Nuisance tripping ASDs
Harmonics	Switch-mode power supplies Nonlinear loads ASDs Single-phase computers	High neutral currents Overheated neutral conductors Overheated distribution and power transformers
Voltage fluctuations	Overheated distribution networks Heavy equipment startup Power line faults Planned and unplanned brownouts Power factor-correction capacitors Unstable generators Motor startups	System lockup Motor overheating System shutdown Lamp burnout Data corruption and loss Reduced performance Control loss
Power outages and interruptions	Blackouts Faulted or overloaded power lines Backup generator startup Downed and broken power lines	System crash System lockup Lost data Lost production Control loss Power supply damage Lost communication lines Complete shutdown
Low voltage electrical noise	Arc welders Electronic equipment Switching devices Motorized equipment Improper grounding Fault-clearing devices, contactors, and relays Photocopiers	Data corruption Erroneous command functions Short-term timing signal variations Changes in processing states Drive-state and buffer changes Erroneous amplifier signal levels Loss of synchronous states Servomechanism control instability In-process information loss Protective circuit activation

Note that common industry practice is to refer to transients as "surges," even though *spike* is more descriptive. Technically, the term *surge* refers to an overvoltage condition of a longer duration.

Whether caused by energization of power factor capacitors, loose connections, load or source switching, or lightning, these transients happen so quickly they are imperceptible to the eye. However, recorded results from a power monitoring analyzer will reveal what types of transients are affecting the system. Transients can cause equipment damage, data corruption, reduced system life, and a host of irreproducible or identifiable problems. Figure 9.5 represents different types of transients and their effect on the normal waveform.

9.4.2 RMS Variations

Normal power follows a near-perfect sine wave. An RMS variation is typically one cycle or longer. Power quality disturbances can be recorded and later identified by comparing the associated waveform deviations with the normal sinusoidal curve. For instance, a swell makes the waveform increase outside of the normal range for a short length of time. A sag is the opposite; it falls below the normal voltage range for a short duration. Overvoltage and undervoltage conditions occur for longer durations than surges and sags, which can cause serious problems. If these events occur often and go undetected, this will stress and eventually deteriorate sensitive equipment. Figure 9.6 shows the key RMS variations.

9.4.2.1 Voltage Sags

Voltage sags usually occur when a large load is energized, creating a voltage dip. The voltage will drop below −10% of normal power for a duration of one-half

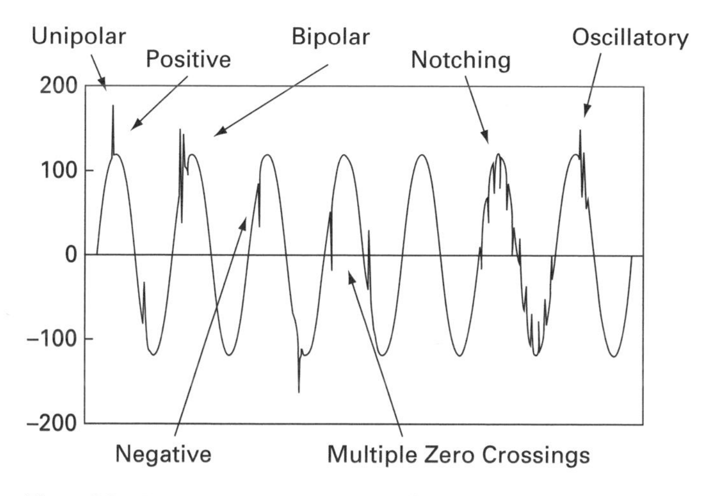

Figure 9.5 Transients shown in waveform. (Courtesy of Dranetz-BMI)

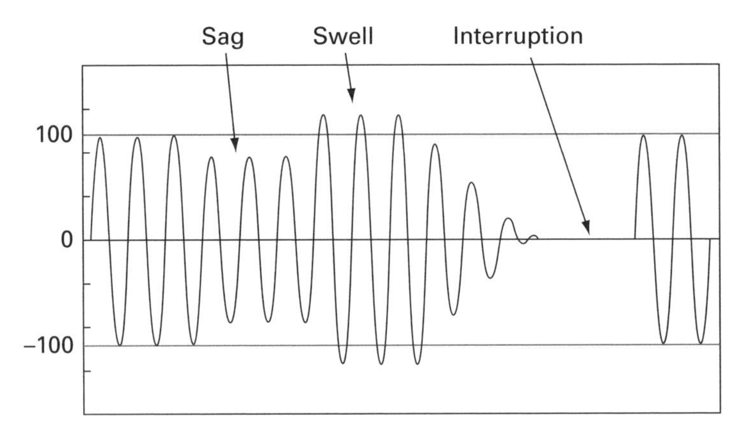

Figure 9.6 Waveforms of RMS variations. (Courtesy of Dranetz-BMI)

cycle to 1 minute. Here, the sine wave is less than its normal range and is not functioning to potential. For digital devices, a sag can be as perilous as an outage. Sags are usually caused by switching on any piece of equipment with a large inrush current, including induction motors, compressors, elevators, and resistive heating elements. A typical motor startup will draw six to ten times the amount of operating power. Figure 9.7 shows the signature sag resulting from a motor start; voltage drops quickly and then steadily increases to nominal.

9.4.2.2 Voltage Swells

Voltage swells usually occur when a large load is turned off, creating a voltage jump, or from a poor transformer tap adjustment. A voltage swell is similar to shutting off a water valve too quickly and creating water hammer. The voltage rises above +10% of normal voltage for one-half cycle to 1 minute. The sine wave exceeds its normal range. Swells are common in areas with large production plants; when a large load goes off-line, a swell is almost inevitable. Figure 9.8 shows a swell generated during a fault.

9.4.2.3 Undervoltage

Undervoltage (long-duration sag as defined by IEEE) occurs when voltage dips below −10% of normal voltage for more than 1 minute. An undervoltage condition is more common than an overvoltage condition because of the proliferation of potential sources: heavy equipment being turned on, large electrical motors being started, and the switching of power mains (internal or utility). Undervoltage conditions cause problems such as computer memory loss, data errors, flickering lights, and equipment shutoff.

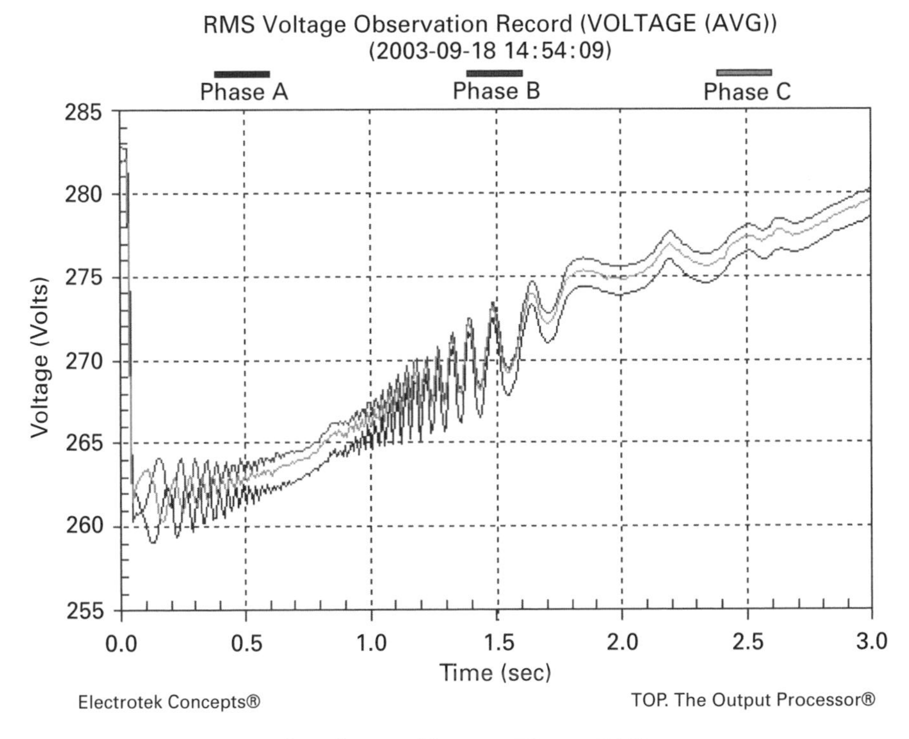

Figure 9.7 Motor-start waveform signature. (Courtesy of Dranetz-BMI)

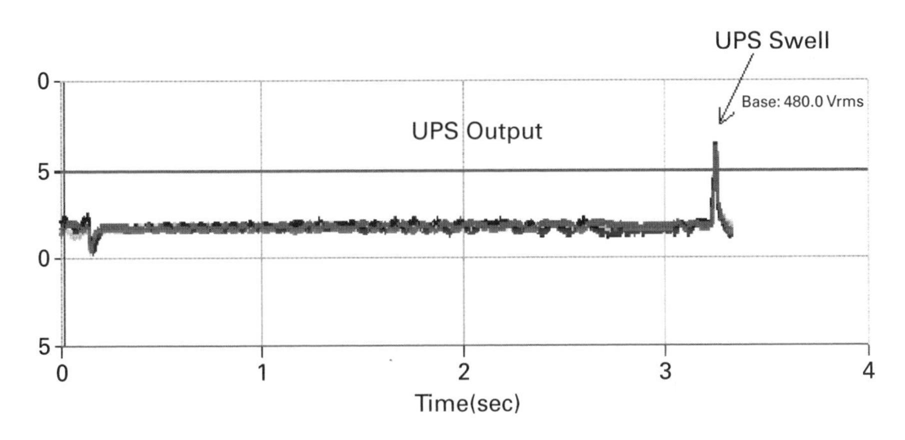

Figure 9.8 Voltage swell timeline. (Courtesy of Dranetz-BMI.)

9.4.2.4 Overvoltage

Overvoltage (long-duration swell as defined by IEEE) occurs when voltage increases to greater than +10% of normal for more than 1 minute. The most common cause is heavy electrical equipment going off-line. Overvoltage conditions also cause problems such as computer memory loss, data errors, flickering lights, and equipment shutoff.

9.4.2.5 Brownouts

A brownout is usually a long-duration undervoltage condition that last from hours to days. Brownouts are typically caused by heavy electrical demand with insufficient utility capacity.

9.4.2.6 Blackouts

A blackout is defined as a zero-voltage condition that lasts for more than two cycles. The tripping of a circuit breaker, power distribution failure, or utility power failure may cause a blackout. The effects of an unanticipated blackout on networked computers include data damage, data loss, file corruption, and hardware failure.

9.4.2.7 Voltage Fluctuations

Voltage fluctuations are any increase or decrease that deviates within +/−10% of the normal power range. The current never remains consistent as the voltage fluctuates up and down. The effect is similar to a load going on- and off-line repeatedly. These variations should be kept within a +/−5% range for the optimal performance of equipment. Voltage fluctuations are commonly attributable to overloaded transformers or loose connections in the distribution system.

9.4.2.8 Voltage Unbalance

Voltage unbalance, as seen in Figure 9.9, is a steady-state abnormality defined by the maximum deviation (from the average of the three-phase voltages or currents) divided by the average, expressed in percentage. The primary source of voltage unbalance is less unbalanced loading in general. Voltage unbalance can also be the result of capacitor bank anomalies, such as a blown fuse on one phase of a three-phase bank. It is most important for three-phase motor loads.

9.4.2.9 Voltage Notches

Voltage notches (Figure 9.10) occur on a continuous basis (a steady-state abnormality), and they can be characterized through the harmonic spectrum of the affected voltage. Three-phase rectifiers (found in DC drives, UPS systems, adjustable speed drives, and other applications) that have continuous DC current are the most important cause of voltage notching.

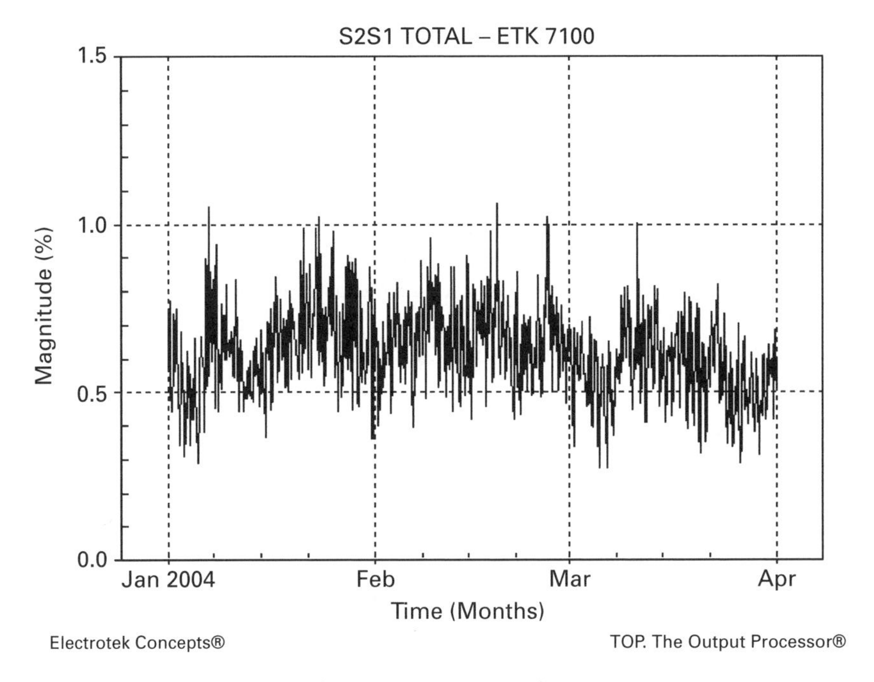

Figure 9.9 Unbalance timeline. (Courtesy of Dranetz-BMI)

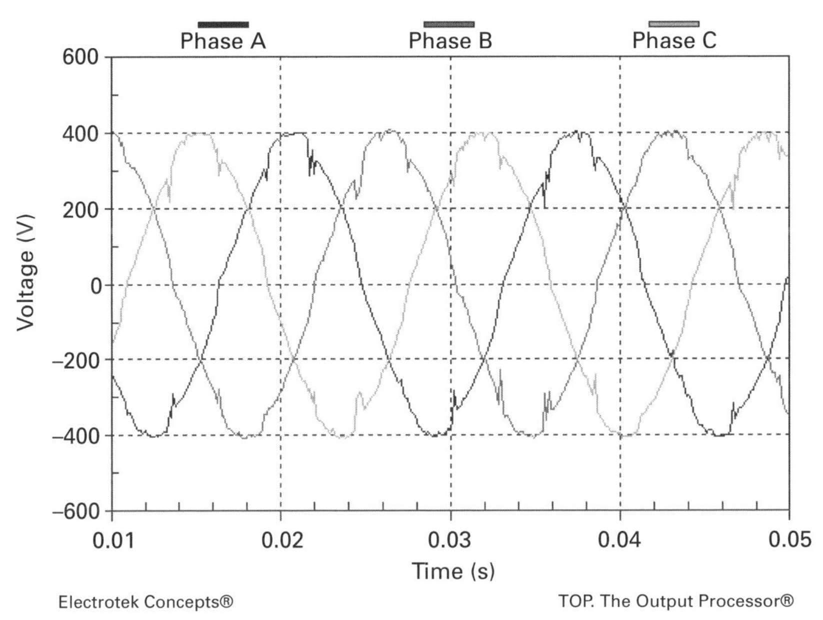

Figure 9.10 Notching waveform. (Courtesy of Dranetz-BMI)

9.4.3 Causes of Power Line Disturbances

There are many causes of power line disturbances; however, all disturbances fall between two categories: externally generated and internally generated. Externally generated disturbances originate from the utility while internally generated disturbances stem from conditions within the facility. The source of the disturbances may determine the frequency of occurrence. For example, if a downed power line initiates a disturbance externally, the disturbance is then random and likely not repeated. However, internal cycling of a large motor may cause repeated disturbances of predictable frequency. Again, facilities engineers must understand the source of the problem to determine a solution to prevent recurrence.

9.4.3.1 Lightning

When lightning actually strikes a system, the lightning causes the voltage on the system to increase, which subsequently induces a current into the system. This transient can instantaneously overload a circuit breaker, causing a power outage. The major problem with lightning is its unpredictability. When and where lightning strikes has a great impact on a facility, its equipment, and personnel safety. A facility manager can only take precautions to damper transients and provide a safe environment.

There is no single way to protect yourself and the equipment from lightning. However, lightning arrestors and diverters can help minimize the effects of a lightning strike. Diverters create a path to a grounding conductor, leading the voltage to a driven ground rod for safety. Many facilities install other devices, called surge protectors, to protect key critical equipment. A surge protector acts as a clamp on the power line, preventing rapid rises in voltage from moving past the choke point.

9.4.3.2 Electrostatic Discharge

Electrostatic discharge (ESD), referred to as static electricity, also produces transients. Static electricity is similar in nature to lightning. When electric charges are trapped along the borders of insulating materials, it builds up until the charge is forcefully separated, creating a significant voltage and sometimes a spark. ESD transients do not originate on the power lines and are not associated with utility power input. ESD requires an electrostatic-generating source (such as carpeting and rubber soled shoes) and is exacerbated by a low-humidity environment.

The high voltages and extremely rapid discharges can damage electronic devices. Static electricity builds on the human body; touching a part of a computer, like the keyboard, allows a discharge into the device, potentially damaging microprocessor components. Minimize ESD by providing paths to the ground, such that charges dissipate before they build to a damaging level. Since humid air is a better conductor than dry air, maintaining a humidity of 40–60% helps control electrostatic buildup.

9.4.3.3 Harmonics

There are two different types of AC electrical loads:

- Linear loads are such things as motors, heaters, and capacitors that draw current in a linear fashion without creating harmonic disturbances.
- Nonlinear loads are such things as switching power supplies and electronic ballasts that draw current intermittently and produce harmonic distortions within the facility's distribution.

Harmonics cause many problems within transformers, generators, and UPSs: overheating, excessive currents and voltages, and an increase in electromagnetic interference (EMI). Major sources of harmonics are variable speed motor drives, electronic ballasts, personal computers, and power supplies. Any device that is nonlinear and draws current in pulses is a risk to an electrical system. Power supplies are important because they draw current only at the peak of the waveform when voltage is at its highest, instead of extracting current continuously. The discontinuous flow produces a great distortion, changing current and adding heating strain.

Figure 9.11 shows how harmonics distort the sine wave. The fundamental waveform and the 5th harmonic are both represented separately. When added together, the result is a disfigured sine wave. The normal peak voltage is gone and replaced with two bulges. What happens to the sine wave affected by three or four harmonics at the same time? Harmonics have tremendous ability to distort the waveforms consumed by sensitive equipment.

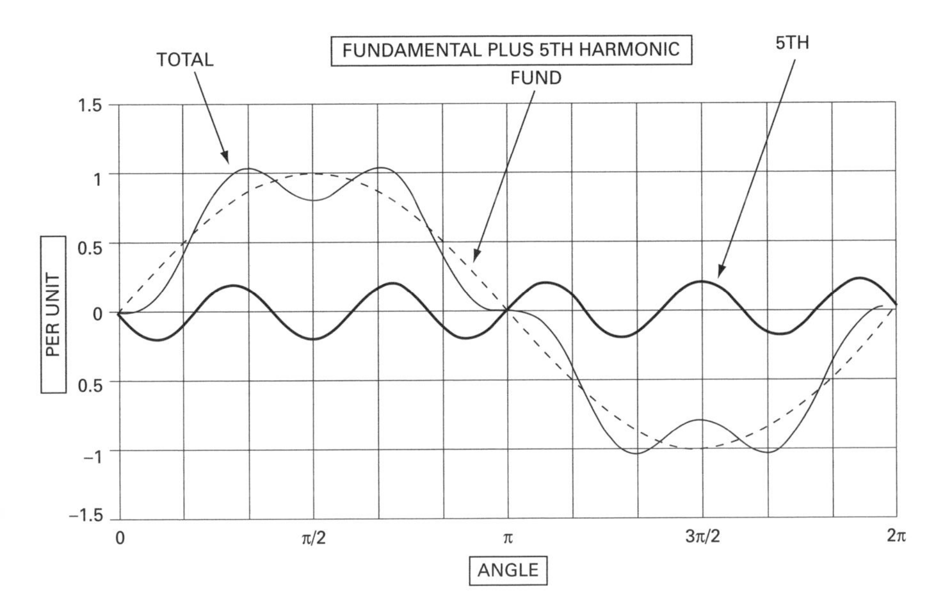

Figure 9.11 The fundamental and the 5th harmonic. (Courtesy of Dranetz-BMI.)

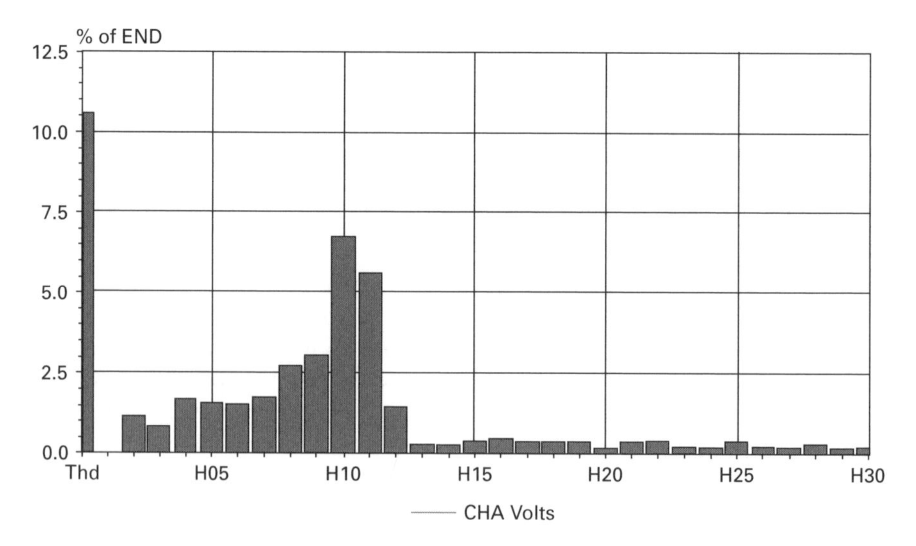

Figure 9.12 The harmonic spectrum indicating a problem. (Courtesy of Dranetz-BMI.)

Harmonic distortion is a disturbance introduced by the computer itself; large computers use power supplies incorporating switching regulators. The critical device may be the very source of harmonic distortion causing the device to malfunction.

A common way to visualize harmonics is the harmonic spectrum (Figure 9.12). IEEE 519 *IEEE Recommended Practices and Requirements for Harmonic Control in Electrical Power Systems* recommends that the voltage total harmonic distortion (THD) should not exceed 5% (this would be all the added frequencies) and no single harmonic should exceed 3% of the fundamental frequency.

Harmonics are generated when voltage is produced at a frequency other than the fundamental. Harmonics are typically produced in whole number multiples of the fundamental frequency (60 Hz). For example, a 3rd harmonic is three times the fundamental frequency, or 180 Hz. The 4th harmonic is four times the fundamental, or 240 Hz, and continues for an infinite number of harmonics. In general, the larger the multiple, the less effect the harmonic has on a power system because the magnitude typically decreases with frequency.

You should be concerned about harmonics when they occur over a long period, because they can have a cumulative effect on the power system. Electronic power supplies, adjustable speed drives, and computer systems, in general, can add harmonics to a system and cause damage over time. Of significant concern are the triplen harmonics (odd multiples of three) because they can add to the neutral phase, causing overheated distribution neutrals and transformers (Figure 9.13).

(1) Phase Sequence. An important aspect of harmonics is the phase sequence, because the phase determines how the harmonic will affect the

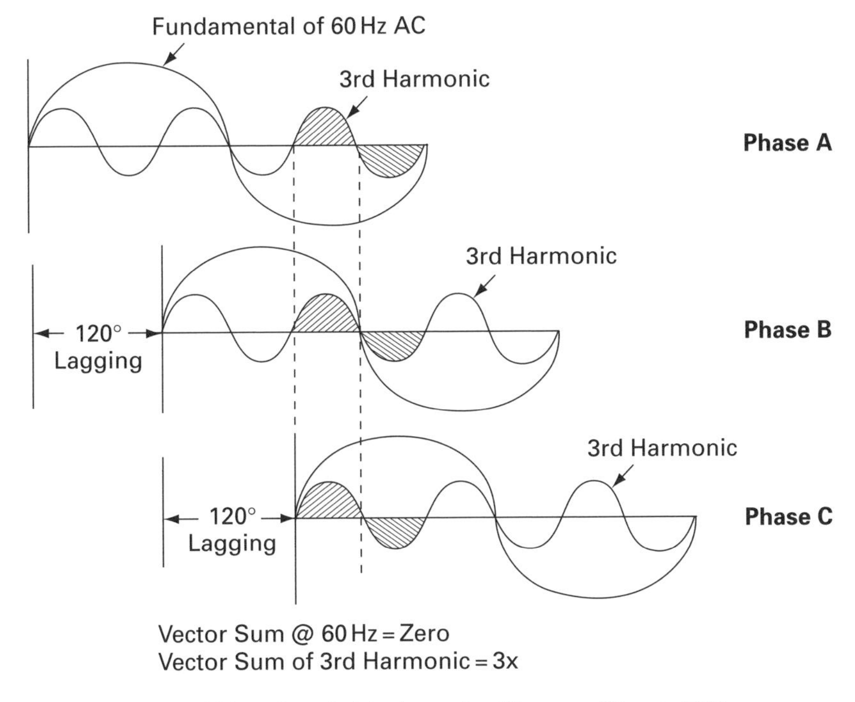

Figure 9.13 The additive effect of triplen harmonics. (Courtesy of Dranetz-BMI)

Table 9.3 The Frequency and Phase Sequence of Whole Number Harmonics

Harmonics	Frequency	Sequence
Fundamental (1st)	60	Positive (+)
2nd	120	Negative (–)
3rd	180	Zero (0)
4th	240	(+)
5th	300	(–)
6th	360	(0)
7th	420	(+)
8th	480	(–)
9th	540	(0)
10th	600	(+)

operation of a load. Phase sequence is determined by the order in which a waveform crosses zero. The result includes a positive, negative, or zero sequence. Table 9.3 shows the frequency and phase sequence of whole number harmonics.

- Positive sequence harmonics (1st, 4th, 7th, etc.) have the same phase sequence and rotational direction as the 1st harmonic (fundamental). The positive sequence increases the amount of heat to the electrical system slightly.
- Negative sequence harmonics (2nd, 5th, 8th, etc.) have a phase sequence opposite of the fundamental's phase sequence. The reverse rotational direction of negative sequence harmonics causes additional heat throughout a system and reduced forward torque of the motor, decreasing the motor's output.
- Zero sequence harmonics (3rd, 6th, 9th, etc.) are not positive or negative. There is no rotational direction, so the zero sequence harmonics do not cancel out, but instead add together in the neutral conductor. The neutral conductor contains little or no resistance, which causes large problems because there is nothing to limit the current flow.

Even numbers (2nd, 4th, 6th, etc.) usually do not occur in properly operating power systems; odd numbers (3rd, 5th, 7th, etc.) are more likely. The most harmful harmonics are the odd multiples of the third harmonic. Table 9.3 shows the zero sequence harmonics (most harmful sequence) to be multiples of three. Because the even harmonics are known not to cause problems, the odd multiples of three are called the triplen harmonics (3rd, 9th, 15th, 21st, etc.). Triplens cause overloads on neutral conductors, telephone interference, and transformer overheating.

Because harmonics are produced in multiples, they can also affect an electric system by combining with each other. Certain multiples of harmonics are easy to identify, but when they come in pairs, like a 3rd and 5th harmonic, the recorded waveform becomes distorted and difficult to identify (Figure 9.14).

9.4.3.4 Utility Outages

An outage or blackout is the most obvious power quality problem and is the most costly to prevent. An outage is defined as the complete loss of power, whether momentary or extended. Utility outages are commonly attributable to rodents and

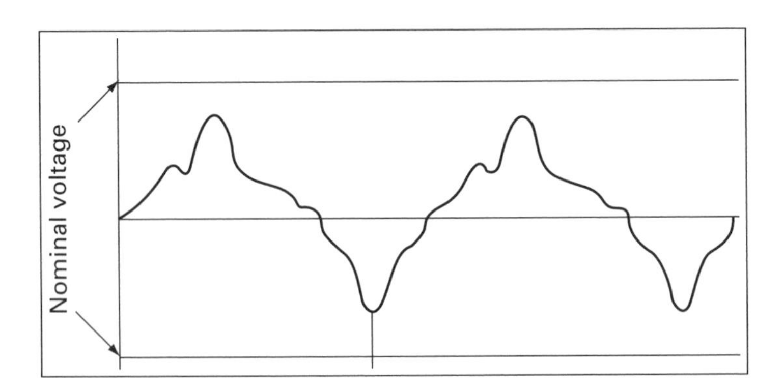

Figure 9.14 Harmonic distortion waveform. (Courtesy of Dranetz-BMI.)

other small animals eating through insulation or nesting in electric distribution gear; downed power lines from accidents or storms; disruption of underground cables caused by excavation or flooding; utility system damage from lightning strikes; and overloaded switching circuits.

In simple terms: In an outage there is no incoming source of power. The consequences are widespread, amounting to more than an inconvenience. Computer data can be lost, telephone communications may be out of service, equipment may fail, products and even lives could be lost (as in hospitals or public transportation). In Figure 9.15 an outage waveform is shown, captured during the August 2003 blackout that hit eight Northeastern states and part of Canada.

9.4.3.5 EMI and RFI

Other, less damaging disturbance initiators are radio-frequency interference (RFI) and electromagnetic interference (EMI). EMI and RFI are high-frequency power line noises, usually induced by consumer electronics, welding equipment, and brush-type motors. Problems induced by RFI and EMI are usually solved by careful attention to grounding and shielding, or an isolation transformer.

9.4.4 Power Line Disturbance Levels

Table 9.4 categorizes power line disturbances by threshold level, type, and duration. Type I disturbances are transient in nature, Type II are considered momentary in duration, while Type III disturbances are classified as power outages. Duration is

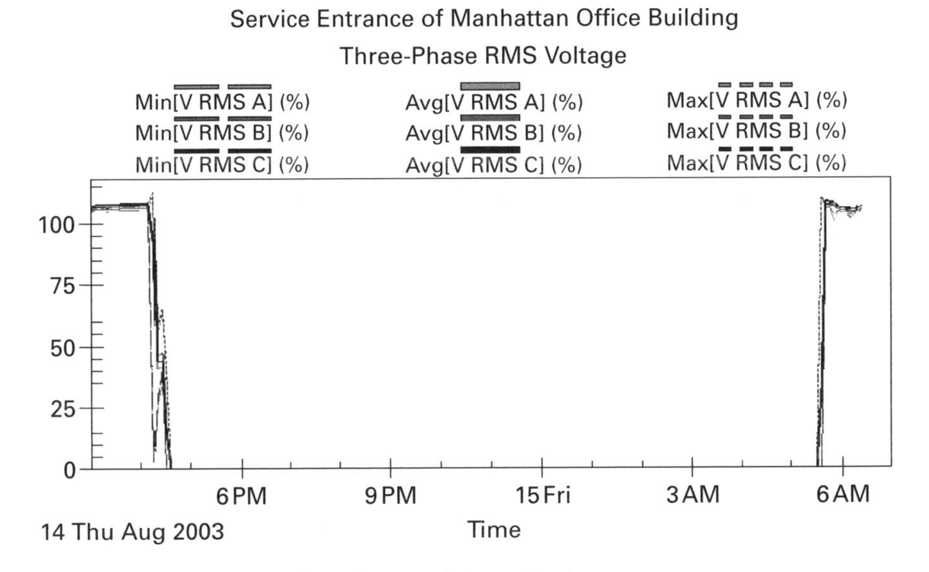

Figure 9.15 Interruption timeline. (Courtesy of Dranetz-BMI)

Table 9.4 Power Line Disturbances and Characteristics

	Type I Transient and Oscillatory Overvoltage	Type II Momentary Undervoltage or Overvoltage	Type III Outage
Causes:	• Lightning power network switching (particularly large capacitors or inductors) • Operation of on-site loads	• Faults on power system • Large load changes Utility equipment malfunctions • On-site load changes	• Faults on power system • Unacceptable load changes • Utility or on-site equipment malfunctions
Threshold Level:	200–400% rated RMS voltage or higher (peak instantaneous above or below rated RMS)	Below 80–85% and above 110% of rated RMS voltage	Below 80–85% rated RMS voltage
Duration:	Spikes 0.5–200 μsec up to 16.7 m sec at frequencies of 0.2–5 kHz and higher	From 0.06 to 1 sec depending on type of power system and on-site distribution	From 2 to 60 sec if correction is automatic; unlimited if manual
Duration in cycles (60 Hz)	0	0.5	120

the important factor because the longer the problem lasts, the worse the effects on an electrical system.

9.5 TOLERANCES OF COMPUTER EQUIPMENT

Computer tolerances of unstable power vary from manufacturer to manufacturer: The most important considerations are the duration and magnitude of the disturbances it can withstand. Understanding the tolerances of computer processing equipment—how well the computer power supply will bear up to transient-induced stress—will assist the facility manager in avoiding information loss and equipment failure. Most solid-state electronic loads will shut down after a certain amount of power instability. If tolerances are known, a facilities manager can take measures to prevent this type of inadvertent shutdown.

The typical RMS voltage tolerance for electronic equipment is +6% and −13% normal voltage, although some electronic equipment requires tighter voltage tolerances. When voltages spike or sag below the computer manufacturer's tolerances, the equipment may malfunction or fail. In practice, sags are more common than spikes. Let's review:

- Starting large electric loads that pull line voltage below acceptable levels usually causes sags; large loads taken off-line usually cause spikes.
- Inadequate wire size causing overloaded circuits and poor electric connections (loose or partial contact) also commonly cause sags.

Consumers often identify sags by noticing momentary dimming in their lighting. A change of 1 V on a 120-V circuit can be seen in the reduced output of an incandescent lamp. More serious sags and surges may cause computer disk read/write errors and malfunctions of other electronic devices. Brownouts (persistent undervoltage condition) cause more general consumer concern due to widespread failures.

A steady state is a range of ±10% of nominal voltage. Voltage and current are monitored using root mean square (RMS) to determine whether it varies slowly or constantly. When a large number of nonlinear, single-phase devices operate from a three-phase (208/120 V) power system, the building wires, transformers, and the neutral conductors are subjected to increased risk from harmonic-induced overloading. A steady state is the goal, and the following curves provide guidelines to prevent mishaps.

9.5.1 CBEMA Curve

The Computer Business Equipment Manufacturers Association (CBEMA) curve represents the power quality tolerances by displaying comparative data. The curve provides generic voltage limits within which most computers will operate (purely a rule of thumb). This curve is used as a standard for sensitive equipment and as a generic guide, because most computers differ in specifications and tolerances.

When analyzing a power distribution system, the recorded data are plotted against the CBEMA curve (Figure 9.16) to determine what disturbances fall outside of the area between the two curves. For example, in Figure 9.16, point A represents recorded data that had no affect on the systems function. The next set of data, point B, occurred within milliseconds of point A; yet fell outside the acceptable range of the CBEMA curve and into the danger zone. Power disturbances that lie outside the "acceptable zone" of the CBEMA curve place the computer processing equipment at risk of losing information or shutting down.

9.5.2 ITIC Curve

CBEMA improved the rule of thumb with a new organization, called the Information Technology Industry Council (ITIC), and updated the comparative data. The ITIC curve is more applicable for computers and to test the performance of electronic equipment. There has been a significant increase in the use of new power conversion technology, so using a more economical curve makes the ITIC curve very useful.

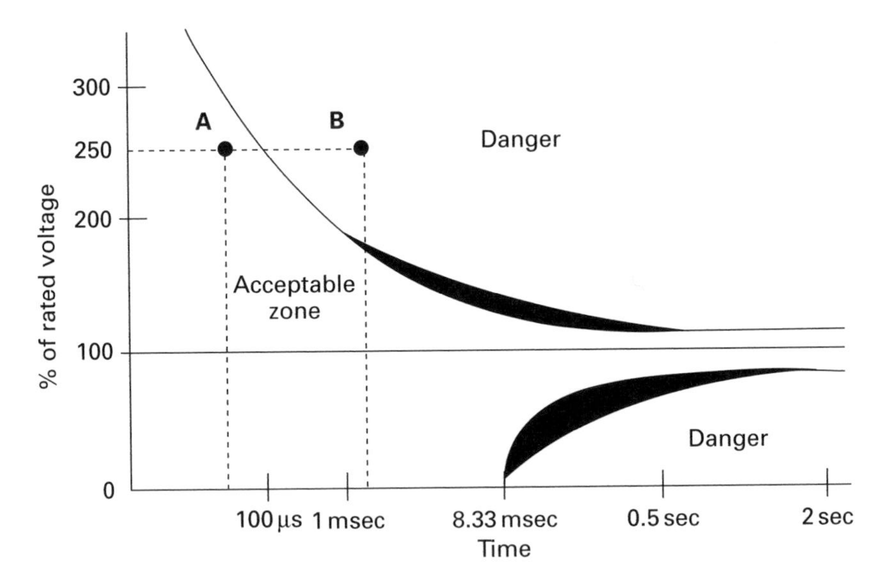

Figure 9.16 CBEMA curve.

The ITIC curve, like the CBEMA curve, has a tolerable zone where a system is generally considered safe from malfunction. Figure 9.17 shows how the ITIC curve is a more technical curve, typically used for higher-performance computer processing equipment.

9.5.3 Purpose of Curves

The CBEMA and ITIC curves may be applied to any electronic equipment that is sensitive to power disturbances, such as computers or programmable logic controllers. When looking at the CBEMA and ITIC curves, you can see the dangerous area versus the "voltage tolerance envelope." If a recorded voltage disturbance fell outside of the tolerable area, the equipment would be susceptible to disruption or an outage. Determining the number and type of disturbances outside the guidelines will assist the designer or facility manager in planning corrective action to prevent loss of critical data and equipment.

9.6 POWER MONITORING

Look at power as a raw material resource, instead of an expense. Today's digital environment is demanding, especially for uptime, nonlinear loads, and power quality assurance. Power monitors provide data for benchmarking the power infrastructure and then sense, analyze, and report disturbances continually. The FM can

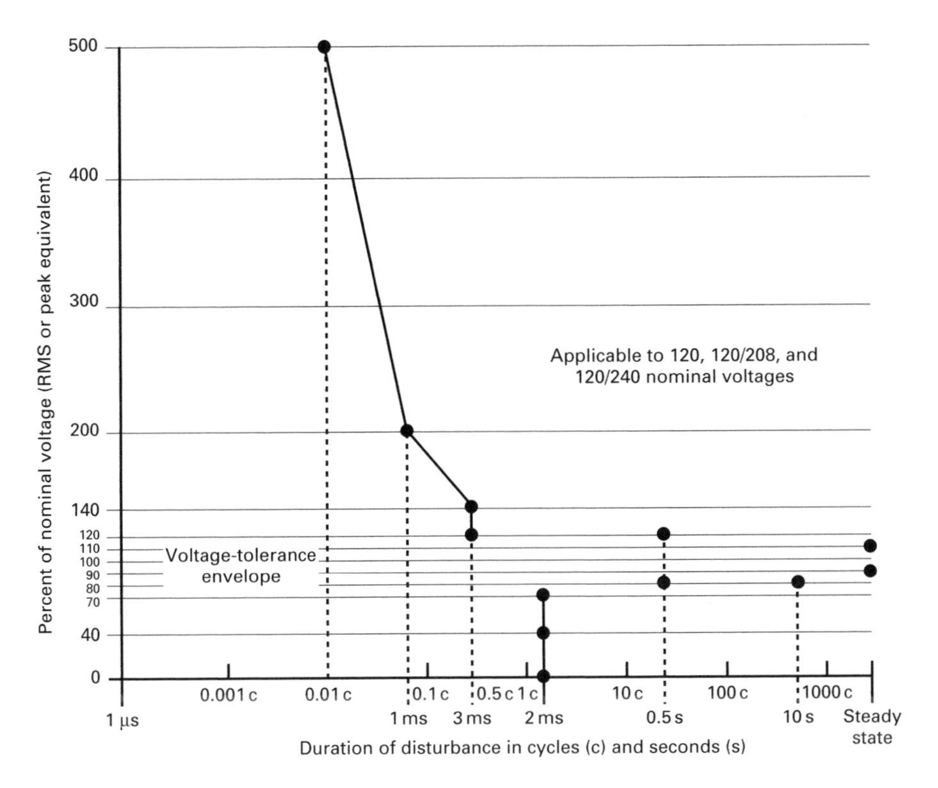

Figure 9.17 ITIC curve.

view voltage and current waveforms, view time sags and swells, study harmonic distortion, acquire transients, and then plot all the data on a CBEMA or ITIC curve.

Instead of reacting to problems individually, the FM should trend the numbers and types of disturbances. Only then can he properly evaluate remediation strategies and develop recommendations for power conditioning technologies. Acting promptly, but thoughtfully, to solve power disturbances will save money in downtime, equipment repair cost, and load analysis. Remember, power distribution systems are only reliable when closely monitored and properly maintained.

The primary purpose of power monitoring is to determine the quality of power being provided to the facility and the effects of equipment loads on the building distribution system. While disturbances may be expected, the source of the disturbance is not always predictable. Power monitoring lets the FM determine whether the power quality event occurred on the utility or facility side of the meter, then pinpoint the cause of disturbances and take countermeasures, hopefully before computer problems arise. For example, a voltage spike is consistently occurring around the same time every day. By monitoring the power, the FM can track down the source of the overload. By correlating disturbances with any equipment performance problems or failures, the FM makes a strong business case for remedial action.

Here we're looking at the importance of power in terms of cost. By managing power, you can maximize the use of power at minimum costs while protecting your sensitive electrical equipment. In computer equipment, improved power quality can save energy, improve productivity (efficiency), and increase system reliability—sometimes extending the useful life of computer systems.

Thanks to power management and analysis software now available, power quality monitoring is effective in a broad range of applications. Power costs comprise electrical energy costs, power delivery costs, and downtime. Common power disturbances affect all three of these cost centers and, thus, can be managed better with monitoring.

- Electrical energy costs come from the types and qualities of services purchased.
  The utility usually handles billing, but any billing errors would likely go undetected without power monitoring.
  Monitoring can save money by evaluating the energy costs to different departments. The FM can then reduce usage in vacant or underproducing cost centers.
  Power monitoring in combination with utility rate analysis can help the FM negotiate utility rates.
  Carefully conserving energy in noncritical areas during peak hours helps to avoid high electric demand charges.
- Power delivery costs comprise equipment, depreciation, and maintenance. Equipment stressed by unseen transients may require increased maintenance for consistent performance. Costs of repairs to maintain sensitive and expensive equipment continually increase.
- Downtime is the easiest way to lose a whole day of profit without taking a day off. Retailing, credit card transactions, and many types of data processing are susceptible to losing millions of dollars for every hour of downtime.

  If the tolerances of the equipment are known, then consistently staying within those limits is cost-effective.

Data from the Electric Power Research Institute shows the duration of interruptions as percentages of all interruptions (Figure 9.18). The most common disturbances occur for only milliseconds to microseconds. Although these brief disturbances are almost invisible on a recording of a 60 Hz sine wave, they are nonetheless damaging to equipment. The most notable fact about this chart is the last column: 3% of interruptions last 30 minutes. With millions of dollars at stake, many companies would find such outages catastrophic.

Utilizing a power monitoring system also allows for improved efficiency and utilization of the existing power system. By having reliable, firsthand data, the FM can determine if and where additional load can currently be added, and then plan for future corporate and technological growth. In an organization that depends upon computerized systems for day-to-day operation, power monitoring data is, well, power.

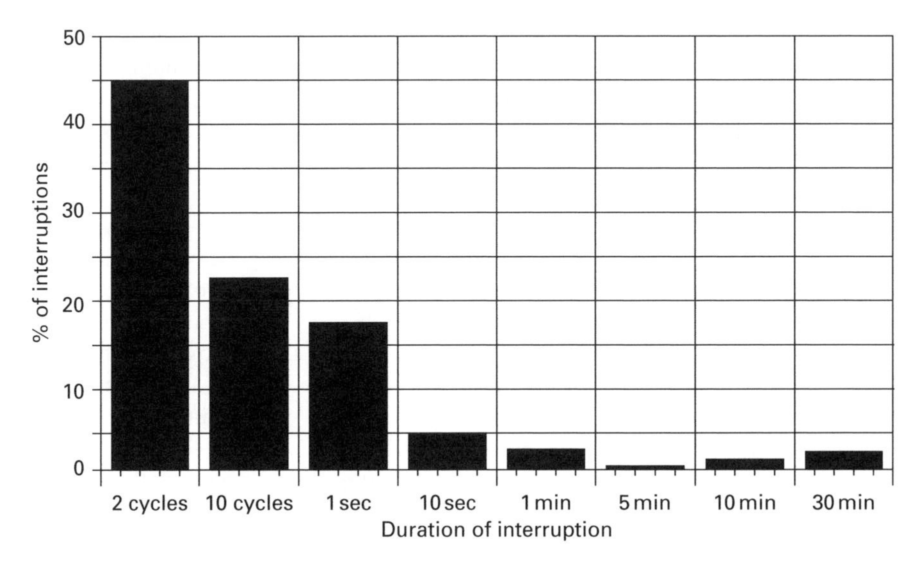

Figure 9.18 A portion of data obtained from the Distribution Power Quality Monitoring Project survey conducted by EPRI and Electrotek Concepts. This survey produced information on voltage sags and swells, transient disturbances, momentary disruptions, and harmonic resonance conditions.

9.6.1 Example Power Monitoring Equipment

- Dranetz-BMI
- Power measurement
- Reliable power meters

Some monitoring devices can be set to specific standards. For example, if current reaches a set level, high enough to generate heat, then the recording devices will detect that and record only the important data. If the monitor recorded data the entire day, then time and money would be wasted.

9.7 THE DEREGULATION WILDCARD

Ask your business these questions: How important is the continuity of your power supply? How will utility deregulation affect the risk of disruption?

Deregulation is the government's way of preventing monopolies in the utility field. By changing some laws, smaller companies are able to supply power to consumers without restrictions imposed by large utilities. The intended benefit of this action is cheaper power. Utilities will compete to serve end-users, and the open market will set the price.

The problem with deregulation is the quality of power supplied. Power produced at a cheap price and sold at a cheap price is probably cheap power. The least-cost utility will have the greatest number of clients, which may cause widespread problems. Making a phone call on Christmas is a good analogy: depending on your provider, you may have to wait for the lines to clear off traffic before you can use the phone. Such factors are crucial to a mission critical facility, where power cannot diminish during peak hours, surges and swells cannot be tolerated, and outages cannot be risked.

With the technology revolution permeating all levels of society, the awareness of power quality becomes a very important issue. Today's sophisticated technology makes even instantaneous disturbances problematic. With little assurance of continuity of power, sophisticated backup power will be a necessity.

Under deregulation, many utility companies need to cut costs and have no choice but to terminate employees and postpone overdue system upgrades. Monitoring your electric distribution system can warn you of impending utility failures and avert disaster. The more knowledge you have about your own system, the better you can service its needs. This becomes important when negotiating with your energy provider.

9.8 TROUBLESHOOTING POWER QUALITY

Power quality problems are common and stem from a variety of sources. Power monitoring can help determine the source and the magnitude of these problems, before incurring damage to expensive equipment. Here is a prescription to find the simplest, least-expensive solutions.

1. Interview the personnel whom observe the operation and take measurements. Use the data to decide what type of problem has occurred.
2. Identify a solution.
3. Implement solution and fix the problem.
4. Observe and measure all data in the problem area to confirm the solution.
5. Institute a procedure to prevent this type of problem in the future.

Power problems are not always easy to find, and a checklist makes troubleshooting much simpler. Figure 9.19 provides a checklist for load problems. A record of the occurrence can help pinpoint load problems by revealing specific times that a problem occurs and what other indications or symptoms were coincident. Figure 9.20 can be used for troubleshooting building distribution problems; Figure 9.21 can be used for facility transformer and main service equipment problems.

POWER QUALITY TROUBLESHOOTING CHECKLIST
LOAD PROBLEMS

Problem Observed or Reported:
☐ Overloads Tripping ☐ Computer Hardware Problems ☐ Erratic Operation ☐ Shortened Life
☐ Other ____________________

Load Type:
☐ Computer ☐ Printer/Copier ☐ PLC ☐ Drive ☐ Motor ☐ Other ____________________

Problem Pattern:
Day(s) of week:
☐ Continuous ☐ Random ☐ Monday ☐ Tuesday ☐ Wednesday ☐ Thursday ☐ Friday ☐ Saturday ☐ Sunday
Time(s):
☐ Continuous ☐ Random ☐ Always Same Time ______ ☐ Morning ☐ Afternoon ☐ Evening ☐ Night

Problem or Load History:
Has problem been observed or reported before? ☐ No ☐ Yes ____________________
Was any corrective action taken? ☐ No ☐ Yes ____________________
Are other loads affected? ☐ No ☐ Yes ____________________
Are nonlinear loads in area, or on same circuit? ☐ No ☐ Yes ____________________
Has there been any recent work or changes made to system lately? ☐ No ☐ Yes ____________________

Possible Problem(s):
☐ Operator Error ☐ Power Interruptions ☐ Sags or Undervoltage ☐ Swells or Overvoltage ☐ Harmonics
☐ Transients or Noise ☐ Improper Wiring/Grounding ☐ Undersized Load ☐ Undersized System
☐ Improperly Sized Protection Devices ☐ Other ____________________

Measurements Taken at Load:
Line Voltage ____ V, Neutral -to-Ground Voltage ____ V, Current ____ A, Power ____ W, VA, ____ VAR
Power Factor ___ PF, ___ DPF, Voltage THD ___, Current THD ___, K Factor ___ Other ____________

Waveform Shape:
Voltage Waveform:
☐ Sinusoidal ☐ Non - Sinusoidal ☐ Flat - Topped ☐ Other ____________________
Current Waveform:
☐ Sinusoidal ☐ Non - Sinusoidal ☐ Pulsed ☐ Other ____________________

Measurements Taken at Load (Over Time):
Normal Voltage _____ V, Lowest Sag ____ V at ____ (Time), Highest Swell ____ V at _____ (Time)
Highest Inrush Current _____ A at _____ (Time)
Number of Transients Recorded ____ over a time period of __________ at a level of _____ % above normal

Possible Problem Solution(s):
☐ UPS ☐ K-Rated Transformer ☐ Isolation Transformer ☐ Zig-Zag Transformer ☐ Line Voltage Regulator
☐ Surge Suppressor ☐ Power Conditioner ☐ Proper Wiring and Grounding ☐ Harmonic Filter ☐ Derate Load
☐ Proper Fuses/CBs/Monitors ☐ Other ____________________

Figure 9.19 Load problems checklist.

POWER QUALITY TROUBLESHOOTING CHECKLIST

BUILDING DISTRIBUTION PROBLEMS

Problem Observed or Reported:
☐ CBs Tripping/Fuses Blowing ☐ Conduit Overheating ☐ Overheated Neutrals ☐ Electrical Shocks
☐ Damaged Equipment ☐ Humming/Buzzing Noise ☐ Other ____________________

Distribution Type:
☐ 1ϕ ☐ 3ϕY ☐ 3ϕΔ ☐ Fuses ☐ CBs ☐ Voltage(s) ____ V ☐ Amperage Rating ____ A ☐ Other __________

Problem Pattern:
Day(s) of week:
☐ Continuous ☐ Random ☐ Monday ☐ Tuesday ☐ Wednesday ☐ Thursday ☐ Friday ☐ Saturday ☐ Sunday
Time(s):
☐ Continuous ☐ Random ☐ Always Same Time ________ ☐ Morning ☐ Afternoon ☐ Evening ☐ Night

Problem or Distribution History:
Has problem been observed or reported before? ☐ No ☐ Yes ____________________
Was any corrective action taken? ☐ No ☐ Yes ____________________
Are other parts of system affected? ☐ No ☐ Yes ____________________
Are nonlinear loads in area, or on same circuit? ☐ No ☐ Yes ____________________
Has there been any recent work or changes made to system lately? ☐ No ☐ Yes ____________________
Are large power loads being switched ON/OFF? ☐ No ☐ Yes ____________________
Is panel properly grounded? ☐ No ☐ Yes ____________________

Possible Problem(s):
☐ Conductors Undersized (Hot) ☐ Neutral Conductors Shared/Undersized ☐ High Number of Nonlinear Loads
☐ Voltage/Current Unbalance ☐ Harmonics ☐ System Undersized ☐ Improper Wiring/Grounding ☐ Other ____

Measurements Taken:
Taken at Panel ______ Located at ____________________
Voltage _____ V, Current _____ A, Power _____ W, _____ VA, ____ VAR, Power Factor _____ PF, _____ DPF
Voltage THD _____, Current THD _____, K Factor _____, Other ____________

Waveform Shape:
Voltage Waveform:
☐ Sinusoidal ☐ Non-Sinusoidal ☐ Flat-Topped ☐ Other ____________
Current Waveform:
☐ Sinusoidal ☐ Non-Sinusoidal ☐ Pulsed ☐ Other ____________

Measurements Taken at Load (Over Time):
Normal Voltage _____ V, Lowest Sag ____ V at ____ (Time), Highest Swell ____ V at _____ (Time)
Number of Transients Recorded ____ over a time period of __________ at a level of _____ % above normal

Possible Problem Solution(s):
☐ Oversize Neutrals ☐ Run Separate Neutrals ☐ Additional Transformer ☐ Separate Loads ☐ Harmonic Filter
☐ Proper Wiring and Grounding ☐ Proper Fuses/CBs/Monitors ☐ Additional Subpanel ☐ Surge Suppressor
☐ Power Factor Correction Capacitors ☐ Other ____________________

Figure 9.20 Building distribution problems checklist.

POWER QUALITY TROUBLESHOOTING CHECKLIST

FACILITY TRANSFORMER AND MAIN SERVICE EQUIPMENT PROBLEMS

Problem Observed or Reported:
☐ CBs Tripping/Fuses Blowing ☐ Conduit Overheating ☐ Overheated Neutrals ☐ Electrical Shocks
☐ Damaged Equipment ☐ Other ______________________________

Distribution Type:
☐ 1ϕ ☐ 3ϕY ☐ 3ϕΔ ☐ Fuses ☐ CBs ☐ Voltage(s) ____ V ☐ Amperage Rating ____ A ☐ Other __________

Problem Pattern:
Day(s) of week:
☐ Continuous ☐ Random ☐ Monday ☐ Tuesday ☐ Wednesday ☐ Thursday ☐ Friday ☐ Saturday ☐ Sunday
Time(s):
☐ Continuous ☐ Random ☐ Always Same Time ________ ☐ Morning ☐ Afternoon ☐ Evening ☐ Night

Problem or Distribution History:
Has problem been observed or reported before? ☐ No ☐ Yes ______________________
Was any corrective action taken? ☐ No ☐ Yes ______________________
Are other parts of system affected? ☐ No ☐ Yes ______________________
Have additional loads been added to system? ☐ No ☐ Yes ______________________
Has there been any recent work or changes made to system lately? ☐ No ☐ Yes ______________________
Are large power loads being switched ON/OFF? ☐ No ☐ Yes ______________________
Is main service panel properly grounded? ☐ No ☐ Yes ______________________
Are any subpanels grounded? ☐ No ☐ Yes ______________________
Has there been a recent lightning storm? ☐ No ☐ Yes ______________________
Has there been a recent utility feeder outage? ☐ No ☐ Yes ______________________

Possible Problem(s):
☐ Conductors Undersized (Hot) ☐ Neutral Conductors Shared/Undersized ☐ Higher Number of Nonlinear Loads
☐ Voltage/Current Unbalance ☐ Harmonics ☐ System Undersized ☐ Improper Wiring/Grounding
☐ Other ______________________________

Measurements Taken:
Taken at Panel ______ Located at ______________________________
Voltage ____ V, Current ____ A, Power ____ W, ____ VA, ____ VAR, Power Factor ____ PF, ____ DPF
Voltage THD ____, Current THD ____, K Factor ____, Other ______________

Waveform Shape:
Voltage Waveform:
☐ Sinusoidal ☐ Non-Sinusoidal ☐ Flat-Topped ☐ Other ________________
Current Waveform:
☐ Sinusoidal ☐ Non-Sinusoidal ☐ Pulsed ☐ Other ________________

Measurements Taken at Load (Over Time):
Normal Voltage ____ V, Lowest Sag ____ V, Highest Swell ____ V
Number of Transients Recorded ____ over a time period of ________ at a level of ____ % above normal

Possible Problem Solution(s):
☐ Oversize Neutrals ☐ Run Separate Neutrals ☐ Additional Transformer ☐ Harmonic Filter
☐ Change to K-Rated Transformer ☐ Add Subpanel ☐ Separate Loads ☐ Proper Wiring and Grounding
☐ Proper Fuses/CBs/Monitors ☐ Power Factor Correction Capacitors ☐ Change Transformer Size
☐ Surge Suppressor ☐ Other ______________________________

Figure 9.21 Facility transformer and main service equipment problems.

Chapter 10

An Overview of UPS Systems: Technology, Application and Maintenance

10.1 INTRODUCTION

Computer systems and applications are growing more important every day. It is difficult to find a business that is not dependent on information technology in its daily operations. From price-scanning cash registers to integrated manufacturing control systems, to 24/7 banking and finance, uninterrupted information flow has become imperative in the operation of every size business.

Today's mission critical infrastructure—infrastructure systems which, if a failure occurs, puts the enterprise at risk of extreme liability or failure—includes communication technology operations such as e-commerce, e-mail, and the Internet that were distant ideas to the average IT department not more than 10 years ago. Take this critical communications infrastructure and combine it with diverse networks, databases, applications, hardware and software platforms that operate just-in-time inventory control, manufacturing scheduling, labor tracking, raw material ordering, financial management, and so on, and the dependence on information flow grows even greater. Facilities management professionals must bridge the platform interface differences to provide an integrated and seamless technology solution that supports the management of all devices and applications. These are large, complex, and real problems faced every day.

Mission-critical power solutions for enterprise management must be available for organizations and systems both large and small. When power flow ceases, so will the flow of information driving the enterprise. Power solutions must be able to provide the exact solution for a given environment, regardless of the size of the application, Enterprise-wide systems demand flexibility and reliability to maximize performance and system availability.

Minimizing the number of systems, subsystems, and components introduced into a given network application simplifies the delivery of reliable power. When it comes to managing these devices, facility professionals face enormous difficulties in keeping up with the complex structure and performance of their networks. The implementation of high availability, continuous software applications at major data centers requires constant processing seven days a week, 24 hours a day. Such continuous operation also means an absence of unscheduled or scheduled interruptions for any reason, including routine maintenance, periodic system level testing, modifications, upgrades, or failures. Various UPS configurations are available today to satisfy these stringent reliability and maintainability requirements.

By deploying the proper power management solution, facility managers gain control of critical power resources while adding functionality and protection to multiple networks. Many power management products offer easy installation, scalable architecture, remote monitoring, shutdown and control, and other convenient features, that can increase total system availability while decreasing maintenance costs and risks for protected equipment. For today's mission critical facility managers, critical power management is an essential tool for providing comprehensive power protection and optimal functionality. With the wide range of choices available, selecting the right power protection architecture for your network can be a daunting decision. On the following pages are the basics of UPS Systems, descriptions of UPS power solutions, configurations you can implement system-wide, and UPS maintenance guidelines

The solution that, arguably, most reliably eliminates all short and long duration power problems less then 15 minutes is an 'on-line' uninterruptible power supply (UPS). On-line UPSs can be designed to operate in various configurations to create systems that vary from small to large capacity and moderate to extremely high availability. On-line UPSs are preferred for all size loads that require a reliable 24/7 operation.

10.2 PURPOSE OF UPS SYSTEMS

In the Industrial Era, our economic system developed a dependency on electric power. Electricity became necessary to provide lighting, energy to spin motors and heat to use in fabrication processes. As our electrical dependency grew, processes were developed that required continuous uninterrupted power. For example, car manufacturers use a step-by-step plan for painting new cars. Every part must be coated precisely and if any errors occur the process must be redone. A power interruption to this process would cost the company a large amount of money in downtime and can delay production deadlines. Manufacturers could not afford to lose power to critical motors or control systems that would shutdown production.

Out of this need for uninterrupted power as the industrial era became the information era, creative electromechanical and electrochemical solutions were developed. The simplest was attaching a backup battery system to the electrical load; however, this arrangement only worked for direct current (DC) powered devices. As alternating current (AC) became the preferred method of transmission/distribution systems, it became more difficult to provide standby power. The added task of remaining in synchronization with the standard frequency (60 hertz in North America, 50 hertz in Europe, most of Asia) had to be accomplished.

The advent of today's Information Age has added the task of providing back-up power without causing an interruption of even 1/2 cycle or 8 milli-seconds (ms) to avoid unplanned shutdown of computer systems. Uninterruptible power sources have become a necessity for powering large or small systems in facilities where personal safety and financial loss is at risk if power is interrupted.

Computers and data communications can be no more reliable than the power from which they operate and the continuous cooling they require. Reliability is a factor that must be accounted for in at least six different areas:

1. Utility / standby power (long-term backup)
2. UPS (short-term backup)
3. HVAC
4. Operations
5. IT power supplies
6. Software

Each different area is reviewed and a suitable UPS system type and configuration is chosen. Utility power varies in reliability from utility to utility, but is generally taken to be no better than available 99.9% of the time. Even within the territory served by a utility, there can be, and often are, wide variances in reliability. Applying the availability of 99.9% to a year, the utility could have interruptions totaling 8 hours and 46 minutes per year. Utility power with emergency generator backup was no longer adequate, thus the uninterrupted power supply (UPS) industry was created.

One advantage of online UPS systems is that they are always operating, guarding against interruptions and a host of other power quality problems. In the event of a utility power outage, UPS systems take power from a backup DC power source (e.g., a battery or flywheel) in order to provide continuous AC output power. The UPS will continue to supply backup power until (a) the utility power returns, (b) an emergency standby power source, such as an engine-generator, can be brought on line, or (c) the backup DC power source is exhausted.

An online UPS system can be generally characterized as a system that normally operates in a forward transferred position with most or all of the power conditioning and output waveform shaping equipment carrying the load most or all of the time. In a full online, double conversion (rectifier-inverter or motor-generator) UPS design, when the input power fails the input of the UPS (rectifier or motor) shuts down but the output (inverter or generator) continues to operate as if nothing happened with no change of state. Therefore the UPS (excluding DC backup) does not have a significant failure rate during an input power failure. If the power conditioning or waveform shaping equipment randomly fails while utility (or generator power) is available, an online UPS will automatically and rapidly reverse transfer the load directly to the utility bypass source. In the event of a UPS failure or overload, the load is saved, the failure is acknowledged and repaired, and the load is forward transferred back onto normal UPS power. An advantage of online UPS designs is that failures may occur randomly while utility bypass power is available to save the load. An advantage of full online, double conversion (rectifier-inverter or motor-generator) designs is that the power is completely conditioned to tight specifications and the output is completely isolated from the input. Disadvantages of online UPS designs include higher cost and lower efficiency, and for very small systems such as plug-in UPS units, the continuous fan noise can be annoying.

There are many UPS designs that can be generally characterized as offline or line-interactive. These designs generally operate normally in a fully or partially bypassed mode most of the time. They pass normal unconditioned or partially conditioned utility (or generator) power to the load up until the normal power fails or degrades outside of acceptable specifications. Then the power conditioning and wave shaping equipment rapidly ramps up or switches on from an off state to save the load. The batteries or other stored energy (flywheels, etc.) of offline or line-interactive UPS systems must be drawn down to correct for source voltage and frequency variations that an online UPS can correct without drawing down the batteries. To minimize battery drawdown or "hits" on the battery, these UPS designs typically allow a wider window of power imperfections to be passed onto the load. Fortunately the vast majority of modern IT equipment can easily tolerate much wider ranges of voltage and frequency, and brief power outages that older IT equipment. Many IT power supplies today are designed for application worldwide and can to operate from roughly 90–250 V and 45–65 Hz. Therefore power availability has become much more important that power quality. Advantages of offline and line-interactive UPS designs include lower cost, higher efficiency, smaller physical size and quieter normal operation. A significant disadvantage of offline and line interactive UPS designs is that the offline or lightly loaded portion of the system must rapidly respond to a power failure. Since the UPS response to a power failure is a significant change of state for the backup equipment, failure rate of the UPS is highest when you need it the most. The classic analogy is the lightbulb—it typically burns out during change of state, when it is turned on from cold, not while it is hot.

An emergency standby power system is off until it is manually or automatically started, when the utility power fails. It may take as little as 3 seconds or as long as several minutes to start a standby generator system, depending whether it's a diesel or turbine engine, the setting of timer delays, and the size of the system. If two or more generators need to synchronize before enough capacity is online to accept load this can add several seconds or more to the startup time. This amount of time can affect systems in a way that they will not be able to function reliably if the UPS and DC backup was not properly designed and integrated into the system.

Equipment can be divided into four categories based on the reliability requirements and the time sensitivity the power supply can operate without electrical power within specifications to the power supply:

1. Data processing equipment that requires uninterrupted power
2. Safety equipment defined by codes that requires power restoration within seconds
3. Essential equipment that can tolerate an interruption of minutes
4. Noncritical equipment that accepts prolonged utility interruptions

10.3 GENERAL DESCRIPTION OF UPS SYSTEMS

10.3.1 What Is a UPS System?

UPS stands for "uninterruptible power supply." Critical loads are usually protected with a backup source of power. When the backup power is constantly available without any interruption it is commonly referred to as an uninterruptible power supply system. Under any condition the power supplied from the UPS would not be interrupted, broken, disturbed or changed in any way that would affect the protected operation. UPS systems maintain a clean electrical supply to a chosen load or group of loads at all times. UPS systems can regulate, filter and provide backup power in the event of a power interruption. They also can reduce or eliminate power quality problems from low voltages to high-speed transients.

A UPS system can be considered to be a miniature power plant. At any sign of a power quality problem, the UPS system protects by providing clean regulated power. A UPS system maintains power to the critical load in the event of a partial or total failure of the normal source of power, typically, the power that is supplied by the electric utility or an on-site prime or standby power system.

10.3.2 How Does a UPS System Work?

The UPS system requires some sort of input power produced at a power plant through a mechanical, physical or chemical process. If something happens to the input power produced by the local utility plant, the UPS uses batteries (or other stored energy), diesel/turbine engine, or a combination of both to provide sufficient power for the system to use. UPS systems can be found in sizes from 50 VA to greater then 4000 kVA. There are three main types of UPS systems: static, rotary (generator), and hybrid.

Static UPS systems, which generate the necessary output power without any rotating power components, condition and/or convert varying input power to dependable, stable output power. Full online, double conversion UPS designs change the alternating current (AC) power supply to direct current (DC) with a rectifier and then change it back to AC with an inverter. Batteries (or flywheels) are used to store energy. The length of time of backup power is determined by the size of the battery plant. For long duration power outage protection, UPS systems use batteries long enough for a standby generator to be reliably started, which can then be used for as long as fuel is available.

10.4 STATIC UPS SYSTEMS

Static UPS systems are electronically controlled solid-state systems designed to provide an continuous source of conditioned, reliable electrical power to critical equipment. This power is provided without rotating power components. Double conversion static systems electronically rectify AC utility power into DC power, which is then paralleled with a DC power storage device, typically a battery. The battery is kept fully charged during normal operation. The DC power is then inverted electronically back into AC power that supplies the critical load. In the event of a loss of utility power, energy from the DC energy storage system is instantaneously available to continue providing power to the critical load. The result is effectively an uninterrupted AC power supply.

It is difficult to differentiate between all the providers of static UPS in terms of reliability because there are a tremendous number of variables. Most providers offer a high degree of sophistication ranging from simple power conversion to a sophisticated synchronization and filtering arrangements. More often, the reliability of a UPS system is driven by the overall system design, operation and maintenance than by the particular type of internal UPS technology. Volume of production of a UPS model also improves reliability.

UPS reliability will typically follows the "bathtub" curve. New units have a higher infant mortality failure rate which rapidly decreases over time. Then the failure rate goes up again near end-of-life. UPS systems greater than 5 years old have increased risk of component failure, including batteries, capacitors, diodes, contactors, motor operators and power switching devices including transistors, thyristors and silicon controlled rectifiers (SCRs). If the UPS is not properly maintained the user should not expect the system to be reliable just as if an automobile was not properly maintained. You can't expect continuous availability and operation without the proper preventative maintenance. Static UPSs are complicated, highly technical pieces of equipment. With proper maintenance and change out of components, larger UPS systems reliably last 15 years or more. Routine maintenance service and technical knowledge about the device are important factors in reliable service of the equipment.

All power protection devices are not created equal, so understanding their capabilities and limitations is very important. Below is an explanation of the leading UPS technologies and the level of protection they offer. There are several types of static UPS designs on the market including *online* double conversion, *online delta conversion, offline* or standby, and line-interactive UPS designs.

10.4.1 Online

Most engineers believe that online UPS systems provide the highest level of power protection and are the ideal choice for shielding your organization's most important computing and processing installations. An online double conversion UPS provides a full range of power conditioning, correcting over and under voltage, voltage and frequency transients, and power-line noise This technology uses the combination of AC-to-DC conversion and DC-to-AC conversion power circuitry and components to continuously power the load, to provide both conditioned electrical power and line disturbances/outage protection.

Online double conversion UPS designs offer complete protection and isolation from all types of power quality problems: power surges, high-voltage spikes (within limitations), switching transients, power sags, electrical line noise, frequency variation, brownouts and short duration blackouts. In addition, they provide consistent computer-grade power within tight specifications not possible with offline systems. For these reasons, they have traditionally been used for mission-critical applications that demand high productivity and systems availability. Today the vast majority of IT equipment can easily tolerate much wider ranges of voltage and frequency, and brief power outages that older IT equipment. Many IT power supplies today are designed for application worldwide and can to operate from roughly 90–250 V and 45–65 Hz. The tight output voltage specifications available with double conversion UPS technology is no longer needed. Simply put, power availability has become much more important that power quality.

10.4.2 Double Conversion

As previously mentioned, this technology uses the combination of a double-conversion (AC-to-DC and DC-to-AC) power circuit, which continuously powers the load, to provide both conditioned electrical power and outage protection. With the rectifier/inverter (double conversion) UPS, input power is first converted to DC. Normal supply of power feeds the rectifier and the AC load is continually supplied by the output of the inverter. The DC power is used to charge the batteries and to continuously operate the inverter under any load conditions required. The inverter then converts the DC power back to AC power. There is an instantaneous static bypass switch path should any components fail in the rectifier/inverter power path [the static bypass switch is part of the UPS system]. Figure 10.1 shows the basic schematic of a double conversion online UPS system.

10.4.3 UPS Power Path

The following figures illustrate the path power takes during each type of operation. Normally AC power is rectified to DC, with a portion used to trickle charge the battery and the rest converted back to AC and supplied to the critical load (Figure 10.1).

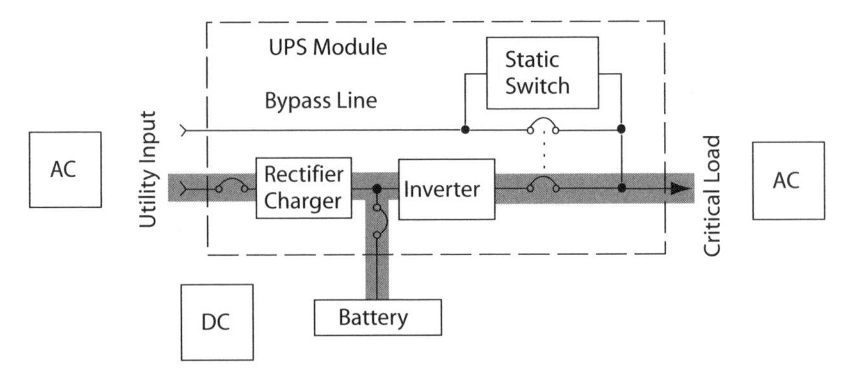

Figure 10.1 Double conversion: In normal operation. (Courtesy of Emerson Network Power.)

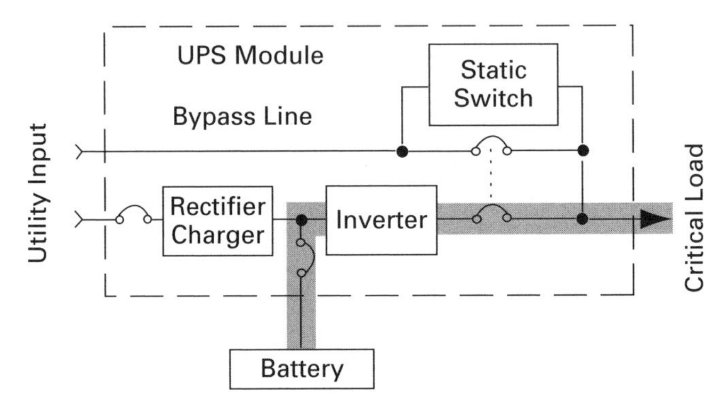

Figure 10.2 Double conversion: Emergency (on battery). (Courtesy of Emerson Network Power.)

When the utility power fails, the battery takes over supplying the power to the critical load after being inverted (Figure 10.2). Since the battery stores DC energy, the power is changed to AC by the Inverter and then supplied to the critical load. This is supplied for only a limited amount of time (depending on the size of the battery plant), typically 5–15 minutes. Much longer (multiple hours) runtimes are possible but not realistic due to the cost and space requirements for batteries and the need to provide cooling to the loads and to the UPS system. It is not feasible to power HVAC by the UPS system.

In the event of an overload or random failure in the rectifier/inverter power path, the utility power is typically available at the bypass input and the UPS system's static bypass switch instantly reverses transfer the critical load to the bypass utility source. Figure 10.3 shows input power fed directly through the static bypass switch to the critical load.

From time-to-time the UPS system will need to be taken offline for maintenance ("bypassed"). In addition to the static (electronic) bypass, UPS systems may have an internal mechanical bypass (as shown), usually a contactor or motorized

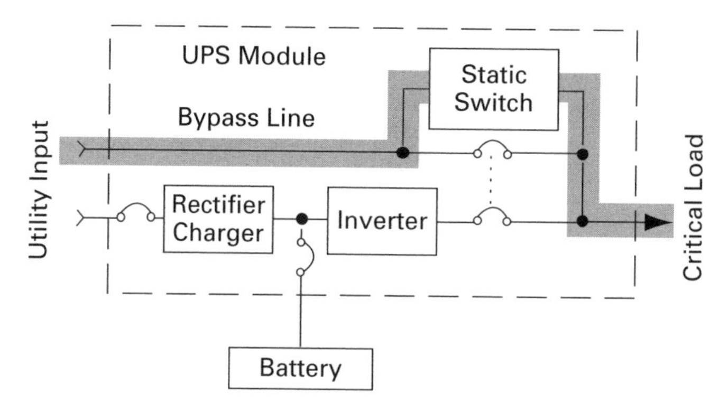

Figure 10.3 Double conversion: Initial bypass. (Courtesy of Emerson Network Power.)

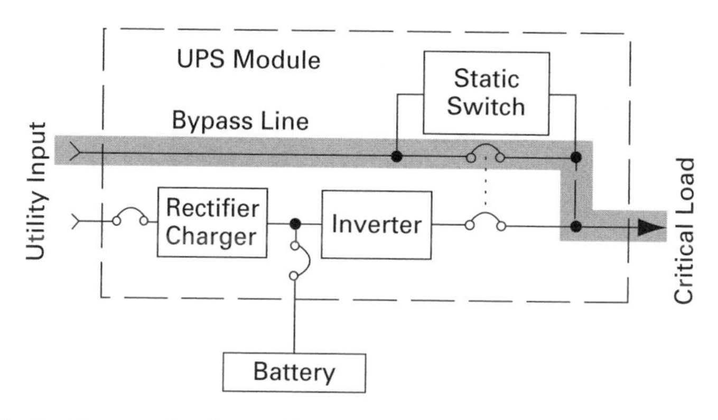

Figure 10.4 Double conversion: Bypass. (Courtesy of Emerson Network Power.)

switch / circuit breaker or external bypass power path. Large UPS systems typically include automatic static, and manual internal and external bypass paths. Because unconditioned utility power may be undesirable to utilize even under maintenance conditions, some users will switch to a standby generator or a redundant UPS system as the bypass source rather than using the utility source for this purpose. Figure 10.4 shows how the bypass power source may feed unconditioned power to the critical load.

10.5 COMPONENTS OF A STATIC UPS SYSTEM

10.5.1 Power Control Devices

10.5.1.1 Silicon Controlled Rectifier (SCR)

An SCR is a solid-state electronic switching device that allows a relatively small current signal to turn on and off a device that is capable of handling hundreds of amperes of current. The switch can be operated at a speed much faster than a mechanical device, into the thousands of hertz range.

The SCR is composed of silicon-based semiconducting materials. In their de-energized state, the materials are excellent insulators and will not allow current to pass through. Yet by applying a slight current via a gate signal, the SCR is instantaneously energized and becomes an excellent conductor of electricity. The SCR remains an excellent conductor until its gate signal is de-energized and the current flow is stopped or reversed. Removal of the SCR gate signal by itself will not turn off an SCR. The ability to precisely time the turn off of an SCR was one of the major challenges for UPS inverter design for decades. Methods to rapidly turn off inverter SCRs include auxiliary commutation using stored energy in a bank of capacitors. To turn off the SCRs the bank of energy is released to back feed into the SCRs and reverse the current flow enough to reach a zero crossing. The result of aux commutation quite often was blown fuses. Modern high power transistors (please see IGBT below) have largely made SCRs obsolete for inverter designs. SCRs are typically still used in the rectifier or battery charger portion of a UPS as AC power control switches. There is no need to rapidly turn off the SCRs in a rectifier / charger. Because the AC power naturally reverses itself 60 times per second, SCRs are an efficient, cost effective device to convert AC power into DC power.

10.5.1.2 Integrated Gate Bipolar Transistors (IGBT)

There are other devices that function similar to the SCR and operate more efficiently in DC applications, such as the insulated gate bipolar transistor (IGBT). A key feature of the IGBT is that it will rapidly turn off with the removal of its gate signal, therefore making it better suited for switching the DC power from the rectifier output or a storage battery. The newer designs for UPS systems utilize the IGBT technology in the inverter section of the UPS.

Using IGBTs in UPS rectifiers can eliminate the majority of input voltage and current harmonic problems associated with silicon control rectifiers (SCRs). With the pulse-width-modulated (PWM)-controlled IGBT UPS rectifier, harmonics can be reduced to between 4% and 7% without custom input filters and resulting leading power factor concerns, and regardless of load from 50% to 100%. However, IGBT-based rectifiers are less efficient and more costly.

10.5.1.3 Rectifier/Battery Charger

In double conversion UPS designs, the rectifier/charger converts alternating current supplied from a utility or generator into a direct current at a voltage that is compatible for paralleling with battery strings. When the UPS is supplied by normal input power, the rectifier provides direct current to power the inverter, batteries are trickle charged and control logic power is provided. The rectifier/charger is sized to carry full UPS load plus battery trickle charge load continuously plus short term battery recharge short term. UPS designs including rotary, delta conversion, line-interactive and offline use a smaller rectifier/charger sized only to recharge the battery, flywheel, or other energy storage. Most major manufacturers of UPS have separate power supplies with dual inputs for internal control power redundancy. They can take power from either the incoming AC source or from the protected inverter output. A few UPS designs have a third input and a DC-to-DC converter to take power from the DC bus.

The rectifier splits the alternating current waveform into its positive elements and negative elements. It does this by using SCRs or a diode bridge in smaller UPS systems to switch the alternating positive and negative cycles of the AC waveform into a continuous positive and negative DC output. All but the largest three-phase UPS rectifiers are typically available with six-pulse rectifiers. UPS designs at 400 kW and higher may be available with twelve-pulse rectifiers, which reduce input harmonics with little or no filtering required. Twelve-pulse rectifiers, where available, cost more and require more space than six pulse rectifiers.

Most UPS manufacturers smooth out the bumps of the AC half cycles by using a DC output filter to prolong battery life. This filter is made of inductors and capacitors or what is referred to as a LC (inductive/capacitive) filter. Most manufacturers additionally use LC filters on the AC input and output for AC harmonic filtering. Alternate DC bus filtering include using DC caps and a DC choke. The choke and DC caps serve the purpose of the filter, but it normally isn't called a DC output filter.

Rectifier/battery chargers are often designed to limit their output to 125% or less of the UPS output continuous rating. Figure 10.5 identifies a typical rectifier schematic utilizing SCR's as the power control devices. No matter what the load placed on the UPS, the rectifier limits how much continuous power the system will produce. The rectifier capacity plus the battery limits how much power could be produced by the inverter. The inverter is typically sized to 100% of the UPS rating continuously, 125% for 10 minutes, 150% for 30 seconds, and 300% with voltage degradation for less than 1 cycle. Rotary UPS designs can typically handle short term overloads better than static designs. Current limiting is an important feature when considering the amount of instantaneous load that could be placed on a UPS

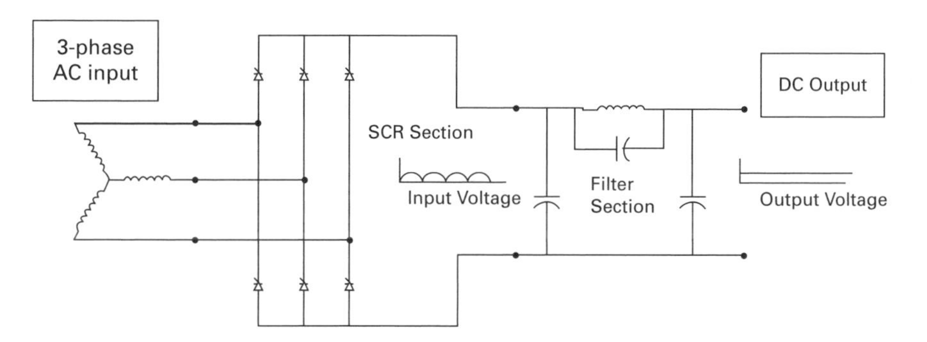

Figure 10.5 Typical three-phase rectifier schematic utilizing SCRs as the power control devices.

system if the output had suddenly experienced an electrical fault. Current limiting also prevents overloading of the upstream power distribution system Input current limits restrict the amount of power that the UPS will consume from the input power distribution system. There is a separate circuit for limiting the output current of the UPS, and it tolerates massive short-term overloads for fault clearing and PDU transformer inrush. Best practices include always switching UPS systems to bypass prior to energizing PDUs and other transformers. Since input current is limited, and since rectifiers cannot change their firing angle instantaneously to accommodate fast-rising step loads, the overload is supplied by the rectifier plus the battery up to 300% of rated load for a tiny time period. The bypass static switch is typically rated up to 1000% of rated load for 40 ms before transferring the load to mechanical bypass with a contactor or circuit breaker.

Another typical design feature is no more than 25% of the available current is supplied to the battery as charging current ensuring the battery recharge does not overload the AC distribution system following a battery discharge. The recharge time on a battery is a factor when there are multiple battery discharges. A rule of thumb is to allow 10 times the discharge rate to recharge the battery or spin a flywheel back up to speed. Therefore, a fifteen minute discharge would result in a 150 minutes of recharge time to restore the battery to 90% of capacity. The remaining 10% of charge for a battery can take several hours to several days.

10.5.1.4 Input Filter

The process of chopping an AC wave into its positive and negative components is much like throwing a stone into a pool of water. The initial wave created by the stone encounters the walls and reflects back to the center. The reflected waves combine and distort the initial wave. In the electronic world, the distortion is referred to as harmonic distortion. In this case, the concern is that the UPS current draw is distorted and the result is a reflection of that distortion in the source voltage. The amount of reflected harmonics in the voltage source is a function of source impedance. A utility source is typically much lower impedance than a standby generator, therefore voltage harmonics from UPS rectifiers are typically higher when measured with the UPS fed from a generator than from utility. The voltage harmonics on the source are not typically a concern for the "offending" UPS system, or the UPS load, but can be a concern for the source and/or other loads fed from the source. Several UPS systems and/or variable speed drives on the same source can be a concern for additive, combined, or resonating harmonics.

To minimize the switching effect a UPS rectifier will have on the building power distribution system, a large three phase UPS is often specified with an *input harmonic filter*. The input filter is designed to trap most of the unwanted harmonics through a tuned inductive-capacitive (LC) circuit.

In brief, the inductor or coil impedes most of the frequencies above a designed value from being fed back into the upstream distribution system, while the capacitive element shunts the frequencies below the designed value. Typical UPS specifications require input current with total harmonic distortion (THD) held to 10% or less at full load. At lighter loads, and especially for redundant UPS applications where the load is often in the 20%–40% range, the THD becomes much higher. However since the higher THD is a percentage of the actual reduced loads, the amount of harmonics at light loads is typically of less concern than at higher loads. Harmonics at light load are typically of so little concern that the capacitive component components of input harmonic filters are typically switched out at light load. The switch can be manual or automatic (with contactors and differential relays), or simply a service visit to briefly shut down the UPS and disconnect and tape off the capacitor leads when the UPS is predominantly operated at light load.

Although well-intentioned, UPS input harmonic filters pose a risk to the upstream electrical system. The concern is leading power factor from the filter capacitors when the UPS is at light load. Risks include nuisance tripping of circuit breakers and unstable operation of standby generators. When a UPS system that includes a large bank of input filter capacitors is energized, the caps "gulp" a lot of current to charge up. The resultant instantaneous high current can nuisance trip a circuit breaker. The leading power factor (current peak ahead of voltage peak in time) can cause instability in generator voltage regulators. The regulators are typically designed for loads with power factor at unity or lagging (current peak follows voltage peak in time). Generator operation can become unstable, and therefore less reliable with a leading power factor load. Designers, specifiers and purchasers of UPS systems must be careful not to overdesign or overspecify UPS input harmonic current reduction. Trying to hard to solve a harmonics problem which may not exist might invite leading power factor problems. There is rarely a need to reduce UPS harmonic THD below 10% at full load and 12%–14% is typically adequate. Six pulse rectifiers with THD filtered to 10%–14% THD at full load, 12-pulse rectifiers without filters (preferred) at 10%–14% THD are adequate. Delta conversion, IGBT rectifiers and rotary UPS designs further reduce input THD without filters. However, a standard six-pulse rectifier with no filter can reflect in excess of 30% current THD to the source and should be avoided.

UPS manufacturers in recent years have recognized the problem with excessive input harmonic filtering. The problem has become more acute in recent years with the increase in redundant UPS applications which UPS systems typically operating well under 50% of capacity. A typical midsize UPS with a six-pulse rectifier and 7% THD at full load harmonic filter draws an input power factor that is at unity at around 40% load. Below 40% load, the input power factor becomes increasingly leading and if not cancelled out by lagging power factor mechanical (HVAC) loads that are always operating anytime the UPS is operating, the overall leading power factor can cause problems. One method to deal with this is to design in an automatic contactor to take out the capacitive element of the UPS input filter at loads below 30%–40%. The contactors and their differential relays and set point adjustments have also become a reliability problem in many cases.

Over time, the capacitors in the input harmonic filter can age and alter the filter's value. The input distortion should be checked annually to assure the filter is functioning within specification and according to benchmark recordings from startup commissioning and previous annual maintenance testing.

10.5.1.5 Inverter

Inverters, utilizing a DC source, can be configured to generate 6 or 12 step-waveforms, pulse width modulated waveforms, or a combination of both, to synthesize the beginnings of an AC output. The *inverter* converts direct current into alternating current utilizing IGBT or equivalent technology. By pulsing on DC current in a timed manner, the resulting square wave can be filtered to resemble a sinusoidal alternating current waveform.

The inverter must produce a sine wave at a frequency that is matched to the utility bypass source while regulating the output voltage to within 1%. The inverter must accommodate load capacities ranging from 0% to 100% of full load or more and have the response to handle instantaneous load steps with minimal voltage deviation to the critical load.

10.5.1.6 Output Filter

A voltage square wave is the sum of all odd voltage harmonics added together. By generating an optimized squarewave at a frequency above 60 Hz for any given load condition, lower order harmonics are minimized. Pulse-width-modulated (PWM) inverters are particularly good in this function. An output filter is used to remove the remaining odd ordered harmonics, resulting in a sine wave, which is then used to power the critical load. As discussed, the inverter generates a square wave by filtering the inverter output. As a result, a clean sinusoidal alternating voltage is produced.

A typical specification for an Output filter is less that 5% THD (total harmonic distortion) with no single harmonic greater than 3%.

10.5.1.7 Static Bypass

In the case of an internal rectifier or inverter fault, or should the inverter overload capacity be exceeded, the static transfer switch will instantaneously transfer the load from the inverter to the bypass source with no interruption in power to the critical AC load.

10.5.2 Line Interactive UPS Systems

Delta Conversion Online UPS are a "line interactive UPS." The delta conversion consists of a delta converter and a main inverter. The delta converter functions as a variable current source. It controls the input power by regulating the DC bus voltage. The main converter is a fixed voltage source. It regulates the voltage to the load. The delta converter controls the UPS input impedance. This is similar to that of a conventional rectifier, to control power across the DC bus and load. Through pulse width modulation (PWM), the main converter recreates AC voltage providing constant output voltage. The delta conversion online consists of a connection between the two converters (from AC to AC). This connection is referred to as the "pure power path." This pure power path allows the delta and the main converter to function as bidirectional power converter. The delta and the main converter has the ability to change AC power to DC power and simultaneously DC power to AC power.

The main inverter is a voltage controlled IGBT (integrated bipolar transistor) PWM inverter. It has the primary function of regulating the input current and input power by acting as a variable current source in the secondary circuit of the transformer, that is, it behaves as a load without power dissipation, except for its switching losses.

Working together, the two converters form a very good power control system. The Main inverter is a voltage controlled IGBT PWM inverter. Its primary function is to regulate the voltage at the power balance point (PBP). The delta converter has an unique approach to compensate loss. It consists of flyback diodes that simultaneously rectify the excess power from the PBP and feed it to the DC bus to account for all the system lossess. The power balance concept control system in a delta conversion UPS uses digital signal processing (DSP). Using DSP gives a very high degree of accuracy.

10.6 ROTARY SYSTEMS

10.6.1 Rotary UPS Systems

Rotary UPS systems can be substituted in-place of any static UPS technology; however, they will be much larger, heavier and less efficient. The rotary power converter uses an electromechanical motor generator combination in lieu of an inverter used in static systems. The motor drive is fed from a combination of the local utility and a rectifier/inverter. Should the utility source become unavailable, the bypass path to the motor become isolated, and the motor is then fed from the DC battery system through the rectifier/inverter path. In normal operation, the critical loads are supplied as regulated computer grade power from the generator. Rather than using electronic technology to create a clean output waveform, the rotary UPS uses rotating AC generator (an "alternator") technology (see Figure 10.6).

The mechanical generator requires fewer parts than their static counterparts. Unlike the static inverter, which requires high frequency switching to form the sine wave and an output filtering to smooth the waveform, the generator portion rotary UPS creates a pure sine wave. The output from the generator is completely isolated from the source. Furthermore, the generator will provide the additional current necessary to clear electrical faults in the downstream distribution without having to transfer to the bypass source as is sometimes required by static UPS systems. The rotary UPS system can sustain the output waveform from short duration power quality problems without transferring to battery because of its motor generator design [it is a basic flywheel—continuing to rotate for some period of time on inertia only]. This feature can preserve the battery system from short duration power quality events thus reducing downtime and maintenance costs.

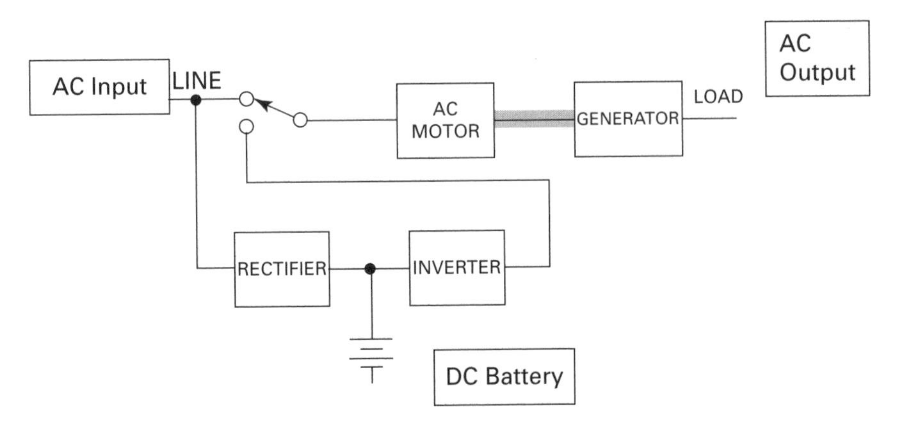

Figure 10.6 Basic rotary UPS. Power is lost and the inverter converts power in the battery to AC.

Rotary UPS systems offer a high initial cost, with lower maintenance cost; reliability is excellent, fewer electronic components mean fewer pieces that fail. M/Gs are durable and can withstand higher room temperatures (less cooling required). Bearing maintenance is minimal with longer life sealed bearings allowing up to ten years between scheduled bearing replacements. Routine vibration analysis can reliably predict the need for overhaul or bearing change out [Rotary systems also offer the possibility of medium voltage outputs. Static UPSs are not economically suitable for this application . . .]

There are three major concerns when considering rotary UPS systems they are:

1. **Initial First Cost** The capital costs of rotary UPS systems are normally 20% to 40% higher than static UPS systems.
2. **Equipment Weight** Regarding equipment weight if these systems are being installed on a slab located on grade the added weight is not a structural concern. However, if this equipment is being installed in a high-rise building then additional reinforcement is required and a lighter-weight static UPS becomes a more attractive alternative.
3. **Service Technician Availability** As with any system, good service is imperative and should be local with response time of 4 hours.

10.6.2 UPSs Using Diesel

Diesel UPSs operate in parallel with the utility supply, delivering clean and continuous power, without the need for batteries. When the utility is operating, the diesel UPS filters out small disturbances and other interruptions such as surges, and voltage drops. This system includes flywheel that stores energy for power during interruptions for duration of a few seconds. When the utility fails, the UPS diesel will start within about 10 seconds and take over as the primary energy source supplying power to the load.

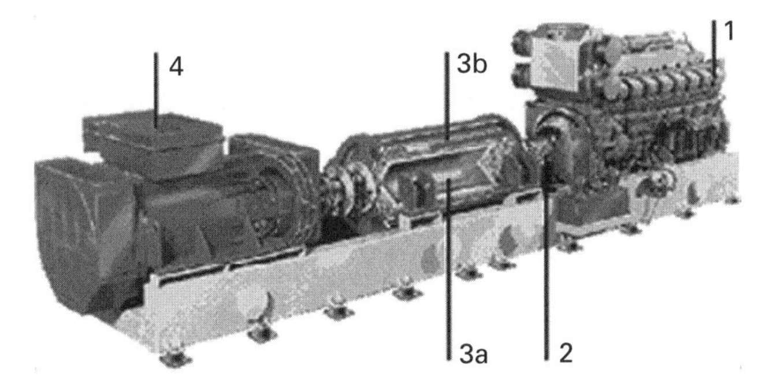

Figure 10.7 Diesel UPS system. (Courtesy of Hitec Power Solutions.)

10.6.2.1 Dual Output Concept

The dual output system is an extension of the diesel UPS system. The dual output system is designed to maintain the quality and continuity of electrical power required by critical loads and noncritical loads. The advantage of this system is that the properties of a UPS and of a generator set are combined, creating a continual supply of power to critical loads as well as protection to noncritical loads.

1. The diesel engine (sized for both critical and noncritical requirements)
2. The free-wheel clutch
3. The induction coupling (sized for critical load)
   a. Inner rotor
   b. Outer Rotor
4. The generator (sized for both requirements)

Description of Operation

- **Normal**: While the utility is supplying power the choke and generator provide clean power to the critical loads while the noncritical loads are fed directly from the utility.
- **Emergency**: When the utility fails the input circuit breaker is opened and the diesel engine is started. At the same time stored kinetic energy in the induction coupling is used to maintain the critical load. When the speed of the engine equals the speed of the induction coupling, the clutch engages and the engine becomes the energy source to support the critical loads. A few seconds after the diesel has started the noncritical loads are connected.

10.7 REDUNDANCY AND CONFIGURATIONS

Multimodule UPS systems (static or rotary systems) are arranged in various ways to achieve greater total capacity and/or improved fault tolerance and reliability.

Multimodule UPS systems allow for redundancy and enable module maintenance by isolating one module at a time.

A single module UPS system must transfer to bypass (commonly referred to as static bypass) in the event of a module failure. Owners must decide whether they are comfortable with the UPS transferring to utility as the backup power source or configure parallel modules to continue carrying the critical load on conditioned UPS power.

10.7.1 Redundancy

When protecting sensitive electronic equipment UPS system configuration is based on the requirements of that critical load. Sometimes a greater factor of protection is needed and more UPSs are added to the system increasing the system reliability. This is called *redundancy*. Redundancy is the placing of additional assemblies and circuits with the condition of an automatic switchover. When an assembly fails, it changes over to its backup counterpart, or is simply isolated without affecting other operations. The purpose of redundancy is when one or more power conversion modules fails or must be taken off line for maintenance or repair, the remaining units are able to support the load.

Standard System ($N+1$): In a standard system, N is equal to the number of UPS modules that are needed to back-up a system at full capacity. ($N+1$) is a combination of UPS modules that meets the criteria of furnishing an essential component plus one additional component as a back up. "N" is typically 1–5 UPS modules, with 1 extra module for redundancy. (See Figure 10.8)

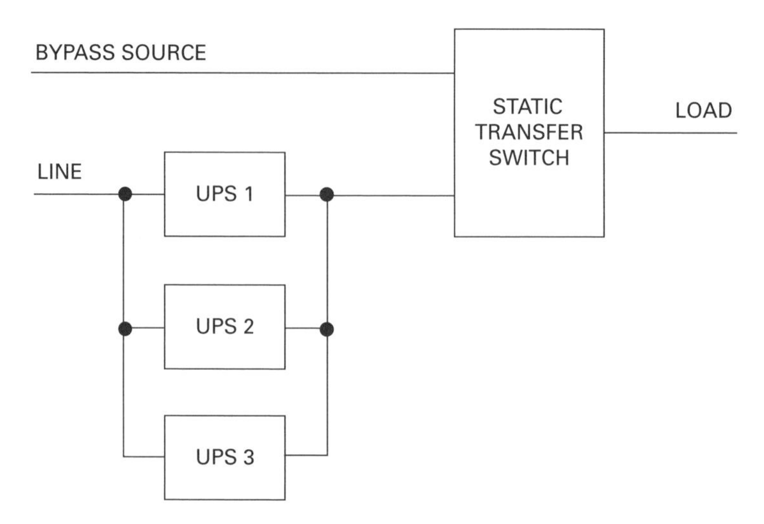

Figure 10.8 Parallel redundant schematic.

- **$N+2$** The $N+2$ design includes an essential component plus two components for backup. When the redundancy is $N+2$, there is the needed amount of systems, plus the addition of two extra components.
- **2N** For a 2*N* redundancy the system is doubled. A 2*N* UPS system includes two separate systems incorporating at a minimum $N+1$ redundancy in each system. There is a backup module, then an exact replica backup module that is added. The 2*N* system can either support a shared load of 50% maximum for each system or operate as an independent system with one system feeding the complete critical load and the other system with 0% or no load.

10.7.2 Isolated Redundant

For this scheme, the primary UPS unit serves the load being fed from the utility. A second UPS unit has its output fed into the bypass circuit of the first. This secondary unit has its input from the utility line but its backup is a reliable source such as a diesel generator set. Thus, the secondary unit provides UPS grade power to the load if the primary UPS unit fails. If the secondary unit fails, both units are bypassed to a standby source and the load will remain served. (See Figure 10.9)

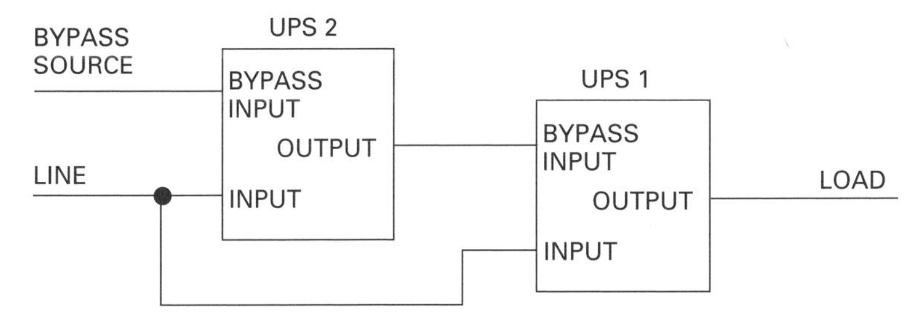

Figure 10.9 Isolated redundant system.

10.7.3 Tie Systems

A tie is made up of two parallel redundant systems with a tie between them. Each system has its own input power feeder and a tiebreaker. The output determines how power is supplied to the load from each supply. If a problem occurs, the tiebreaker closes and the remaining system continues to give power to the load, without the load noticing a change. Each parallel redundant system must be able to handle the load.

Some manufacturers have safeguards to lessen the risk of transferring a fault from one side to the other. Tie operations are usually performed for maintenance reasons, transferring system loads from one side to the other so one side at a time can be completely de-energized. Some manufacturers, transfers are manually initiated (specific sequence of button pushing) but automatically executed (motorized breakers executing the transfer when conditions are within parameters). The tie breaker never connects automatically in response to a system problem. However, downstream static transfer switches if installed perform automatic transfers between sources.

10.8 BATTERIES AND ENERGY STORAGE SYSTEMS

10.8.1 Battery

Lead-acid batteries are the most commonly used power storage devices to provide the UPS inverter with energy when the normal AC power supply is lost. Battery systems can be designed to provide enough capacity for a predetermined duration of normal power loss. This is typically 15 minutes or less, although there are some applications which may require hours of run time on a battery.

Depending on the design load the battery capacity will vary. Some systems are designed to accommodate momentary power interruptions. Applications such as these typically utilize a few minutes of reserve time on battery. Data center applications typically utilize 15 minutes of battery, providing enough time for several attempts to start a genset or to provide enough time for an orderly system shutdown of the major computer hardware. Although today with the advent of high density data centers it is almost impossible to shutdown a data center of significant size in 15 minutes. For example a $10,000\text{-ft}^2$ data center can take a minimum of 2 hours to shutdown. There are data centers that exist that are $100,000\,\text{f}^2$ so the questions becomes why is 15 minutes still the industry standard for larger data centers? While applications, like telephone switches or broadcasting may design enough capacity for many hours of uninterrupted service.

Common batteries in use today are lead calcium batteries that have a float voltage between 2.25 volts per cell. 2.35 Vpc is extreme and in the gassing range and for VRLA would be forbidden due to the dry-out and heat issues. Depending on the application, over 200 cells may be required for a 480-V UPS system. The footprint and floor loading of large battery systems are a major consideration when designing a UPS system.

Battery cells are connected in series to provide enough voltage to supply the UPS system. Some applications run parallel strings to provide enough capacity for larger systems, and add one additional string should redundancy be required.

Large UPS applications use flooded or wet cell lead acid batteries that require a significant level of monitoring and maintenance. The sturdy construction of wet cells is designed for up to 20 years of service life and is capable of repetitive high discharge rates. Smaller UPS applications and increasingly large applications use valve regulated lead acid (VRLA) batteries VRLA batteries are easier to maintain but have a life span of approximately 5 to 10 years. Monitoring can extend that useful period 1–2 years. It's not unusual to begin to replace VRLA batteries after 4 years.

Many managers make the mistake of buying a battery system and neglecting the maintenance, often times as a means of cost savings. It is important to understand that one faulty battery cell can render an entire UPS system useless if it only has a single string of batteries. While battery life is not entirely predictable, the vast majority of defective batteries can be detected through a good maintenance and load-testing program. The addition of monitoring adds a layer of security for its trending and predictive failure analysis ability.

Batteries supply power to the inverter when input power has been lost. A battery is a group of single cells connected in series and or parallel. The need for reliable and maintained batteries is imperative to a UPS system. Batteries need to be able to ride through short-term outages and have enough time to do orderly shutdowns during long-term outages. There are two types of batteries used in most UPS applications.

1. **Flooded or Wet Lead Acid**: Flooded lead acid batteries have been in use for more than a century. Flooded or "wet" design refers to those batteries that require periodic replenishment of water or acid, which make maintenance costs higher. The battery must be filled with distilled water and vented to keep the hydrogen gas that is emitted during charging, from accumulating. Lead acid batteries must be taken care of and inspected to assure proper productivity and longevity. The expected life of this battery can be up to 20 years, with proper care, and is quite reliable with the proper maintenance program. The disadvantages of flooded batteries are that they require large floor space, ventilation systems, rack or stands, acid containment and regular maintenance.

   The design of flooded batteries is quite simple. Negative plates made of lead or a lead alloy is sandwiched between positive plates of lead or lead alloy with an additive of calcium. Sheets of nonconductive material separate the plates. All positives and negatives plates are welded with their like and attached to a terminal post. The assemblies are placed in a container and filled with a dilute sulfuric acid solution whose specific gravity can be changed to increase capacity.

2. **Valve-Regulated Lead Acid (VRLA)**: VLRA batteries should have no need for special ventilation or acid containment. However it is important to note that while the acid is held in an absorbent material and there is no "free" electrolyte, they can and do still have the potential to leak or to vent hydrogen, which is why the term sealed is no longer used. The valves are designed to emit small bursts of gas when pressures reach predetermined levels. This battery needs no watering for its life, however due to its design it is subject to dry-out and its temperature must be maintained around 77°F. Operating a battery 10° above ambient will cut a battery's life in half. High temperatures will also hasten dry-out and eventually cause the battery to fail catastrophically. The negative aspects about this type of battery are their shorter life, only 1/4-max the life of flooded lead acid batteries, and much lower reliability with a much higher chance to fail open. And they actually

require more care as we cannot see the condition of the plates and must rely on voltage and internal measurements to provide the data necessary to determine its relative health. Monitoring is becoming a must if the critical equipment must stay up at all times.

The design of VRLA batteries includes an absorbent glass mat (AGM) that holds the needed electrolyte (water-acid mixture) between the plates. The thicker the mat is the longer the battery will last without drying out.

Batteries can be sized for any specific period. The discharge limit can range between 5 minutes and 1 hour, but the most common length of time used is around 15 minutes. The longer the duration, the larger the amount of batteries are required, which also means the larger the storage space. With generators reaching synchronization in far less time than engines of the past, it is not unusual to size batteries for 10 minutes and less. Again, if the generator does not start, the extra time in the battery does nothing to help provide orderly shut down or any other benefits for that matter. An example of a flooded cell battery room is shown in Figure 10.10.

Safety operations need to be considered when planning on the size and space needed for batteries. Natural disasters, such as earthquakes, ventilation (depending on type of battery), and even operating temperatures (preferred battery room temperature is 25°C/72°F) are important considerations to take in the installing of large UPS systems that requires a significant amount of batteries. Maintaining proper room temperature and cell voltages can best optimize the life of a battery.

Fail Modes:

The most significant difference between Flooded and VRLA batteries is how they manifest in failure. The flooded battery tends to fail capacity, which generally means that with proper maintenance and monitoring you have plenty of time to react to a degrading battery. The VRLA on the other hand tends to fail catastrophically and *can be almost without notice* making proper maintenance and monitoring much more critical for the life of the battery.

Figure 10.10 Flooded cell battery room.

10.8.2 Flywheel Energy

Essentially, a flywheel is a device for storing energy or momentum in a rotating mass, where high power is required only intermittently and briefly. Flywheels can be high-mass, low-speed (as low as a few hundred RPM) or low-mass, high-speed (greater than 50,000 RPM). The greater the inertia (mass $\times$ RPM^2) the higher the stored kinetic energy. Table 10.1 discusses the characteristics of a flywheel.

Compared to batteries, flywheels generally have a higher initial capital expense, lower installation costs, lower operational costs, and an extended working lifetime. Flywheels are used with motor-generators to increase the ride-through time of the set. Flywheels can also be used to store DC energy that can be used in parallel with a battery, to improve its overall reliability and life. A third application is using flywheels in place of a battery, to bridge the few seconds needed to start a standby genset.

Standalone flywheel cabinets can be used in place of either VRLA or flooded batteries with double-conversion UPS products. For mission-critical applications, flywheel systems should always be employed with properly maintained backup gensets capable of quickly starting and assuming load.

Some companies have designed integrated flywheel UPS systems. These typically combine a medium-speed flywheel with single-conversion UPS electronics, all in a single cabinet (see Figure 10.11). These integrated UPS designs are surprisingly compact for their rated power output, requiring less than half the floor space of a double-conversion UPS plus batteries.

The footprint advantages of flywheels are especially helpful to system designers coping with the power density of some modern data centers. By replacing battery racks with flywheel cabinets, a system designer can reclaim enough floor space to add another complete UPS-plus-flywheel system. For new construction, using the integrated flywheel/UPS products can enable the system designer to fit as much as 3–4 times the UPS capacity into a given amount of equipment space (see Figure 10.12).

Table 10.1 Flywheel Characteristics

Advantages	Disadvantages
Long life (15–20 yr.) unaffected by number of charge/discharge cycles	Low energy (low reserve time)
Lower lifecycle cost	Higher initial cost
High Efficiency—save power	"Batteryless" operation requires use with a genset
High charge/discharge rates and no taper charge required	Expensive bearings (some manufacturers)
Inherent bus regulation and power shunt capability	
Reliable over a wide range of ambient temperatures	
High power density compared to batteries	

Figure 10.11 A medium-speed flywheel. (Courtesy of Active Power.)

10.9 UPS MAINTENANCE AND TESTING

Today's UPS systems are so well designed and manufactured that they continue to produce an acceptable sine wave despite a remarkable degree of degradation to internal components. They are no less than the Sherman tank of the electronics world.

If we count the number of failures of modern UPS systems, they are statistically insignificant when compared with the millions of successful run hours. Quite often, we hear terms like six sigma's of availability or 10,000 hours mean time between failure attached to the success of electronic engineering in providing computer grade power. Nevertheless, when you fall into the ranks of the statistically insignificant your perspective on reliability changes forever. Suddenly that last micro percentage point is more important than the previous year of online operation.

Figure 10.12 An integrated 300-kVA flywheel/UPS. (Courtesy of Active Power.)

Such is the case when manager's find after 10 years of successful uninterrupted operation a system will fail, stranding a data center without power for only minutes, but a recovery process that takes months. In the post mortem search for a smoking gun, a technician may find buckets of failed components; diodes, capacitors, printed circuit boards, SCRs, fuses, and even essential logic power supplies.

What is remarkable is not that a box fails, but that it can continue to operate despite a high degree of degradation.

Still the failure leaves its mark and we come to realize that there are thousands of boxes out there continuing to run, but waiting for that last component to fail before the system comes unglued. Consequently, our testing procedures need to include functionally testing of UPS equipment along with the associated batteries.

Because of the complexity of a UPS system, managers tend to leave the UPS maintenance in the hands of a vendor service contract. For 10 years, the technicians may provide good service but often they are not allowed to functionally test the equipment because of business constraints. Additionally service varies with technicians.

In order to assure functionality of the UPS equipment, a test procedure must be instituted. One that is centered on accurately recording and verifying real life conditions. The test is primarily based on standard factory acceptance testing and it is broken down into three phases:

1. **Physical PM**: Testing of a UPS system as described is a complex and dangerous evolution if not performed correctly. The skills needed to run a thorough UPS functional test are beyond the infrequent experience of most facility managers. It is important to contract with a reputable test engineering group that has strong experience in UPS maintenance and testing of critical systems to oversee the procedure and analyze the test results and make recommendations.

   The physical PM includes infrared testing prior to shutting down the unit. Many vendors must rent an expensive camera, consequently the machine is infrared scanned only once a year, and quite often, the vendor neglects this as well. However, by purchasing a relatively inexpensive grayscale camera, we are able to perform an independent scan whenever the box is open or there is need to check an alarm cause. Check for loose connections on power supplies, diodes, capacitors, inverter gate, and drive boards, and check for swollen capacitor cans. In general, a swollen or leaking capacitor is an indication that the capacitor has failed or is beginning to fail.

2. **Protection Settings, Calibration, and Guidelines**: The next aspect is checking of protection settings and calibration. By simulating an alarm condition, the technician can verify that all alarms and protective settings are operational. Using independent monitoring the system metering can be checked to verify its operation. If the UPS system metering is out of calibration, it may affect the safety and operation of the UPS system.

   For instance if a meter that monitors DC bus voltage is out of calibration, it may affect the DC over voltage or under voltage alarm and protection. If the UPS were to transfer to battery, the ensuing battery discharge might be allowed to continue too long. As the battery voltage dropped, the current must increase to continue supporting the same power. When current becomes too high the temperature in the conduction path increases to the point of destruction.

   The following is a basic list of typical UPS alarms:

   - AC over voltage
   - AC under voltage
   - DC over voltage
   - DC under voltage

3. **Functional Load Testing**: Functional load testing can be broken down into four distinct operations: steady-state load test, transient response test, filter integrity, and battery rundown. For systems with multiple modules, a module fault test restoration should be performed in addition.

If you think of a UPS system as a black box, the best way we can determine the health of that box is by monitoring the input into the box and the outputs from the box. In most UPS systems, power is drawn from the utility to create a DC power supply that may be paralleled with the battery. That power is drawn equally across all phases with all modules sharing in the transmission of power.

On the output side: The power distribution is dependent upon the actual load that is drawn from the UPS box. In many instances the three phases will not be equally loaded, consequently current draw is not equal across all three phases. On the other hand, output phase voltage should be equally balanced across all phases and across all modules (for multi-module systems).

10.9.1 Steady-State Load Test

Under a steady-state test all input and output conditions are checked at 0%, 50%, and 100% load. The test provides a controlled environment to verify key performance indicators like input currents and output voltage regulation.

The steady-state test checks that input currents are well-matched across all phases of a module and that all modules are equally sharing the load. Output currents are load dependent. That is, there may be a heavier load on an individual phase due to load distribution. The input currents, however, should be closely matched as each module is sharing in the generation of a DC power source from which the load will draw from.

Voltage readings are also taken during the steady-state conditions to check for good evenly matched voltage regulation on the output side. Since power is equal to voltage times current, a degraded voltage from a single module or phase means that other modules or phases must produce more current to accommodate the voltage drop. Multiple modules are intended to run closely matched in output voltage and frequency.

Steady-State Load Test at 0%, 50%, and 100% load:

- Input voltage
- Output voltage
- Input current
- Output current
- Output frequency
- Input current balance
- Output voltage regulation

10.9.2 Harmonic Analysis

The input and output of the UPS should be monitored during the steady-state load testing for harmonic content. Observing the harmonic content at 0%, 50%, and

100% allows the engineer to determine the effectiveness of the input and output filters.

It is important to note that most UPS filtering systems are designed for greatest efficiency at full load. Consequently, the harmonic distortion is greatest when the system is least loaded and smoothes out as load is increased. As long as the distortion is consistent across phases and modules, this is no reason for alarm.

Total harmonic distortion (THD) is essentially a measure of inefficiency. A unit operating with high THD uses more energy to sustain a load than a unit with low THD, that additional energy is dissipated as heat. Consequently, a UPS operating with high THD at low load, while inefficient is in no danger of damaging conduction components. A UPS operating with high THD at high load is extremely taxed since the unit is producing energy to sustain the load along with additional heating due to a poor efficiency.

Another way of explaining this concept is a system rated for 1000 amps operating at 100 amps of load with 20% THD is producing 120 amps, which is easily accommodated on a 1000 amp rated system. That same system operating at full load or 1000 amps with a 20% THD is producing 1200 amps or in an overloaded condition.

Harmonic Analysis:

- Input voltage
- Output voltage
- Input current
- Output current

10.9.3 Filter Integrity

For most UPS systems there are three filters: input filter, rectifier filter, and output filter. The filters commonly use an arrangement of inductors (coils) and capacitors to trap or filter unwanted waveform components. Coils are relatively stable and seldom breakdown. They are quite simply wound metal. Indications of a coil failure are typically only picked up on thermal scans.

Capacitors are more prone to failure under stress. Like batteries, capacitors are subject to expansion and leakage or electrolyte drying up. AC capacitors are subject to rapid expansion or popping when over stressed. The resultant consequence is acidic electrolyte sprayed onto sensitive electronic components. Under good operating conditions, AC capacitors should last the life of the UPS.

DC capacitors have a definitive life span. The electrolyte in DC capacitors is prone to drying out over time, which is difficult to detect short of a capacitance test. DC capacitors are subject to swelling as well, but are normally equipped with a relief indicator that can be readily spotted.

In large three-phase UPS systems, there are virtually hundreds of capacitors wired in series and arranged in banks. Identifying a single failed capacitor can be difficult and time consuming. A simplified means of checking filter integrity is by performing a relative phase current balance. By checking phase current through each leg of the filter, a quantitative evaluation may be made. A marked difference in phase currents drawn through a filter assembly is an indicator that one or more capacitors have degraded. Further analysis may be directed to a filter assembly drawing abnormal current relative to corresponding phases.

Filter Testing:

1. Input filter capacitance

   Phase current balance

   Individual capacitance test

2. Rectifier filter capacitance

   Phase current balance

   Individual capacitance test

3. Output filter capacitance

   Phase current balance

   Individual capacitance test

10.9.4 Transient Response Load Test

The transient response test simulates the performance of a UPS given large instantaneous swings in load. The UPS is designed to accommodate full load swings without a distortion in output voltage or frequency. Periodically testing the response to load swings and comparing them with the original specifications may ascertain the health of the UPS.

In order to observe the response to rapid load swings, a recording oscilligraph and load banks are connected to the UPS output. The oscilligraph monitors three phases of voltage along with a single phase of current. (The current trace acts as a telltale as to when load is applied and removed.) As the load is applied and removed in 50% increments, the effect is observed on the voltage waveform. The voltage waveform should not sag, swell or deform more than 8% with the application and removal of load.

Similar tests are performed simulating the loss of AC input power and again on AC input restoration.

Module Output Transient Response (Recording Oscilligraph)
Load Step Testing:

- 0%–50%–0%
- 25%–75%–25%
- 50%–100%–50%
- Loss of AC input power
- Return of AC input power

10.9.5 Module Fault Test

Multimodule systems should also be functionally tested to verify the system would continue to maintain the critical load in the event of an inadvertent module failure. By applying full system rated load (via load banks), a single module should be simulated to fail. The system should continue to maintain the load without significant deviation of voltage or frequency as verified by recording oscilligraph. Each module should be tested in such a manner:

10.9.6 Battery Run Down Test

The final test of a system should be a battery rundown. Quite often, the battery system may be subject to failures that go undetected. The most meaningful test of a battery is to observe the temperature, voltage, and current under load conditions. Disparate voltages from cell to cell or string to string provides the best indication a battery or string is degrading. The load test depends on which type of batteries is connected to the UPS system. Wet cell batteries are more robust and are capable of undergoing a full battery run down test once a year. Valve regulated batteries are many times over stressed by a full battery rundown and must be tested to a shorter duration test.

Regardless of the battery design or duration, a battery should undergo a discharge that will record the "knee of the curve." That is the point when battery voltage with respect to time begins to decay at an exponential rate. While undergoing the discharge test, battery cell voltages should be continuously checked either with a cell monitor or manually. Batteries that exhibit lower than average cells should be marked for potential replacement.

When the battery string is no longer able to provide designed discharge time (e.g., 15 minutes), then it is time to consider full battery replacement.

Battery temperatures should be continuously monitored during the discharge test, particularly at the plates and terminal posts, to determine abnormally high temperatures. High temperature is an indication that it is working too hard to provide the designed output.

10.10 STATIC UPS AND MAINTENANCE

Static is static, it is difficult to differentiate between all the providers of Static UPSs in terms of reliability. Most providers offer a high degree of sophistication ranging simple power conversion to sophisticated synchronization and filtering arrangements.

A UPS that is greater than 10 years old begins to show signs of age via component failure. Capacitors, SCRs and diodes have in most instances been in service continuously and will begin to fatigue and fail. Maintenance awareness must be increased. The standard maintenance routine of tightening screws and dusting off the circuit boards is not adequate to determine failed components. It may be worthwhile to contract with a reputable power-engineering firm to evaluate your current maintenance program.

Static UPS is a complicated, highly technical piece of equipment. It is difficult to hold a maintenance vendor or manufacturer responsible without of fair degree of internal technical knowledge. It is imperative you obtain reliable service followed with a full service report and follow up recommendations. Skipping routine maintenance service will inevitably lead to disaster and it doesn't take too many data center crashes to realize the importance of UPS maintenance.

10.10.1 Semi-Annual Checks and Services

1. **Perform a temperature check on**:
   - All breakers, connections and associated controls
   - Report and if possible, repair all high-temperature areas
   - Greater then 10°C; corrective measures required
   - Greater then 30°C; corrective measures required immediately
2. **Perform a complete visual inspection of the equipment**:
   - Including subassemblies, wiring harnesses, contacts, cables and major components
3. **Check module(s) completely for**:
   - Rectifier and inverter snubber boards for overheating
   - Power capacitors for swelling or leaking oil
   - DC capacitors vent caps that have extruded more than 1/8 in.
4. **Current sharing balance capacitor assemblies**:
   - Record all voltage and current meter readings on the module control cabinet or the system control cabinet
   - Measure and record 5th, 7th, and total harmonic trap filter currents
5. **Visual inspection for the battery room to include**:
   - Check for grease or oil on all connections (if applicable)
   - Check battery jars for proper liquid level (if flooded cells)
   - Check for corrosion on all the terminals and cables
   - Examine the cleanliness of the battery room and jars

10.11 UPS MANAGEMENT

The UPS may be combined with a software package to automatically shut down a server or workstation when a power failure occurs. The server/workstation programs are typically shut down in predetermined sequence to save the data before the UPS battery is out of power. This package is operated from the power-protected workstation or computer and functions as if it were in a stand-alone environment, even when the computer is networked. A UPS connected in a power management function usually provides UPS monitoring and computer shutdown functions.

In addition, the manager also can control the operations of the UPS from a remote location, including turning the UPS on and off, configuring communications capability, and making performance adjustments. This control is typically executed either within an SNMP environment, where it uses not only "traps" but also "gets and sets" (the remote control functions), or through software, which is supplied by the UPS vendor. This software operates directly with the network and typically is an attached module in the network management software.

10.12 ADDITIONAL TOPICS

10.12.1 Offline (Standby)

Offline or standby UPSs consist of a basic battery/power conversion circuit and an internal power transfer switch that senses irregularities in the electric utility. The computer is usually connected directly to the utility, through the sensing power switch that serves as the primary power source. Power protection is available only when line voltage dips to the point of creating an outage. A standby power supply is sometimes termed "off-line UPS" since normal utility power is used to power the equipment until a disturbance is detected and the power switch transfers the load to the battery-backed inverter. The transfer time (typically, from a few milliseconds to a few seconds, depending on the required performance of the connected load) from normal source to the battery-backed inverter is important.

Some offline UPSs do include surge suppression circuits, and some possess optional built-in power line conditioners to increase the level of protection they offer. In the case of power surges, an offline UPS passes the surge voltage to the protected load until it hits a predetermined level, around 115% of the input voltage. At the surge limit value, the unit then goes to battery. With high-voltage spikes and switching transients, they give reasonably good coverage, but not the total isolation needed for complete input protection. For power sags, electrical line noise and brownouts, offline UPSs protect only when the battery is delivering power to the protected system. A similar limitation exists in the case of frequency variation. An offline UPS protects only if the inverter is operating and on battery. If the input frequency varies outside the device's range, the unit is forced to go to battery to regulate the output to the computer loads. In very unstable utility power conditions, this may drain the battery making it unavailable during a blackout or "0" voltage condition. Standby UPSs are cost effective solutions particularly suitable for single-user PCs and peripheral equipment that requires basic power protection.

Small critical electric loads, like personal computers, may be able to use standby power supply (SPS) systems. These systems have the same components as on-line UPS systems but only supply power when the normal source is lost.

The SPS senses the loss and switches the load to the battery inverter supply within 1 cycle. Some computer equipment may not tolerate this transfer time, potentially resulting in data loss. In addition, SPS systems do not correct other power quality problems. They are, however, less expensive than online UPS systems.

Chapter 11

Data Center Cooling: Systems and Components

Don Beaty

11.1 INTRODUCTION

Different areas of the datacom industry have different terms, training, and focus that inhibit or handicap collaboration. Compaction and increasing loads are creating significant cooling challenges requiring increased collaboration and a holistic approach to cooling. This holistic approach includes considering the interdependence of space geometry, obstructions, and IT equipment, as well as the implications of "adds, moves, and changes."

Although there is no reason to believe that the IT and facilities industries are intentionally limiting their interaction and collaboration, it is no secret that there needs to be a concerted effort for increased collaboration and integration. ASHRAE is an engineering society whose specialty includes cooling as well as writing standards, guidelines, and model codes.

Although ASHRAE is best known for the facilities/buildings sector, ASHRAE Technical Committee TC9.9 (specializing in the Data Processing and Communication industry) has strong representation from the IT sector including IBM, HP, Intel, and Cisco. So inherent to TC 9.9 is a working form of this integration.

11.2 BUILDING COOLING OVERVIEW

The intent of this chapter is to provide basic information about various aspects of the industry directly or indirectly associated with or impacting cooling.

Maintaining Mission Critical Systems in a 24/7 Environment By Peter M. Curtis

There are a number of options for specific building cooling systems, and there are types of cooling equipment that are both common to all systems and specific to certain ones. The chapter will start with some background on the basic components of a cooling system such as chillers, cooling towers, condensers, dry coolers, heat exchangers, air handlers, fan coil units, and computer room air-conditioning units (CRAC units).

This equipment can be labeled and/or classified in numerous ways, but these names are common ones used by the cooling equipment manufacturers as well as engineers and contractors. The intent is for this background to be in layman's terms so that it will be useful for areas of the industry that do not have a cooling engineering background and/or a facilities cooling background (e.g., datacom equipment engineers).

Some of what will be discussed will refer to guidelines and standards that are good sources of information for understanding the cooling of computer equipment, such as ASHRAE's Thermal Guidelines for Data Processing Environments.

11.3 COOLING WITHIN DATACOM ROOMS

The fundamental premise behind a cooling system is the transportation of heat. In the broadest sense, heat is generated by the computer equipment and is transported away from the room where the computer equipment resides and released into the atmosphere. Along the way, one or more methods of heat transportation may be utilized.

It is simple to understand that if a confined space houses a heat-generating piece of equipment (e.g., computer equipment) and there is no means of regulating the temperature of that room, then as soon as the piece of computer equipment is operational, the temperature in the room will increase steadily until it exceeds the maximum operational temperature of that equipment and ultimately cause the equipment to fail.

In order to counteract the heat gained by the room as a result of the equipment, we need to introduce a source of cooling. However, the magnitude of the cooling source we introduce is still in question. If we provide less cooling than the amount of heat being produced, the temperature will still steadily increase over time, just not as quickly as with no cooling.

Therefore, in order to establish an equilibrium or steady-state temperature within the room, the amount of cooling we need to provide should equal the amount of heat being produced. In other words, if we have a desired room temperature of 75 °F and the computer equipment with no cooling raises the temperature in the room by 5 °F per hour, we have to introduce a source of cooling in that room that, if there were no heat source, would lower the temperature of the room by 5 °F per hour. In simplistic terms:

$$75\,^\circ\text{F (room temp.)} + 5\,^\circ\text{F (heat gain per hr)} - 5\,^\circ\text{F (cooling per hr)}$$
$$= 75\,^\circ\text{F (room temp.)}$$

The actual design temperature (and other attributes such as humidity, etc.) of the room needs to be within the manufacturers operating conditions, but if you do not know the manufacturer of the equipment that is being used or if you have multiple manufacturers of equipment that have different operating conditions, you can use the environmental classifications and specifications listed by ASHRAE in their publication "Thermal Guidelines for Data Processing Environments."

11.4 COOLING SYSTEMS

11.4.1 Airside

The most common source of cooling of datacom rooms is condition and recirculate the air within the room. The types of equipment that are used to condition the air could be an air handling unit (AHU), a packaged rooftop unit (RTU), a fan-coil unit, or a computer room air-conditioning (CRAC) unit. These types of units all work in a similar manner, with the key components being a fan and a coil.

The fan in these units induces air movement over the coil that is made up of a serpentine configuration of finned tubes that carry a fluid within them. Heat is exchanged by the warm air from the room passing over the coil that has a cool fluid inside it, resulting in cold air being distributed back to the room and warm fluid being transported away from the coil.

Figure 11.1 shows this heat transfer in a simple graphical format.

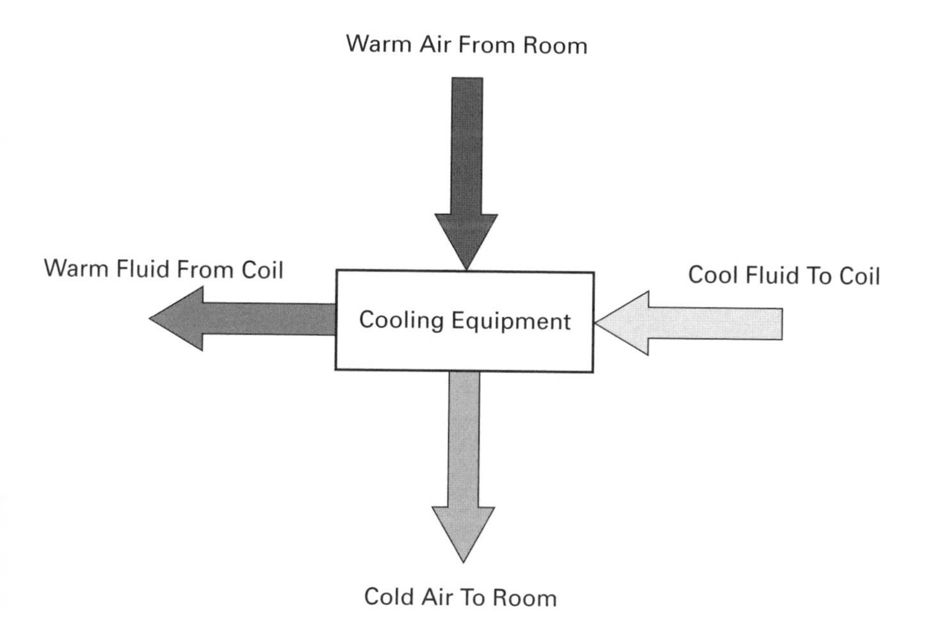

Figure 11.1 Airside cooling system heat transfer for datacom rooms. (Courtesy of DLB Associates.)

11.4.2 Waterside

From the above system, the two products that are "made" by the cooling equipment are cold air and warm fluid. As mentioned prior, the cold air is recirculated back to the datacom room so that half of the equation is known. However, the warm fluid needs to be returned back to the coil as cool fluid so another means of heat transfer needs to take place in order to have this happen.

The type of fluid that is used dictates its thermal properties and also determines the type of equipment that is used to perform the heat transfer required to remove the heat from the warm fluid and turn it into cool fluid. Some examples of the type of liquids used include water, refrigerant, glycol, or a glycol–water mixture.

The fluid is transported via piping to a piece of equipment that relies on a further heat transfer taking place, with the byproduct being either another fluid or air. In the case of air, the piece of cooling equipment essentially performs the reverse of the airside cooling that was described in the previous section, with warm fluid and cooler air entering the equipment and cool fluid and warmer air leaving the equipment.

The cool fluid goes back to the airside equipment described earlier, and the warmer air is released into the atmosphere; therefore, the type of cooling equipment that performs this type of heat transfer is sometimes referred to as air-cooled (since the fluid is cooled by air) and also as heat rejection equipment (since the heat is rejected to the atmosphere).

Given its operation, air-cooled heat rejection equipment is located outside the building envelope, either on the roof or at ground level. The types of equipment that fall into this category include air-cooled chillers, air-cooled condensing units, and drycoolers. Figure 11.2 is a graphical representation of the heat transfer that occurs with this type of equipment.

The second type of waterside cooling equipment that we will discuss falls under the category of water-cooled. There are two facets to water-cooled equipment. One type involves a piece of equipment that transfers heat from one fluid stream to another fluid stream and an example of that is a water-cooled chiller. The warm fluid from the air handling equipment shown in Figure 11.1 enters the water-cooled chiller where it is mechanically cooled and returned forming a closed loop.

The heat from the entering warm fluid in that process is transported away from the chiller using another fluid loop. The fluid between the airside cooling equipment and the chiller is typically referred to as the chilled water. The fluid that is used to transport the heat away from the chiller is referred to as condenser water. The condenser water leaves the chiller warmer than it enters and is returned to the chiller after being cooled by water-cooled heat rejection equipment such as a cooling tower.

Figure 11.3 shows the operation of a water-cooled chiller and Figure 11.4 shows the continuation of the condenser water loop by showing the heat rejection that occurs at a cooling tower.

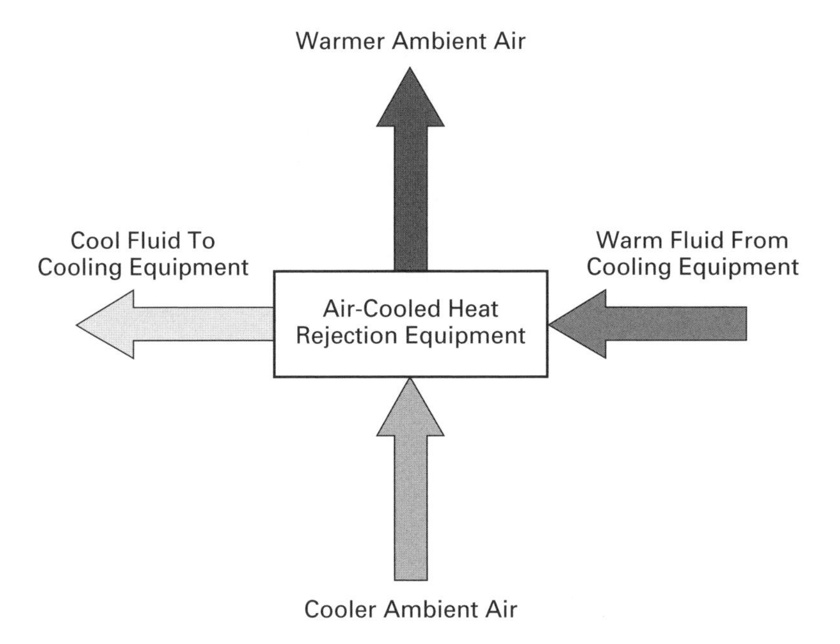

Figure 11.2 Heat transfer for air-cooled heat rejection equipment. (Courtesy of DLB Associates.)

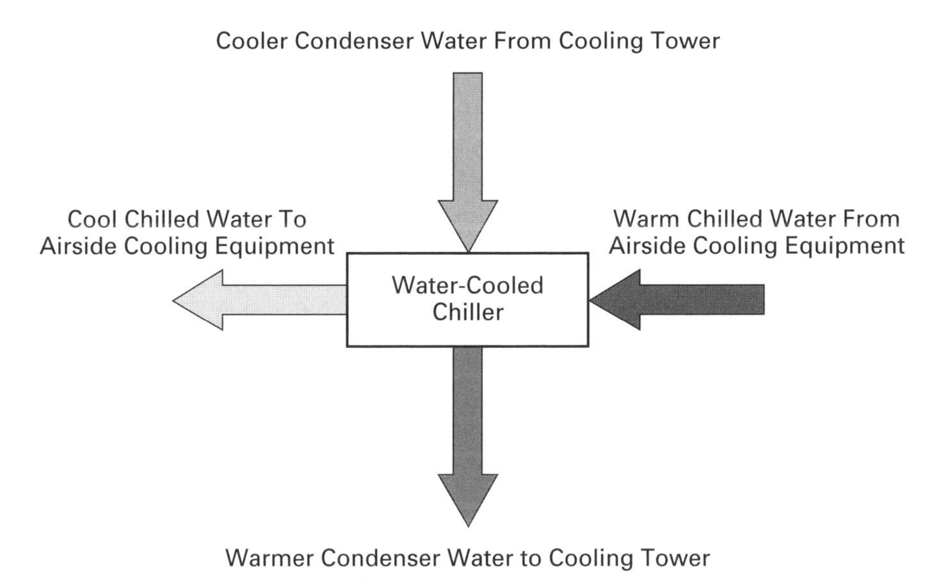

Figure 11.3 Heat transfer for a water-cooled chiller. (Courtesy of DLB Associates.)

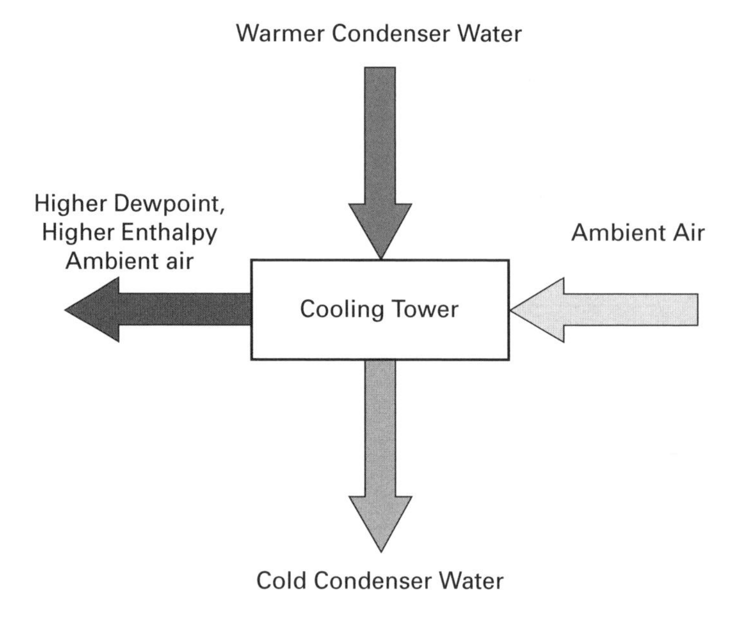

Figure 11.4 Heat transfer for a cooling tower. (Courtesy of DLB Associates.)

11.4.3 Air- and Liquid-Cooling Distribution Systems

11.4.3.1 Introduction

The terms air cooling and liquid cooling are often used to describe the two major cooling distribution methods. It is commonly accepted that air cooling or liquid cooling is terminology that refers to the cooling media that is used within the cooling system. However, the definition of what constitutes an air-cooled or liquid-cooled system differs based on the common terminology used by the different industries.

In this section, we shall attempt to paint a comprehensive picture of the thermal architecture paths and list some options for the cooling distribution media for each path segment. As will be shown, there are air-cooling and liquid-cooling options for almost all segments, and your choice regarding which particular segment you deem to be the critical one from your standpoint may ultimately dictate whether you define a system as being air-cooled or liquid-cooled.

a. Thermal Architecture Path. There are basically six segments to the heat flow path of any cooling system dealing with electronic equipment. The six segments all involve the transportation of heat away from the source to its eventual released to the atmosphere. Not all six segments are used in every cooling system, and a given system may bypass or even eliminate a segment.

The six segments listed will each be described in more detail below.

- Segment 1: Heat generating source to the extents of the packaging.
- Segment 2: Packaging to extents of equipment framework (rack/cabinet).

- Segment 3: Extents of framework to terminal cooling equipment.
- Segment 4: Terminal cooling equipment to mechanical refrigeration equipment.
- Segment 5: Mechanical refrigeration equipment to heat rejection equipment.
- Segment 6: Heat rejection equipment to atmosphere.

Segment 1: Heat Generating Source to the Extents of the Packaging. There are multiple heat generating sources within electronic equipment. The two major ones are the central processing unit (CPU) and the hard drive. This segment of the thermal architecture path deals with how the heat from those sources makes it to the extents of the packaging of the equipment. Packaging is a term used by the IT equipment industry to describe the enclosure or case that the electronic components and boards are housed in.

Packaging should not to be confused with framework or rack as will be described in the next segment, but this would be the smaller box that typically houses and protects the more fragile computer components.

Heat from the hard drive is usually cooled by a fan that is internal to the packaging. Cooler air is drawn into the packaging by the fan and is drawn over the hard drive and expelled from packaging. In many cases there are multiple fans used sometimes on the inlet and outlet of the packaging. For this scenario, segment 1 is essentially air-cooled since air is the medium that transports the heat from the source to the extents of the packaging.

The CPU operates at a much higher temperature than the hard drive and can exceed 212 °F (100 °C). Heat can be transported away from the CPU in a number of ways. For servers and desktop equipment, it is common to utilize a heat sink.

A heat sink is a metal extrusion that is bonded to the CPU using a thermal paste. The heat sink is typically made from a thermally conductive metal such as copper or aluminum, and its design is such that it maximizes the surface-area-to-volume ratio and allows for airflow through the packaging to interact more intimately within the fins of the heat sink and therefore cool it.

Often, heat sinks have small fans mounted directly to them in order to induce more airflow interaction between the air passing through the packaging and the heat sink itself. This heat sink as described, fan-assisted or not, represents an air-cooled system for segment 1 since air is the medium that is transporting the heat from the source—in this case the CPU—to the extents of the packaging.

The role that the heat sink plays is simply to extend the source from the actual CPU but not extend it beyond the extents of the packaging. Alternate methods of achieving this include the use of a liquid-cooled technology such as a heat pipe.

A heat pipe is filled with a suitable liquid; one end is placed near the CPU while the other is placed at a remote location within the packaging. The liquid at the hot end (i.e., at the CPU) is heated to where it vaporizes and moves to the colder end. The vapor at the colder end condenses and is returned to the hot end by capillary action within the wall construction of the heat pipe.

Almost all laptop computers utilize heat pipes to cool their CPUs since the volume required for a heat sink is not conducive to the volumetric compaction sought by the laptop market.

Yet another method for transporting heat from the CPU is to use a cold plate (sometimes also known as a liquid heat sink). A cold plate is a liquid-filled chamber that is bonded to the CPU. The cold plate has physical hose connections constantly flowing cold liquid through the chamber. The hot and cold hoses are extended to the extents of the packaging and sometimes beyond. This would be considered a liquid-cooled system for Segment 1.

Segment 2: Packaging to Extents of Equipment Framework (Rack/Cabinet). Once the heat has made it to the extents of the packaging, it now has to travel from the packaging to the extents of the framework that it is housed in. This framework is referred to as a rack or an enclosure or a cabinet. Its construction ranges from something as simple as two vertical mounting posts to something as complex as an enclosed cabinet with built-in cooling, and there are multiple variations in between.

In the case of the two-post or four-post rack, there is very little interface in terms of this affecting the thermal architecture of this segment. Whatever cooling method is utilized for Segment 1 is typically extended through Segment 2. The same can be said of enclosed cabinets that have perforated doors or panels.

More sophisticated cabinet enclosures include air management systems. These air management systems are designed to more precisely channel air to the air inlet locations at the packaging extents and then to manage the exhausted hot air so as to mitigate the possibility of a short cycle with the hot air being drawn through an inlet.

Such air management cabinets represent air-cooled systems for Segment 2. There are other cabinets that utilize a liquid piping infrastructure to transport the liquid piping from a liquid-cooled Segment 1 to the extents of the rack. Thermosyphons are one liquid-cooled technique that has been developed at a rack level.

Further complicating this segment is that in some configurations, the extents of the rack also house the terminal cooling equipment. For example, a rack may have a fan and cooling coil built into it. The fan recirculates air within the extents of the rack and the cooling coil cools the returning warm, air.

This would mean that either (a) the segment could be considered air-cooled since the only medium that is transporting the heat within the extents of the rack is air or (b) it could be considered water-cooled since the only interface or connection to the rack consists of the liquid connections to the cooling coil.

Segment 3: Extents of Framework to Terminal Cooling Equipment. So far, the two scenarios that we have been following involve either (a) air being exhausted from the extents of the rack in some manner or (b) liquid piping connections that need to be made either to the rack or directly to the individual electronic equipment packaging within the rack.

The liquid piping connections are typically connected to terminal cooling devices that are integrated in the rack (i.e., rack housed fan and cooling coil), So Segment 3 is incorporated into Segment 2 with liquid cooling.

For the air-cooled system, there is an airflow path to the terminal cooling device from the extents of the rack. The terminal cooling equipment can be a computer room air-conditioning (CRAC) unit or a more localized cooling unit that may be mounted on top of the rack or in the aisle space adjacent to the rack.

The terminal cooling equipment will typically consist of a fan and a coil. The fan draws in the warmer air that has been exhausted by equipment via the racks and then passes it over a cooling coil before reintroducing cold air into the room, preferably directed toward the inlets of the rack/IT equipment packaging.

Segment 4: Terminal Cooling Equipment to Mechanical Refrigeration Equipment. Segment 4 is predominantly a liquid-cooled segment. In the multiple scenarios described above, we are dealing with either the piping connections directly from the IT equipment packaging within the rack, the piping connections to the cooling coil that is integrated within the rack, or the piping connections from the cooling coil that is located in the terminal cooling equipment.

The cooling coils in both the integrated and the remotely located terminal cooling equipment will have liquid connections supplying cooler liquid to the coil and returning warmer liquid from the coil. In some cases the liquid medium is a refrigerant that vaporizes at atmospheric pressure, but for the intent of this chapter, we shall consider that to also represent a liquid-cooled medium for this particular segment.

These liquid piping connections, via a pumping system, will either go directly to the heat rejection equipment (i.e., bypass this segment) or undergo a mechanical refrigeration process. The determination of whether the heat exchanger is involved depends on the type of cooling system being employed at a central plant level.

For some liquid-based cooling systems used in Segment 3, the liquid piping can be directly routed to exterior heat rejection equipment (i.e., bypassing this segment and the next). This would be in line with the air-cooled system that was introduced earlier.

Alternatively, this segment describes the scenario whereby the liquid piping is directed from the terminal cooling equipment to the equipment (e.g., a chiller) that houses the mechanical refrigeration components. The mechanical refrigeration components may also be housed in certain terminal cooling equipment such as a CRAC unit which would make this segment represent a shorter physical path and the terminal cooling unit would then be described as self-contained.

The mechanical refrigeration equipment can be thought of as a heat transfer device with warmer liquid from terminal cooling equipment entering and cooler liquid leaving. The heat transfer within the mechanical refrigeration equipment would then be transported away as described by the next segment.

Segment 5: Mechanical Refrigeration Equipment to Heat Rejection Equipment. The heat that is transferred to the mechanical refrigeration equipment is then transported to heat rejection equipment. As will be discussed in more detail later in this chapter, the heat rejection equipment is typically located outside of the building envelope. The transport medium is typically a liquid and this separate liquid loop is piped between the two devices.

In some cases the mechanical refrigeration equipment and the heat rejection equipment is a part of the same piece of equipment, such as an air-cooled chiller. In other cases they are separate pieces of equipment, as with a water-cooled chiller and a cooling tower.

Segment 6: Heat Rejection Equipment to Atmosphere. This final segment typically utilizes air as a medium, whether directly or indirectly. The piping that has made its way to the heat rejection equipment utilizing some or all of the five segments of the thermal architecture path is rejected to the atmosphere in an open or closed system using the latent heat of vaporization (e.g., cooling towers) or simply by the cross-flow of air via propeller fans over a piping coil (air-cooled condensers, drycoolers).

b. Summary. The definition of air and liquid cooling by a given industry or individual basically applies the term to the medium associated with Segment 1, Segment 2, or Segment 3. There is no correct definition, and it is better to develop an understanding of the entire thermal architecture path segment structure and options to be able to classify a system.

The following sections of this chapter will discuss in more detail the individual components that make up some of the terminal cooling, mechanical refrigeration, and heat rejection equipment that have been introduced within the thermal architecture path segments.

11.5 COMPONENTS OUTSIDE THE DATACOM ROOM

11.5.1 Refrigeration Equipment—Chillers

In the building cooling industry, a chiller is the term used to describe mechanical refrigeration equipment that produces chilled water as a cooling output or end product (Figure 11.5). The term chiller is not normally used in the building cooling

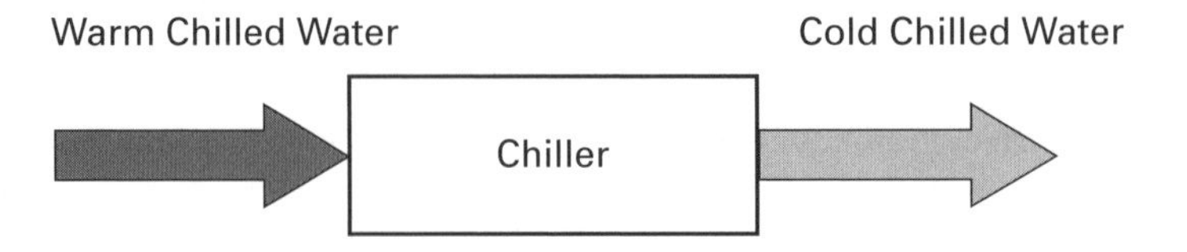

Figure 11.5 Generic chiller diagram. (Courtesy of DLB Associates.)

industry to describe a unit that produces chilled air (e.g., a CRAC unit). In fact, chilled air is not a common term in the building cooling industry; a normal term is conditioned air or supply air.

It is extremely rare for the chiller itself to be located in the same room as the computer equipment for a variety of reasons including room cleanliness, maintenance, spatial requirements, acoustics, and so on. Depending on the type of chiller, it is typically located in a dedicated mechanical room or outside the building envelope at ground level or on a roof.

The chilled water produced by a chiller for building cooling is commonly 42–45 °F but can be selected and designed to produce temperatures that are higher or lower than this range. The chilled water can be 100% water or a mixture of water and glycol (to prevent the water from freezing if piping is run in an unconditioned or exterior area) or other additives such as corrosion inhibitors. The cooling capacity of the chiller is influenced or de-rated by whether glycol or additives are included.

In a typical chilled water configuration, the chilled water is piped in a loop between the chiller and the equipment that supplies conditioned air to the electronic equipment room or space (e.g., a CRAC unit). The heat that the chilled water gains from the hot air in the electronic equipment room is rejected in the chiller itself. The method of the chilled water heat rejection represents the first delineation in the way that chillers are classified. Using this approach, chillers are described as being either air-cooled chillers or water-cooled chillers.

11.5.1.1 Air-Cooled Chillers

For air-cooled chillers, the chilled water loop rejects the heat it gains to the atmosphere using fan assisted, outdoor air (Figure 11.6). Air-cooled chillers are often described as packaged equipment meaning that the heat rejection component is an integral part of the chiller package and visually looks like a single assembly (Figure 11.7).

The majority of air-cooled chillers are located outside of the building envelope to facilitate heat rejection to the atmosphere. However, due to spatial or other constraints the air-cooled chiller components can be split with the heat rejection component (typically an air-cooled condenser) located remotely from the chiller.

11.5.1.2 Water-Cooled Chillers

Water-cooled chillers employ a second liquid loop in addition to the chilled water loop mentioned above. The two loops exchange thermal properties between them within the chiller before being transported (i.e., piped) to different components of the facility cooling system. The chilled water media described above leaves the chiller colder than it arrived. The other liquid loop is called the condenser water loop and leaves the chiller warmer than it arrived (Figure 11.8).

A separate piece of heat rejection equipment, typically a cooling tower, is used to cool a water-cooled chiller's condenser water loop. The heat rejection equipment is typically located outside of the building envelope while the water-cooled chiller

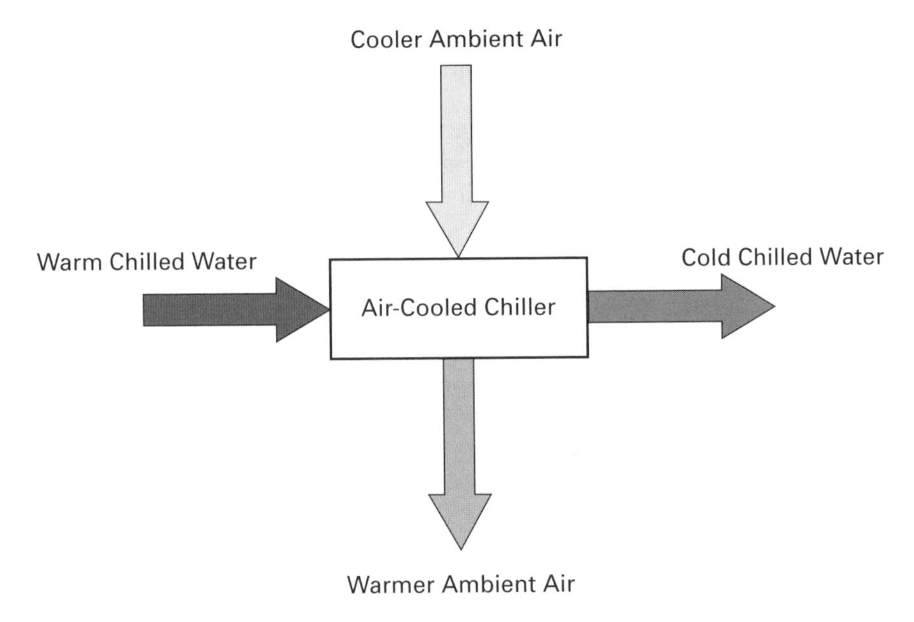

Figure 11.6 Air-cooled chiller diagram. (Courtesy of DLB Associates.)

Figure 11.7 Typical packaged air-cooled chiller. (Courtesy of Trane.)

unit itself (Figure 11.9) is located inside the building and the condenser water loop piping connects the two. How the condenser water is cooled can vary greatly, depending on the region (climate), availability of water, quality of water, noise ordinances, environmental ordinances, and so on. Water-cooled chillers are typically more efficient than air-cooled chillers. Based on electrical consumption per quantity of cooling produced and depending on the type of configuration used, water-cooled chillers can be as much as 25–75% more efficient than the air-cooled equivalent.

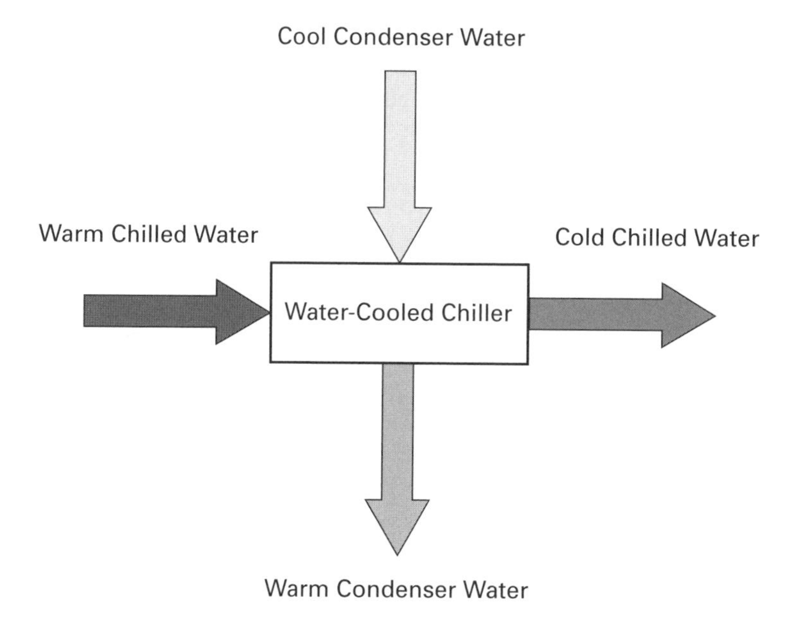

Figure 11.8 Water-cooled chiller diagram. (Courtesy of DLB Associates.)

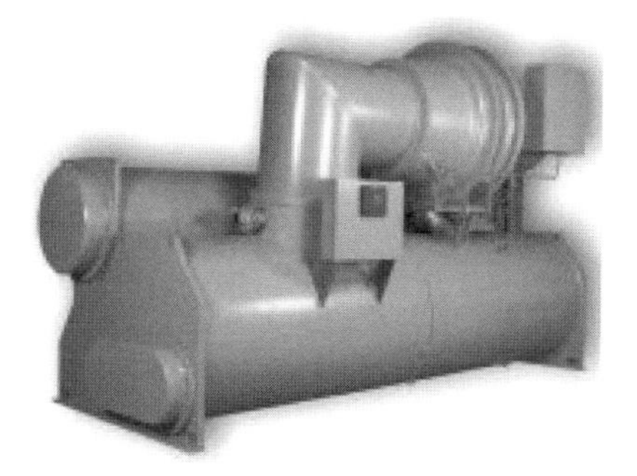

Figure 11.9 Typical water-cooled chiller. (Courtesy of Trane.)

11.5.1.3 Chiller Classification and Typical Sizes

Another way to classify chillers is based on the type of compressor they utilize. Compressor types include reciprocating, screw, scroll, or centrifugal, and the type of compressor selected is commonly based on the cooling capacity needed (e.g., centrifugal compressors are a common choice for large capacities where reciprocal chillers are a common choice for small capacities). Another parameter influencing the chiller selection is efficiency and operating cost at part load versus full load conditions.

In the building cooling industry, common commercial size chillers range from 20 tons to 2000 tons (or 60–6000 kW of electronic equipment power cooling capacity), with chiller sizes outside of this general range being available but not as common. Air-cooled chillers are typically only available up to 400 tons (1200 kW). Chiller configurations also often utilize additional units for redundancy and/or maintenance strategies, and this can play an integral part in determining the quantity and the capacities of chiller units for a given facility.

11.5.2 Heat Rejection Equipment

11.5.2.1 Cooling Towers

Introduction. A cooling tower is a heat rejection device that rejects waste heat to the atmosphere though the cooling of a water stream to a lower temperature.

The type of cooling that occurs in a cooling tower is termed "evaporative cooling." Evaporative cooling relies on the principle that flowing water, when in contact with moving air, will evaporate. In a cooling tower, only a small portion of the water that flows through the cooling tower will evaporate. The heat required for the evaporation of this water is given up by the main body of water as it flows through the cooling tower.

In other words, the vaporization of this small portion of water provides the cooling for the remaining un-evaporated water that is flowed through the cooling tower. The heat from the water stream transferred to the air stream raises the air's temperature and its relative humidity to 100%, and this air is discharged into the atmosphere. Figure 11.10 provides a simple schematic overview of a generic cooling tower flow diagram.

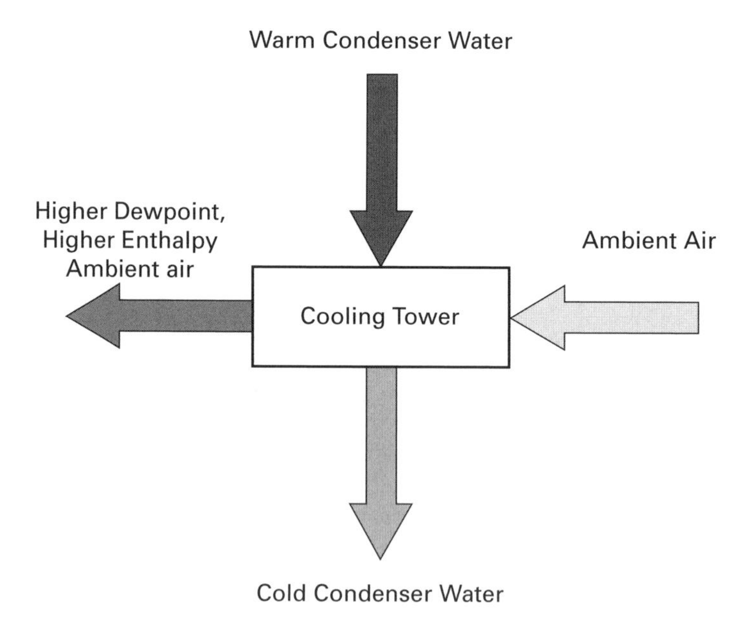

Figure 11.10 Schematic overview of a generic cooling tower flow. (Courtesy of DLB Associates.)

Evaporative heat rejection devices, such as cooling towers, are commonly used to provide significantly lower water temperatures than are achievable with "air-cooled" or "dry" heat rejection devices. "Air-cooled" or "dry" heat transfer processes rely strictly on the "sensible" transfer of heat from the fluid to the cooling air across some type of system barrier and do not depend on the evaporation of water for cooling. Generally, in "air-cooled" or "dry" heat transfer processes, the fluid is separated from the cooling air by a system barrier, and it is not in direct contact with the cooling air.

A radiator in a car is a good example of this type of system and heat transfer process. "Sensible" cooling is limited by the temperature of the air flowing across the system barrier; the temperature of the fluid can never be cooled to a temperature that is less than the temperature of the air flowing across the system barrier. Evaporative cooling, on the other hand, can cool the fluid to a temperature less than the temperature of the air flowing through the cooling tower, thereby achieving more cost-effective and energy-efficient operation of systems in need of cooling.

Common applications for cooling towers are to provide cooled water for air-conditioning, manufacturing, and electric power generation. The smallest cooling towers are designed to handle water streams of only a few gallons of water per minute supplied in small pipes (e.g., 2 in.), while the largest cool hundreds of thousands of gallons per minute supplied in pipes as much as 15 ft (about 5 m) in diameter on a large power plant.

Cooling towers come in a variety of shapes, sizes, configurations, and cooling capacities. For a given capacity, it is common to install multiple smaller cooling tower modules (referred to as cells) rather than one, single, large cooling tower.

Since cooling towers require an ambient air flow path in and out, they are located outside, typically on a roof or elevated platform. The relative elevation of the cooling tower with the remainder of the cooling system needs to be considered when designing the plant because the cooling tower operation and connectivity relies on the flow of fluid by gravity.

Classification. The generic term "cooling tower" is used to describe both open-circuit (direct contact) and closed-circuit (indirect contact) heat rejection equipment. While most think of a "cooling tower" as an open, direct contact heat rejection device, the indirect cooling tower is sometimes referred to as a "closed-circuit fluid cooler" and is, nonetheless, also a cooling tower.

A direct or open-circuit cooling tower (Figure 11.11) is an enclosed structure with internal means to distribute the warm water fed to it over a labyrinth of packing or "fill." The fill may consist of multiple, mainly vertical, wetted surfaces upon which a thin film of water spreads (film fill) or several levels of horizontal splash elements that create a cascade of many small water spreads (splash fill). The purpose of the "fill" is to provide a vastly expanded air-water interface for the evaporation process to take place. The water is cooled as it descends through the fill by gravity while in direct contact with air that passes over it.

Figure 11.11 Direct cooling towers on an elevated platform. (Photo courtesy of DLB Associates.)

The cooled water is then collected in a cold water basin or sump that is located below the fill. It is then pumped back through the process equipment to absorb more heat. The moisture-laden air leaving the fill is discharged to the atmosphere at a point remote enough from the air inlets to prevent its being drawn back into the cooling tower (Figure 11.12).

An indirect or closed-circuit cooling tower (Figure 11.13) involves no direct contact of the air and the fluid being cooled. A type of heat exchanger defines

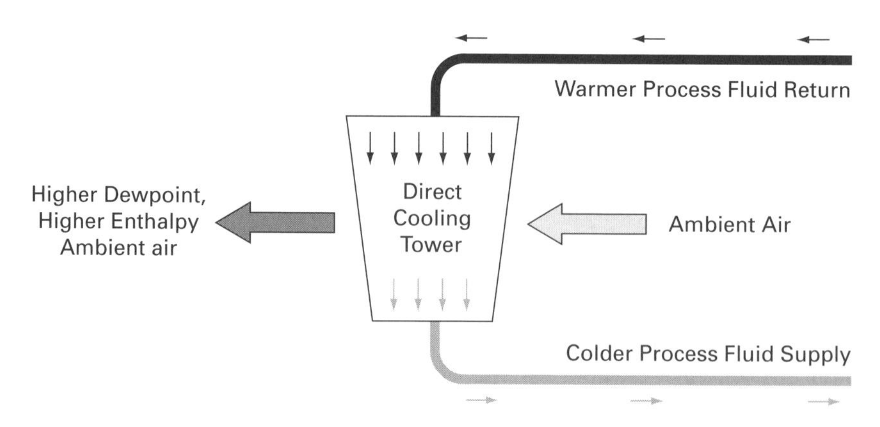

Figure 11.12 Direct or open circuit cooling tower schematic flow diagram. (Courtesy of DLB Associates.)

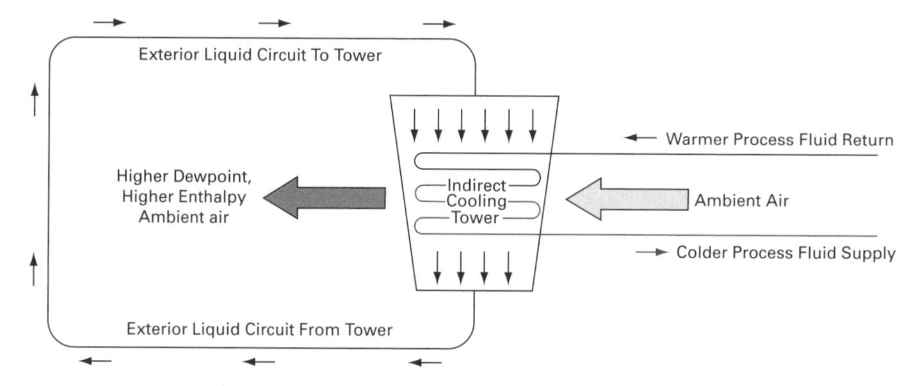

Figure 11.13 Indirect cooling tower schematic flow diagram. (Courtesy of DLB Associates.)

the boundaries of the closed circuit and separates the cooling air and the process fluid. The process fluid being cooled is contained in a "closed" circuit and is not directly exposed to the atmosphere or the recirculated external water. The process fluid that flows through the closed circuit can be water, a glycol mixture, or a refrigerant.

Unlike the open-circuit cooling tower, the indirect cooling tower has two separate fluids flowing through it; one is the process fluid that flows through the heat exchanger and is being cooled while the other is the cooling water that is evaporated as it flows over the "other" side of the heat exchanger. Again, system cooling is provided by a portion of the cooling water being evaporated as a result of air flowing over the entire volume of cooling water.

In addition to direct and indirect, another method to characterize cooling towers is by the direction the ambient air is drawn through them. In a "counter-flow" cooling tower, air travels upward through the fill or tube bundles, directly opposite to the downward motion of the water. In a cross-flow cooling tower, air moves horizontally through the fill as the water moves downward.

Cooling towers are also characterized by the means with which air is moved. Mechanical-draft cooling towers rely on power-driven fans to draw or force the air through the tower. Natural-draft cooling towers use the buoyancy of the exhaust air rising in a tall chimney to provide the air flow. A fan-assisted natural-draft cooling tower employs mechanical draft to augment the buoyancy effect. Many early cooling towers relied only on prevailing wind to generate the draft of air.

Make-Up Water. For the recirculated water that is used by the cooling tower, some water must be added to replace, or make up, the portion of the flow that evaporates. Because only pure water will evaporate as a result of the cooling process, the concentration of dissolved minerals and other solids in the remaining circulating water that does not evaporate will tend to increase. This buildup of impurities in the cooling water can be controlled by dilution (blow-down) or filtration.

Some water is also lost by droplets being carried out with the exhaust air (a phenomenon known as drift or carry-over), but this is typically reduced to a very small amount in modern cooling towers by the installation of baffle-like devices, appropriately called drift eliminators, to collect and consolidate the droplets.

To maintain a steady water level in the cooling tower sump, the amount of make-up water required by the system must equal the total of the evaporation, blow-down, drift, and other water losses, such as wind blowout and leakage.

Terminology. Some useful cooling tower terms commonly used in the building cooling industry are as follows:

- **Drift**. Water droplets that are carried out of the cooling tower with the exhaust air. Drift droplets have the same concentration of impurities as the water entering the tower. The drift rate is typically reduced by employing baffle-like devices, called drift eliminators, through which the air must travel after leaving the fill and spray zones of the tower.
- **Blow-out**. Water droplets blown out of the cooling tower by wind, generally at the air inlet openings. Water may also be lost, in the absence of wind, through splashing or misting. Devices such as wind screens, louvers, splash deflectors, and water diverters are used to limit these losses.
- **Plume**. The stream of saturated exhaust air leaving the cooling tower. The plume is visible when water vapor it contains condenses in contact with cooler ambient air, like the saturated air in one's breath fogs on a cold day. Under certain conditions, a cooling tower plume may present fogging or icing hazards to its surroundings. Note that the water evaporated in the cooling process is "pure" water, in contrast to the very small percentage of drift droplets or water blown out of the air inlets.
- **Blow-down**. The portion of the circulating water flow that is replaced with "clean" water in order to reduce the concentration of dissolved solids and other impurities at an acceptable level.
- **Noise**. Sound energy emitted by a cooling tower and heard (recorded) at a given distance and direction. The sound is generated by the impact of falling water, by the movement of air by fans, by the fan blades moving in the structure, and by the motors, gearboxes, or drive belts.

Water Storage. For data centers or other mission critical facilities, on-site water storage is a consideration to avoid the loss of cooling capability due to a disruption in water service. For large data centers, 24–72 hours of on-site make-up water storage can be in the range of 100,000 to well over 1,000,000 gallons.

Typically, the selection of cooling towers is an iterative process since there are a number of variables resulting in the opportunity for tradeoffs and optimization. Some of those variables include size and clearance constraints, climate, required operating conditions, acoustics, drift, water usage, energy usage, part load performance, and so on.

11.5.2.2 Condensers and Drycoolers

Introduction. Heat rejection equipment is the term commonly used by the building cooling industry for the equipment that ultimately discharges the heat from the conditioned area to the atmosphere.

Cooling towers (both open and closed circuit types) are a part of a water-cooled system and are evaporative cooling devices using the cooling effect of the phase change of water to provide cooling. This equipment is a predominantly wet-bulb-centric device.

The heat rejected from chillers can utilize either air or water media. Water-cooled chillers utilize the wet-bulb-centric heat rejection devices (i.e., cooling towers) mentioned above. Air-cooled chillers, on the other hand, utilize a different type of heat rejection equipment.

Air-cooled chillers have an integral air-cooled condenser section. This section contains condenser fans that induce airflow over a coil that contains refrigerant. The refrigerant inside this coil cools and changes phase from a gas to a liquid as it passes through this coil. Air-cooled condensers are basically dry-bulb-centric devices that use the movement of ambient air passing through a coil to reject heat to the atmosphere by condensing the refrigerant vapor in the coil to a liquid state (see Figure 11.14).

Types of Dry-Bulb-Centric Heat Rejection Devices. There are a number of different types of dry-bulb-centric heat rejection devices. Similar to the

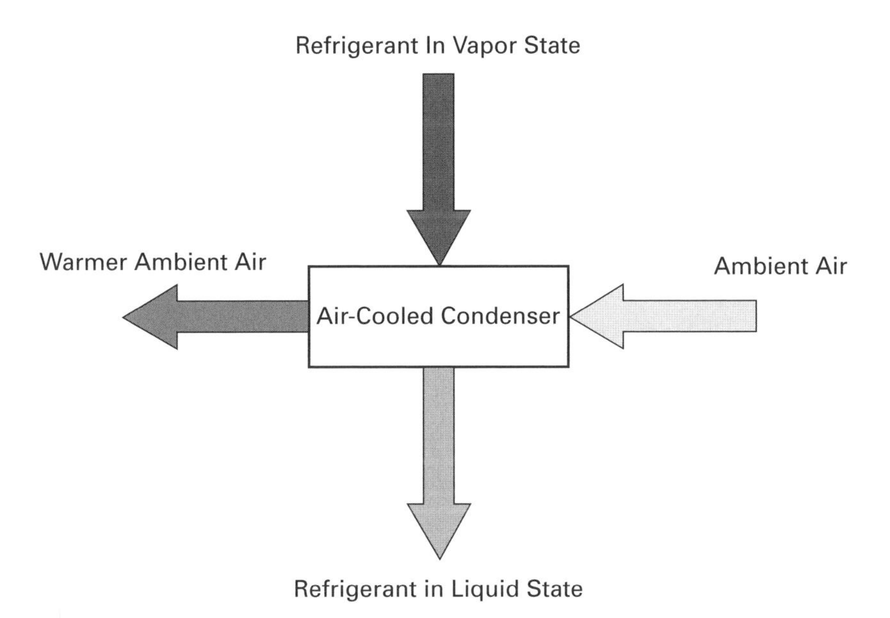

Figure 11.14 Schematic overview of a generic air-cooled condenser. (Courtesy of DLB Associates.)

wet-bulb-centric devices (i.e., cooling towers), these devices are all located outdoors (since they all utilize ambient air as a means of heat rejection). Depending on the specific type of heat rejection device and the application, their location varies from being at grade or on a roof. Other common design influences include:

- Noise ordinances and acoustics
- Clearances for both air inlets and discharges to provide adequate air flow to the coil
- A configuration that avoids the short-circuiting of the hot discharge air with the air inlets

Too frequently, proper clearances are not maintained. This is especially true when multiple units are installed in the same area. Also, it is important to consider the source and impact of flying debris, dirt, vegetation, and so on.

The three main types of dry-bulb-centric heat rejection devices that we will be discussing are:

- Air-cooled condensers
- Self-contained condensing units
- Drycoolers

Air-Cooled Condensers. The air-cooled condenser section in an air-cooled chiller is, in some cases, a completely separate unit that can be located remotely from the compressor section of the chiller or other compressorized piece of equipment for certain products and situations. This stand-alone piece of equipment is simply called an air-cooled condenser.

The two main components of an air-cooled condenser are a fan to induce the air movement through the unit and a coil (typically with fins) to provide the heat transfer surface area that is required for the refrigerant to condense to a liquid state. The combination of a fan and a coil are the components that provide the means for the refrigeration circuit to reject its heat to the air.

Air-cooled condensers come in various shapes, sizes, and orientations. Figures 11.15 and 11.16 show examples of air-cooled condensers with different air inlet and outlet (discharge) configurations. Figure 11.15 is an air-cooled condenser that is typically paired with a small wall- or ceiling-mounted air handling unit used in many smaller computer server rooms for office applications (i.e., one or two racks).

Figure 11.16 is an air-cooled condenser that is more typically used in larger server rooms in conjunction with air-cooled CRAC units. The multiple condenser fan quantities and larger footprints allow it to have a greater heat rejection capacity than the air-cooled condenser shown in Figure 11.15.

Self-Contained Condensing Units. In some cases, the compressor and air-cooled condenser sections are combined into a single piece of equipment that is called a self-contained condensing unit or simply a "condensing unit"). A condensing unit is the type of equipment that is most commonly used in residential

Figure 11.15 Side inlet and side outlet, air-cooled condenser. (Courtesy of Sanyo.)

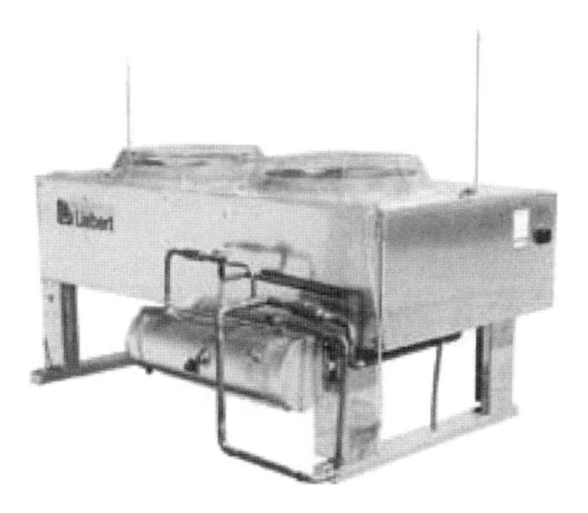

Figure 11.16 Bottom inlet and top outlet, air-cooled condenser. (Courtesy of Liebert.)

applications and looks like a cube with a round propeller fan horizontally mounted on top discharging air upwards (Figure 11.17). It is also used in larger systems where the noise generated by the compressors must be outside the occupied space.

Drycoolers. The air-cooled condenser shown in Figure 11.16 looks very similar to another type of dry-bulb-centric heat rejection device, a drycooler. Whereas an air-cooled condenser cools a refrigerant, a drycooler utilizes water as its cooling media (typically with a mixture of glycol in the water solution to prevent freezing). The drycoolers themselves are used with a glycol-cooled CRAC unit.

One of the biggest advantages of using drycoolers is that the mechanical cooling provided by the compressors can be eliminated during "low ambient" temperatures, providing energy savings to the owner. This is accomplished by the installation of a supplementary cooling coil inside the CRAC unit. This "economizer" coil can provide either all or a portion of the cooling required by the space.

Figure 11.17 Typical self-contained condensing unit. (Courtesy of Carrier.)

Performance. As is typical with all coil selections, a heat rejection device's capacity is impacted by coil specific parameters such as:

- Coil or tube size
- Number of rows of coils
- Fin density (number of fins per inch)
- Materials of construction
- Thickness and shape of the tubes and fins

In the building cooling industry, there is a standardized way to rate the capacity of a heat rejection device. The Air-Conditioning and Refrigeration Institute (ARI) sets a fixed dry and wet bulb condition that all manufacturers test to and base their published capacity data on. This allows for easier comparison of products.

The performance of all outdoor heat rejection devices is largely based on climate. Therefore, the maximum output for a given device will differ, depending on the climate of that region. Furthermore, its energy usage will vary along with the weather. Depending on the requirements, a heat rejection device can have one or more fans and the controls can vary from a single control per condenser to staged control of a multi-fan, multi-section condenser. Variable speed is also possible.

Geothermal Heat Rejection Systems. In a geothermal or earth-coupled system, coils are installed horizontally or vertically in the ground and take advantage of the earth's temperature. Since its heat rejection does not utilize ambient air, it is not considered to be either dry-bulb-centric or wet-bulb-centric.

An earth-coupled heat rejection system has the advantage that it does not require fans to reject heat. Having no fan is an energy saver and eliminates the noise associated with it. A disadvantage of an earth-coupled system is that typically it is a much higher installation cost. In some cases, utility rebates are available to make the TCO (total cost of ownership) more attractive. Its main advantage, however, is that it is generally a very efficient system.

11.5.3 Energy Recovery Equipment

11.5.3.1 Economizers

Introduction. The term "economizer" can be used to describe a piece of equipment (or a component that is part of a larger piece of equipment), but it can also refer to an energy savings process. Building cooling systems utilize energy provided by the utility company (gas or electric) to drive mechanical components that reject heat from a transport media to produce a cooling effect. If the heat transferred by this equipment is ultimately rejected to the atmosphere, then the performance of this equipment is dependent on the ambient weather conditions.

Weather conditions or climate also are the main drivers in determining the economic feasibility of using an economizer or "free cooling." An economizer basically takes advantage of favorable outdoor conditions to reduce or eliminate the need for mechanical refrigeration to accomplish cooling.

Design Weather Conditions. The design parameters used to select and size heat rejection equipment are typically based on ambient weather conditions that will occur, along with the frequency of these conditions, in the project location. Depending on the form of heat rejection utilized, either the (dry bulb) temperature or a value that relates to the amount of moisture in the air (wet bulb or dew point) will be used. Sometimes a "mean coincident" value will be combined with either a dry bulb or wet bulb temperature to more accurately state the "maximum, actual" condition (or total energy of the ambient air) that will most likely be experienced. A common source for this information is the ASHRAE Fundamentals Handbook, which lists weather data for over 750 locations nationwide (and another 3600 locations worldwide). The weather data are listed in columns with "0.4%," "1%," and "2%" as the headings of these columns. The values provided under these percentages represent the amount of time in one calendar year that the conditions listed for a particular location will be exceeded.

For example, in Boston, MA, the dry bulb temperature stated in the 0.4% "dry bulb/mean coincident wet bulb" bin is 91 °F. This means that the temperature in Boston will only exceed 91 °F 0.4% of the year (which is about 35 hours). The 2% value (about 175 hours) for this same heading for Boston is 84 °F.

Although cooling equipment is typically selected based on the data found in one of these categories (the "design conditions" for the project) the vast majority of time, the actual ambient conditions are below these "design conditions." (The specific data used to select the equipment depends on the particular emphasis of the application and the client's preferences.) Generally, the "design conditions" are used to select the maximum capacity of the cooling equipment. As a result, the full capacity of the cooling equipment is not utilized for the majority of the year.

Economizer Process. Fundamentally, the economizer process involves utilizing favorable ambient weather conditions to reduce the energy consumed by the building cooling systems. Most often this is accomplished by limiting the amount of energy used by the mechanical cooling (refrigeration) equipment. Since the use of an economizer mode (or sequence) reduces the energy consumption while maintaining the design conditions inside the space, another term that is used in the building cooling industry for economizer mode operation is "free cooling."

ASHRAE Standard 90.1 (Energy Code for Commercial and High-Rise Residential Buildings) is a "standard" or "model code" that describes the minimum energy efficiency standards that are to be used for all new commercial buildings and includes aspects such as the building envelope and cooling equipment. This standard is often adopted by an Authority Having Jurisdiction (AHJ) as the code for a particular locale. Sometimes it is adopted in its entirety, while in other instances it is adopted with specific revisions to more closely meet the needs of the AHJ. Within that standard, tables are provided stating the necessity of incorporating an economizer into the design of a cooling system, depending on design weather conditions and cooling capacity.

In simplified terms, ASHRAE 90.1 mandates that unless your building is in an extremely humid environment or has a cooling capacity equivalent to that of a single family home, an economizer is required to be designed as a part of any cooling system for your building.

Economizers can utilize "more favorable" ambient air directly to accomplish the cooling needs of a building or process. This is known as an "airside economizer." If a water system is used to indirectly transfer heat from the building or process to the atmosphere, as in a condenser water/cooling tower system, this known as a "waterside economizer."

Airside Economizer. In order to simplify the following discussion, the moisture content of the air streams will be considered to be of no consequence. However, this is not a practical approach when building systems are designed.

During normal cooling operation on a hot, summer day, in order to maintain a comfortable temperature inside the building in the mid-70s (°F), a cooling system for an office space might typically supply air at temperatures around 55 °F. The temperature of the air returned back (recirculated) to the cooling system from the space would approximate the temperature maintained inside the space before any ventilation (outdoor) air in introduced into the air stream. Therefore, it requires less energy by the cooling system to cool and recirculate the air inside the building than it would to bring outdoor air into the building (i.e., cooling the air from 75 °F down to 55 °F uses less energy than going from, say, 90 °F to 55 °F).

However, building codes mandate that a minimum quantity of ventilation (outdoor) air be introduced into a building to maintain some level of indoor air quality. This minimum quantity of ventilation air is often based on the expected occupancy load (i.e., the number of people per 1000 square feet) in a space multiplied by a prescribed quantity (CFM) of outdoor air per person. This formula can yield a very high percentage of outdoor air (as high as 75%) under certain "densely populated" situations. The percentage of outdoor air is the ratio of the amount of outdoor air (in CFM) divided by the amount of supply air (in CFM). For purposes of this discussion, let's assume that the outdoor air percentage is 25% of the total amount of air supplied by the cooling system. Therefore, this minimum quantity of outdoor is mixed with the recirculated (return) air, producing a mixed air temperature closer to 75 °F than 90 °F prior to passing through the cooling coil in an air handling unit. This scenario is graphically depicted in Figure 11.18.

For days in the spring or fall, the outdoor air temperature may drop to below 75 °F (the temperature of the return air). During those days, it would take less energy to use 100% outdoor air in the cooling air stream for the same reasons highlighted above. Figure 11.19 shows the economizer mode operation of an air handling unit.

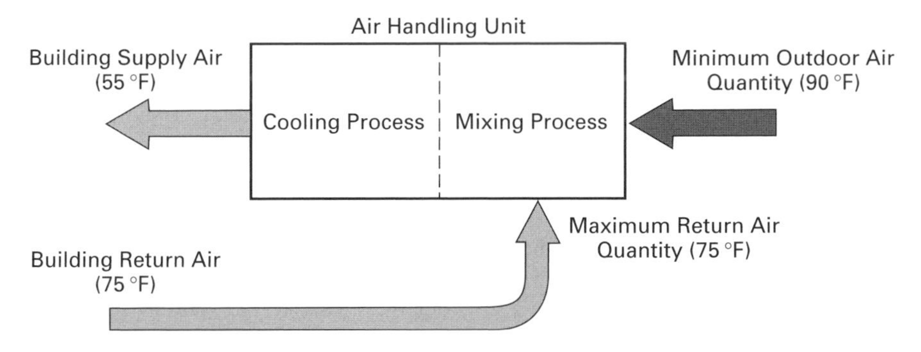

Figure 11.18 Hot summer day cooling operation. (Courtesy of DLB Associates.)

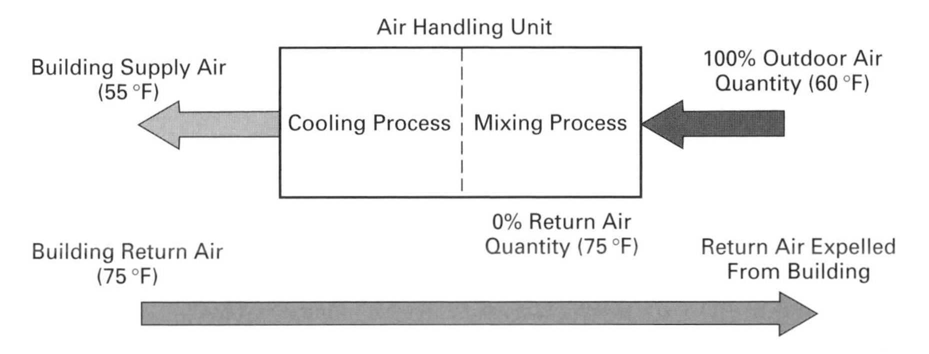

Figure 11.19 Spring/fall day cooling operation—economizer mode. (Courtesy of DLB Associates.)

In an airside economizer setup, the air handling equipment can vary the amount of outdoor air introduced into the building through the use of a mixing box and dampers that are configured as a part of the air handling equipment itself. Outdoor air temperature sensors and motorized controllers are used to optimize the energy consumption for any weather condition. Keep in mind that in addition to temperature, humidity as well as contaminants contained in the outdoor air are also a concern and should influence the operation of an airside economizer.

For datacom (data processing and communications) facilities the use of airside economizers has been a controversial topic. Some reasons behind the controversy include:

- There is a lack of consensus about the number of hours that the economizer can actually be used. In addition to the number of "economizer hours" being based on geography, the operation of datacom facilities typically needs to adhere to a stricter tolerance of environmental conditions as mandated by legally binding Service Lease Agreements (SLAs) executed between the facility owner and the end-user. Therefore, anytime the economizer mode is enabled, the facility operator needs to be confident that the resulting interior environment is within the limits set forth in the SLA. This situation narrows the window (number of hours) deemed suitable for economizer operation (i.e., the tendency will be against using economizer operation on borderline days).
- There is a lack of consensus about the amount of energy savings that can be realized. Since outdoor air typically has a greater quantity of contaminants, especially in "industrial" areas, the filtering process to ensure the correct indoor air quality is more substantial. If the quantity of contaminants inside the space is a concern, then the use of more exacting filtration methods can be utilized to reduce/remove these contaminants. The downside of adding this filtration is the increase in fan power that is required to overcome the pressure drop of this filtration. This added static pressure drop requires additional fan horsepower, resulting in the use of more energy both during economizer operation and during normal operation. Therefore, any energy savings achieved through the elimination of the mechanical cooling process needs to be offset against the energy increases through the operation of a larger fan. The air in coastal areas is known to be very corrosive as a result of the salt carried by the air currents. Removing this salt could be a costly operation. The ongoing maintenance costs associated with the replacement of filters also needs to be considered as part of the overall cost savings calculation.
- Humidity control inside the space also needs to be addressed when the topic of airside economizers is discussed during the planning stages of a datacom project. Even though the accepted ranges of relative humidity inside the space have widened over the past few design cycles, there typically is some requirement to control the relative humidity in datacom facilities. "High"

relative humidly levels could make electronic equipment prone to hygroscopic dust failures (HDF), conductive anodic failures (CAF), tape media errors, and corrosion while "low" relative humidity levels might leave electronic equipment susceptible to the effects of electrostatic discharge (ESD). The cost of removing moisture from outdoor air is not insignificant. While there are various ways to remove moisture from air, such as refrigeration or desiccants, each method represents "substantial" investments in capital and operating expenditures. Adding moisture to a space can cost as much as trying to remove it. Adding moisture to increase the humidity level inside the building requires the evaporation of water, and there are several methods available to accomplish this. Some of them require evaporating or boiling water so that the moisture can be more easily introduced into the air. This method has the substantial side benefit of allowing just pure water vapor to be introduced into the space, because the dissolved minerals and other impurities that could settle on electronic components and compromise their reliability are left behind in the vessel used to boil the water. Another method to introduce moisture into the space uses compressed air and high-pressure water to help atomize the water, decreasing the scale of the water "plume" generated by the nozzle. If there is concern about the introduction of minerals and impurities into the space, then water treatment is an option with this method.

- Some facility owners may have first cost concerns. A lot of data center owners focus in on the smallest first cost possible for the construction of their data centers in order to use the revenue from customers to fund capacity upgrades as needed after the initial construction period. The addition of an economizer operation results in a greater first cost of equipment and also requires more sophisticated controls. Although payback studies may prove a lower total cost of ownership (TCO), the emphasis of building the facility, attracting the customers and then expanding as needed differs from an emphasis of achieving the lowest possible TCO. The TCO calculation may also be skewed since many owners of datacom facilities have negotiated "very favorable" utility rates so the cost of energy may be substantially less than what is available to commercial customers.
- The effectiveness and use of an airside economizer in datacom facilities is therefore dependent on a number of factors ranging from geographical climate to owner preferences. Since each datacom facility is unique with regard to these factors, each project needs to be evaluated individually to determine whether an airside economizer is right for a specific application.

Waterside Economizer. A more common economizer application in datacom facilities is the use of a waterside economizer. In order to understand the operation of a waterside economizer, we should recap some of the information presented earlier in this chapter. Previously, we introduced two liquid media loops that are used to transport heat:

- **Chilled or Process Water Loop**. Transports heat from the cooling coil within the air handling equipment (e.g., CRAC unit) to the mechanical cooling equipment (e.g., chiller).
- **Condenser Water (or Glycol) Loop**. Transports heat from the mechanical cooling equipment (e.g., chiller) to the heat rejection equipment (e.g., cooling tower, drycooler, etc.).

The mechanical cooling process involves an input of energy to realize a transfer of heat between these two liquid loops. The net result is that the temperature of the condenser water increases and the temperature of the chilled or process water decreases as it passes through the mechanical cooling equipment.

In some air handling or CRAC unit configurations, it is possible for the mechanical cooling process to be internal to that piece of equipment (also called self-contained). In that configuration, the condenser water loop runs from the air handling or CRAC unit directly to the exterior heat rejection equipment (typically a drycooler).

Again, the drycooler is typically selected and sized based on the hot, summer day design conditions; therefore, there is an opportunity to use more favorable ambient weather conditions to save energy, this time through the use of a waterside economizer. Similar to the process described for airside economizers, the lower ambient air temperature can result in a lower temperature condenser water being produced and the temperatures may be such that the energy required for the mechanical cooling of the process water loop can be reduced or, in some cases, eliminated entirely.

In order to utilize the condenser water as a part of a waterside economizer process, the necessary components need to be included as part of the mechanical design. If the cooling equipment utilizes a refrigerant (as opposed to water) to transfer heat from the cooling coil to the heat rejection equipment (drycooler or cooling tower), then a second cooling coil will be required in the air handling equipment. When the ambient conditions are "favorable" and "cold" condenser water can be produced by the heat rejection equipment, then the condenser water can be pumped through this coil, providing all or part of the required space cooling. If water, instead of refrigerant, is used for the cooling medium, then the designer has two options available. If the condenser water system has been provided with "above average" water treatment, then the "cold" condenser water can be pumped directly through the chilled/process water piping system. The second option would be for the designer to include a heat exchanger in the system design to allow for the transfer of heat between the systems while keeping the 2 water streams separate. Please keep in mind that producing cold water with heat rejection equipment can require a considerable amount of fan (both in the heat rejection and air handling equipment) and pumping energy, so the "free" cooling may not be as "free" as one would expect.

Additional components required include motorized control valves and condenser water temperature sensors in order to control the routing of the condenser water dependent on its temperature.

11.5.3.2 Heat Exchangers

Introduction. Building cooling systems rely on thermal transportation in order to condition the air within an enclosed space. This thermal transportation typically involves multiple media loops between various pieces of building cooling equipment. Heat exchangers represent equipment that is supplementary to the cooling system, and they are installed in order to either attain some energy recovery from the thermal transportation media flows in order to reduce the amount of energy required to perform the building cooling by providing a nonmechanical cooling process when ambient conditions allow it (similar to a waterside economizer but at a larger scale).

As its name suggests, a heat exchanger is a device that relies on the thermal transfer from one input fluid to another input fluid. One of the fluids that enter the heat exchanger is cooler than the other entering fluid. The cooler input fluid leaves the heat exchanger warmer than it entered, and the warmer input fluid leaves cooler than it entered. Figure 11.20 is a simplistic representation of the process.

Heat Exchanger Types. Heat exchangers can be classified in a number of ways ranging from the media streams involved in the transfer process to the direction

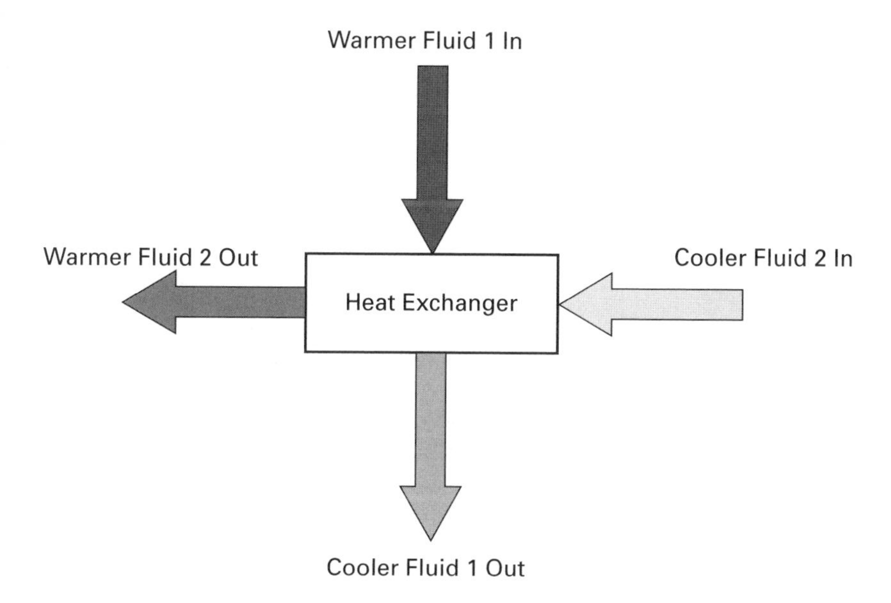

Figure 11.20 Simple overview of the heat exchanger process. (Courtesy of DLB Associates.)

of the flow of the media within them to their physical configuration. We will discuss the types of heat exchangers in turn and then classify them using the criteria listed above.

Shell and Tube Heat Exchanger. A shell and tube heat exchanger is a water-to-water type of heat exchanger where one liquid fills a larger cylindrical vessel (i.e., the shell) and the other liquid is transported through the filled vessel in multiple small tubes. The tubes are typically made from a thermally conductive material such as copper, whereas the shell material is more durable and is made from stainless steel or brass.

The configurations of shell and tube heat exchangers can be expressed in terms of the number of passes, with each pass indicating the transportation of the liquid in the tubes across the length of the shell. For example, a one-pass configuration would have the tube liquid input at one end of the shell and then the output at the other end. A two-pass configuration would have the tubes double back, and the input and output of the tube liquid would be at the same end.

Four-pass configurations are also available; also, the longer the passage for the tube liquid in the submerged shell, the greater the heat transfer that occurs (depending on the relative temperatures of the liquids involved). Shell and tube types of heat exchangers are typically used for applications that involve hot water systems. Figure 11.21 shows a typical shell and tube heat exchanger.

Plate and Frame Heat Exchanger. A plate and frame type of heat exchanger is a water-to-water type of heat exchanger that consists of a series of flat hollow plates (typically made from stainless steel) sandwiched together between two heavy cover plates. The two liquids pass through adjacent hollow plates such that one liquid in a given plate is sandwiched between two plates containing the other liquid. In addition, the direction of the liquid flow is opposed for the two liquids and hence from one plate to the next (Figure 11.22).

All of the liquid connections are in the same side, but one liquid travels from the bottom of the plate to the top while the other liquid travels from the top of the adjacent plates to the bottom, thus maximizing the heat transfer rate by always

Figure 11.21 Shell and tube heat exchanger. (Courtesy of DLB Associates.)

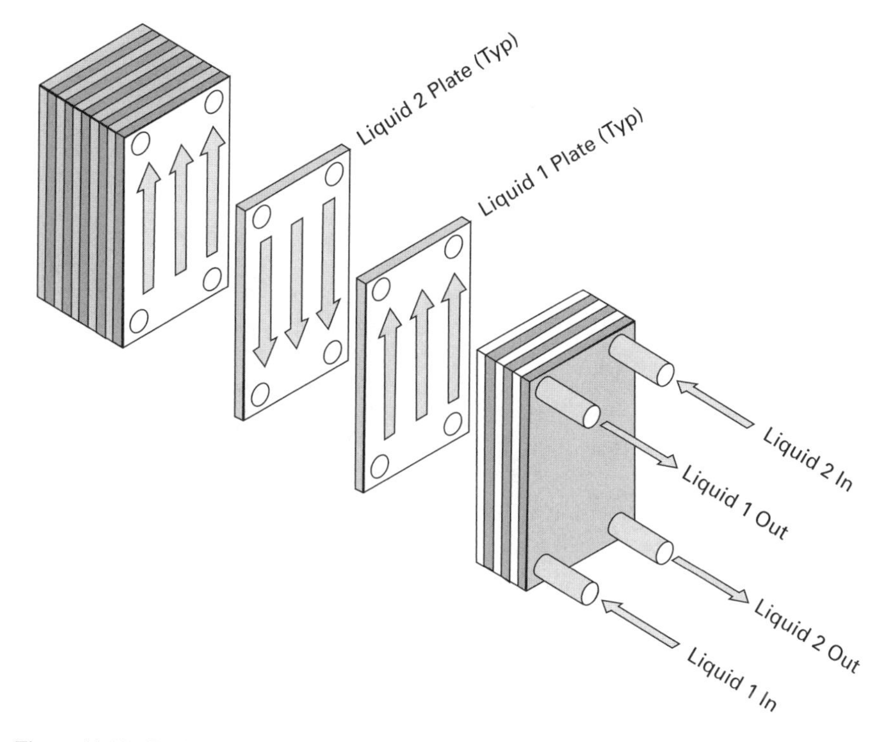

Figure 11.22 Exploded isometric showing plate and frame heat exchanger operation. (Courtesy of DLB Associates.)

having the difference in liquid temperatures be as large as possible through the process. Figure 11.23 shows a typical installation of plate and frame heat exchangers.

One of the main advantages of plate and frame heat exchangers over shell and tube heat exchangers is that the overall heat transfer coefficients are three to four times higher, and this type of heat exchanger is more common in water-cooled chilled water applications.

Other Types of Heat Exchangers. There are many other types of heat exchangers for more specialized applications that are beyond the scope of this chapter and will not be covered in this publication. The two types mentioned above are the most common types used for the majority of applications.

Heat Exchanger Operation for Datacom Facilities. For datacom (data processing and telecom) facility operations that use water-cooled chilled water systems, heat exchangers are typically used when condenser water conditions are low enough to allow for the returning chilled water to be cooled via the heat exchangers' passive and energy efficient heat transfer process and without the mechanical operation of a chiller.

The two liquid streams normally connected to the chiller are now re-routed to one or more heat exchangers that are installed in parallel to the chillers. The

Figure 11.23 Installed plate and frame heat exchangers. (Photo courtesy of DLB Associates.)

switchover from the heat exchangers to the chillers can be as simple as a set day of the year to switch over, or it can be dynamically controlled depending on the ambient conditions.

11.6 COMPONENTS INSIDE DATACOM ROOM

11.6.1 CRAC Units

Introduction. The final component of building cooling equipment we will introduce is a computer room air-conditioning unit—or as it is more commonly known, a CRAC unit. CRAC units are typically used to cool large-scale (20 kW or more) installations of electronic equipment. They are sometimes called by other names such as CAHU, CACU, and so on.

Although these units all look very similar from the outside (Figure 11.24), the components within the CRAC unit can be configured to integrate with a wide variety of building cooling systems including chilled water, direct expansion (DX), glycol, air-cooled, water-cooled, and so on, and the units themselves are available in capacities ranging from 6 to 62 "nominal" tons (or 20–149 kW of electronic equipment cooling).

For the majority of electronic equipment installations, the heat generated by the equipment is discharged into the computer room environment in the form of hot air through small fans that are integral to the electronic equipment packaging. This hot air from the electronic equipment is known as return air (i.e., air that is being

Figure 11.24 Typical computer room air conditioning unit. (Photo courtesy of Liebert.)

returned to the cooling equipment to be cooled). The return air is transported to the CRAC units, where it is cooled and redelivered to the room environment adjacent to the air inlets of the electronic equipment.

The cold air delivered to the electronic equipment room environment from the CRAC unit is known as supply air or conditioned air. In the electronic equipment industry, this air is sometimes referred to as chilled air, but the term chilled air is not used in the building cooling industry.

Location. More often than not, the CRAC units themselves are located in the same room as the electronic equipment. The CRAC unit itself has a fairly large physical footprint (the larger CRAC units are 8 to 10 ft wide by 3 ft deep) and also requires unobstructed maintenance clearance areas to the front and the two sides of the unit, making the overall impact to the floor area of an electronic equipment room significant.

Where spatial constraints result in the electronic equipment room not being able to accommodate any CRAC units, they may be located in adjacent rooms with ducted connections to supply and return air from the electronic equipment room (Figure 11.25). However, the CRAC units cannot be located too far from the computer equipment that they are meant to cool due to limitations of the CRAC unit fans.

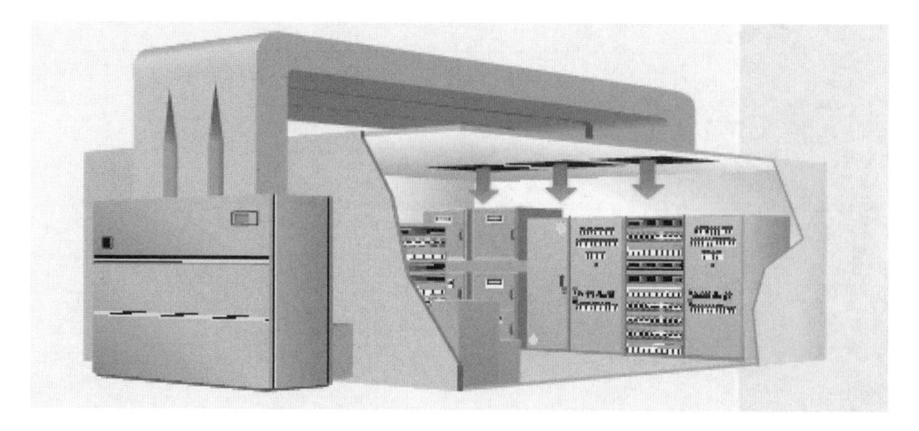

Figure 11.25 CRAC unit located outside the electronic equipment room. (Image courtesy of Liebert.)

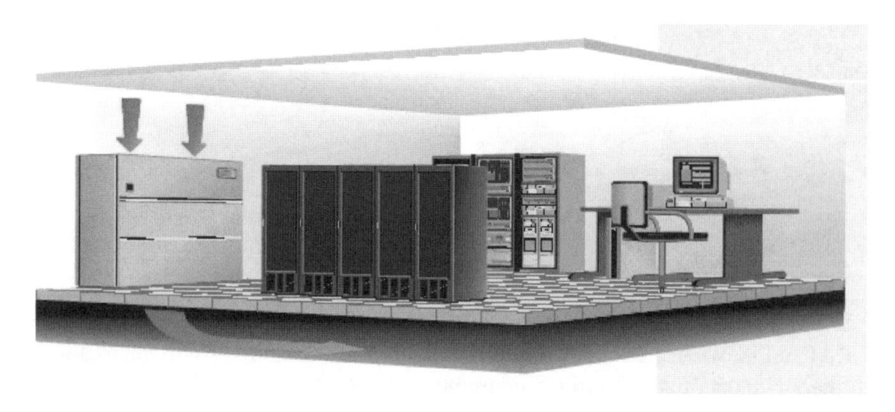

Figure 11.26 Downflow CRAC unit airflow path. (Image courtesy of Liebert.)

Airflow Paths. The airflow paths to and from a CRAC unit can be configured a number of different ways. The cold supply air from the unit can be discharged through the bottom of the unit, through the front of the unit or through the top of the unit. The warmer return air can also be configured to enter the CRAC unit from the front, the rear, or the top.

The most common airflow path used by CRAC units is referred to as "downflow" (Figure 11.26). A downflow CRAC unit has the combination of return air entering the top of the unit and supply air discharged from the bottom of the unit, typically into a raised floor plenum below the electronic equipment racks. The supply air pressurizes the raised floor plenum and is reintroduced into the electronic equipment room proper through perforated floor tiles that are located adjacent to the air inlets of the rack-mounted electronic equipment.

Upflow CRAC units have return air entering the CRAC unit in the front of the unit, but toward the bottom. The supply air is either directed out of the top or out the front but near the top of the unit. A frontal supply air discharge is accomplished through the used of a plenum box, which is basically a metal box with no bottom panel that is placed over and entirely covers the top of the unit. The plenum box has a large grille opening in its front face to divert the air out in that direction (Figure 11.27).

Instead of using a plenum box, an alternate airflow distribution method for an upflow CRAC unit is to connect the top discharge to conventional ductwork. The ductwork can then be extended to deliver supply air to one or more specific locations within the electronic equipment room (Figure 11.28).

Upflow CRAC units are typically used either in areas that do not have raised floors or in areas that have shallow raised floors used only for power/data cable distribution.

Figure 11.27 Upflow CRAC unit airflow path. (Image courtesy of Liebert.)

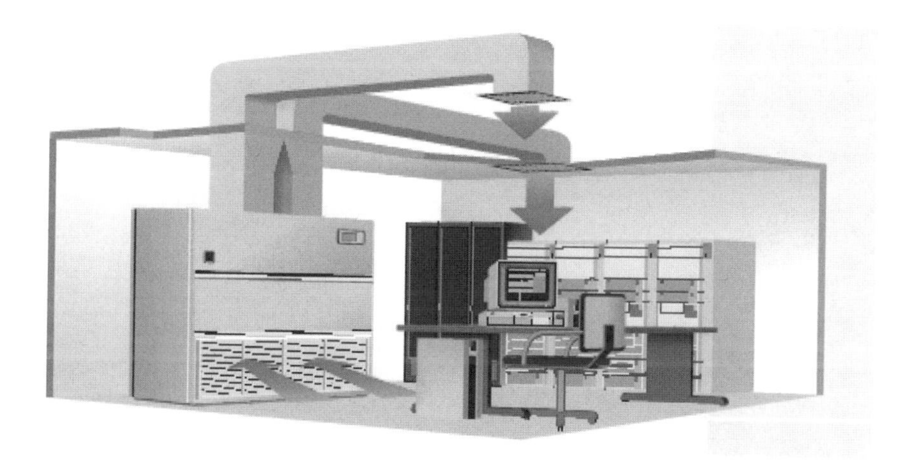

Figure 11.28 Ducted upflow CRAC unit airflow path. (Image courtesy of Liebert.)

Configurations. The common components within all cooling configurations of CRAC units are a fan and a coil. Other optional components that are available to all CRAC unit configurations include a humidifier and a heating element, the combination of which help control the humidity of the electronic equipment room environment. The specific type of building cooling system that is being utilized determines the additional components located within the unit as well as the type of liquid media that is circulated in the coil.

Classification. In addition to the upflow and downflow approach to classifying the CRAC units, the units can also be classified by whether or not the refrigeration process occurs within their physical enclosure. These fall into two main categories called compressorized (e.g., air-cooled, water-cooled, glycol, etc.) and chilled water systems.

A CRAC unit is said to be compressorized or self-contained if it contains the compressors within its housing. Compressorized CRAC units require a remote piece of equipment to perform the heat rejection, and a liquid (condenser water, glycol, etc.) is piped between the CRAC unit and the heat rejection device.

In a chilled water CRAC unit, the mechanical refrigeration takes place outside of the physical extents of the CRAC unit enclosure in a remotely located chiller. No compressors are required in the CRAC unit, and only the chilled water piping from the chiller is required to provide chilled water to the coil. Chilled water CRAC units have a greater capacity (up to 60 nominal tons or 149 kW) than compressorized systems (up to 30 nominal tons or 88 kW).

Chapter 12

Raised Access Floors

Dan Catalfu

12.1 INTRODUCTION

12.1.1 What Is an Access Floor?

An access floor system is literally an elevated or "raised" floor area upon another floor (typically a concrete slab in a building).

This raised floor (Figure 12.1) consists of 2-ft × 2-ft floor panels, generally manufactured with steel sheets—welded construction filled with lightweight cement to provide a strong, quiet accessible floor structure—that are supported by adjustable understructure support pedestals.

The access floor system (floor panels and support pedestals) create a space between the floor slab and the underside of the access floor. This space is where you can cost-effectively run any of your building services. These services consist of electric power, data, telecom/voice, environmental control and air conditioning, fire detection and suppression, security, and so on. The access floor panels are designed to be removable from their support so that "access" to under the floor services is fast and easy. Access holes are made in the floor to allow services to terminate at floor level using hinged lid termination boxes, or pass through the floor panels into the working space of the building area.

Raised access floors are an essential part of mission critical facilities due to the need for efficient and convenient air and electrical distribution. Access floors were in fact developed during the beginning of the computer age in the 1950s to provide the increased airflow required to keep systems and control rooms at safe operating temperatures. While increased airflow is still a major benefit, as data centers have evolved, so have access floors. There are now a variety of panel grades and types to accommodate practically every situation. There have also been advancements in electrical and voice/data distribution which increase flexibility, adding to the

Maintaining Mission Critical Systems in a 24/7 Environment By Peter M. Curtis

Figure 12.1 Typical raised floor. (Courtesy of Tate Access Floors.)

benefits of an access floor. Overall, the use of an access floor will help you to maximize the use of your services.

12.1.2 What Are the Typical Applications for Access Floors?

Specific requirements of the space will determine the potential areas in which raised access flooring can be considered. Areas in which raised access flooring can be used:

- General office areas
- Auditoriums
- Conference rooms
- Training rooms
- Executive offices
- Computer rooms
- Telecom closets
- Equipment rooms
- Web farms
- Corridors
- Cafeterias/break rooms
- Bathrooms/restrooms

- Storage areas
- Lobby/reception areas
- Print room

12.1.3 Why Use an Access Floor?

In short, because it is a more cost-effective service distribution solution from a first-cost and operating-cost perspective. Easy-to-move service terminals (power/voice/data) and air at any point on floor plate.

Furthermore, as technology advances, it is critical that your facility can be easily and economically modified with minimal interruption. An access floor gives you that flexibility. The interchangeability of solid and airflow panels allows you to concentrate your airflow wherever needed to cool sensitive equipment, which can produce an enormous amount of heat.

To summarize, an access floor is:

- Cheaper to build
  - Speeds up construction cycle
  - Reduces slab-to-slab height by up to 1 ft per story
  - Greatly reduces HVAC ductwork
- Cheaper to operate
  - Ongoing cost saving per HVAC zone addition
  - Efficient equipment relocation or addition with minimum interruption

12.2 DESIGN CONSIDERATIONS

- When selecting the appropriate access floor system for your facility, you must take several factors into consideration. These factors will be in terms of structural performance, finished floor height, understructure support design type, and the surface finish of the floor panels.

12.2.1 Determine the Structural Performance Required

It is important at an early stage in the consideration of a raised access floor that a detailed assessment is made of the likely loadings that will be imposed on the floor surface. These loadings need to be assessed in terms of the following:

12.2.1.1 Concentrated Loads

A concentrated load (Figure 12.2) is a single load applied on a small area of the panel surface. These loads are typically imposed by stationary furniture and equipment which rest on legs. In the laboratory test, a specified concentrated load is applied to the panel on a one-inch-square area while deflection is measured under

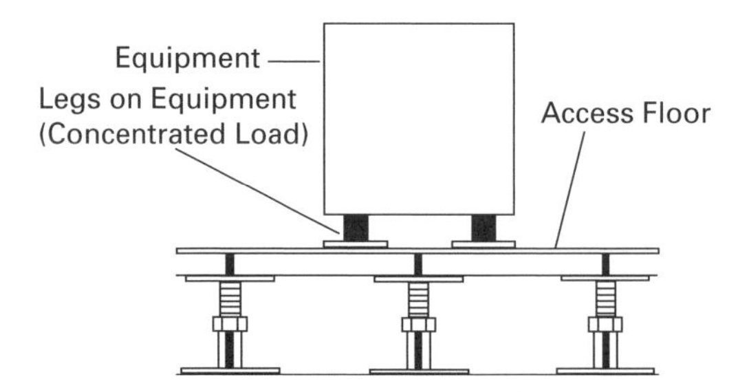

Figure 12.2 A concentrated or static load is a single load applied on a small area of the panel surface. (Courtesy of Tate Access Floors.)

the load. Permanent set (rebound) is measured after the load is removed. The rated concentrated load capacity indicates the load that can be concentrated on a single floor panel without causing excessive deflection or permanent set. Concentrated loads are also referred to as static loads.

12.2.1.2 Rolling Load

Rolling loads (Figure 12.3) are applied by wheeled vehicles carrying loads across the floor. Pallet jacks, electronic mail carts, and lifts are some examples. The rated rolling load capacity indicates the load allowed on a single wheel rolling across one panel.

Rolling loads are defined by the amount of weight on the wheels and number of passes across the panel. The 10-pass specification is used to rate a panel for rolling loads that will occur *infrequently* over the life of the floor (usually referred to as move-in loads). The 10,000-pass specification is used to rate panels for rolling loads that may occur many times over the life of the floor.

12.2.1.3 Uniform Loads

Uniform loads (Figure 12.4) are applied uniformly over the entire surface of the panel and are typically imposed by items such as file cabinets, pallets, and furniture

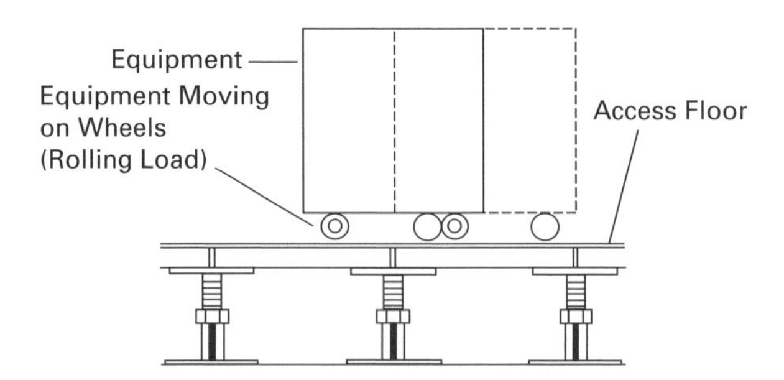

Figure 12.3 Rolling loads are applied by wheeled vehicles carrying loads across the floor. (Courtesy of Tate Access Floors.)

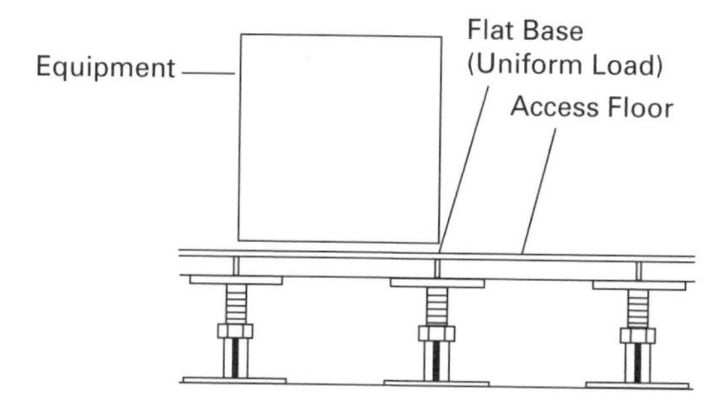

Figure 12.4 Uniform loads are applied uniformly over the entire surface of the panel, typically imposed by file cabinets, pallets, and furniture or equipment without legs. (Courtesy of Tate Access Floors.)

or equipment without legs. The rated uniform load capacity indicates the panel's load capacity stated in pounds per square foot.

12.2.1.4 Ultimate Loads

The ultimate load capacity of a floor panel is reached when it has structurally failed and cannot accept any additional load (Figure 12.5). The ultimate load capacity is not an allowable load number—it merely indicates the extent of the panel's overloading resistance. A panel loaded to this degree will bend considerably and will be destroyed. You should never intentionally load a floor panel anywhere near its ultimate load rating. It is recommended that the panel's ultimate load or safety factor be rated at least 2.5 times its concentrated or working load.

12.2.1.5 Impact Loads

Impact loads (Figure 12.6) occur when objects are accidentally dropped on the floor. In the laboratory test, a load is dropped 36 inches onto a steel plate that is resting on top of a one-inch-square steel indentor placed on the panel. Because the test load is

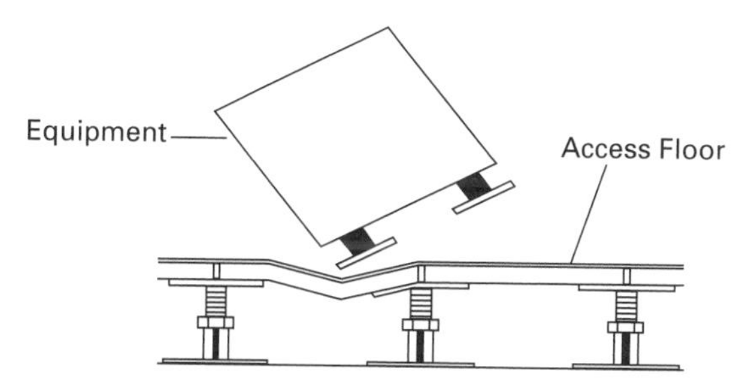

Figure 12.5 The ultimate load capacity of a floor panel is reached when it has structurally failed. (Courtesy of Tate Access Floors.)

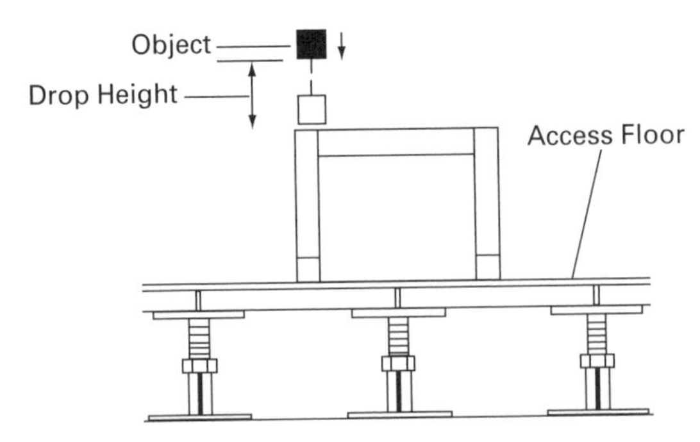

Figure 12.6 Impact loads occur when objects are accidentally dropped on the floor. (Courtesy of Tate Access Floors.)

concentrated onto a small area, it generates a severe shock—which makes the test a safe indication of the panel's impact capacity. Heavy impact loads, which most often occur during construction, move-in, and equipment/furniture relocations, can cause permanent panel damage.

12.2.2 Determine the Required Finished Floor Height

Finished floor height is defined by the distance from the concrete slab subfloor to the top surface of the access floor panel. Finished floor heights can range from as low as 2.5 inches to as high as 6 ft and above with special understructure supports, but typical mission critical facilities will be in the 12- to 36-inch range most of the time, depending on the services required under the floor. When selecting the appropriate finished floor height, take the following into consideration:

12.2.2.1 Consider the Service Requirements

These services will typically include electrical power, data, telecom/voice, environmental/air conditioning, fire detection/suppression, security, and so on. An evaluation of each of these underfloor services will give an indication of the space required for each service run. Due regard needs to be taken for any multiple service runs, crossover of services, separation requirements, and so on. This information can then be used to determine the cavity depth required under the raised floor and hence determine the finished floor height which will then be used in specifying the raised access floor system. For access floors using underfloor electrical distribution, a 12-in. finished floor height would be typical. For an access floor utilizing underfloor air, 18- to 24-in. finished floor heights are more typical. Although these finished floor heights are typical, each facility will need to be evaluated based on the amount of cable and air handling equipment to determine the appropriate finished floor height.

12.2.2.2 Minimum Floor to Ceiling Requirements

Within any building design the top of floor to underside of ceiling minimum dimension is one aspect that is outside the designer's area of influence as it is governed by building regulations. However, the use of a raised access flooring system in conjunction with under floor modular cable management and/or under floor HVAC will allow this dimension to be optimized.

12.2.3 Determine the Understructure Support Design Type Required

Various understructure design options are available to accommodate structural performance requirements, floor height, and accessibility needs. It is important that at an early stage in the consideration of a raised access floor a detailed assessment is made of the likely loadings that will be imposed on the understructure. These loadings need to be assessed in terms of the following:

12.2.3.1 Overturning Moment

Overturning moment is a lateral load applied to the pedestal due to rolling load traffic, underfloor work due to wire pulls, or seismic activity. It is determined by multiplying the lateral load applied to the top of the pedestal by the pedestal's height. A pedestal's ability to resist overturning moment is a function of its attachment method to the structural slab, size and thickness of the base plate, and pedestal column size.

12.2.3.2 Axial Load

Axial load is a vertical load applied to the center of the pedestal due to concentrated, rolling, uniform, and other loads applied to the surface of the access floor panel.

12.2.3.3 Seismic Load

Seismic load is a combination of vertical (both up and down) on the surface of the access floor as well as lateral loads.

12.2.3.4 Understructure Design

Ninety percent of mission critical data centers utilize bolted stringer understructure for lateral stability and convenient access to the underfloor plenum (Figure 12.7). Bolted stringers can be 2 ft, 4 ft or a combination of both. The 4-ft basketweave system will provide you with the most lateral support, since it ties multiple pedestals together with a single stringer. It also makes aligning the grid easier at installation and helps keep the grid aligned during panel removal and reinstallation. The 2-ft system gives you flexibility, because fewer panels need to be removed if stringers

Figure 12.7 Typical bolted stringer understructures and cable tray. (Courtesy of Tate Access Floors.)

need to be removed to perform service or run additional cabling under the floor. Cables can be laid directly on the slab, or can be cradled in cable trays positioned between the slab and the bottom of the access floor panels. The combination 2-ft/4-ft system will give you some benefits of both of the previous options.

12.2.4 Determine the Appropriate Floor Finish

There are many options for factory-laminated floor finishes, but most are not appropriate for mission critical data centers. You will most likely want to select a finish with some degree of static control or low static generation.

Floor tile for rooms requiring control of static electricity is often specified incorrectly because the tile names and definitions are easily misconstrued. There are two basic types of static-control floor tile: low-static-generating tile and static-conductive tile. Static-conductive tile has a close cousin named static-dissipative tile.

12.2.4.1 Low-Static-Generation Floor Tile

High-pressure laminate tile (HPL) is a low-static-generating tile informally referred to as anti-static tile. It typically provides the necessary static protection for computer rooms, data centers, and server rooms. Its purpose is to provide protection for equipment by limiting the amount of static charges generated by people walking or sliding objects on the floor. It offers no protection from static charges generated in other ways. The static dissipation to ground is extremely slow with HPL because its surface-to-ground electrical resistance is very high. In addition to using tile to control static generation and discharge, generation should also be controlled by maintaining the proper humidity level, as well as by the possible use of other static-control products.

The standard language for specification of HPL electrical resistance is the following: *Surface to Ground Resistance of Standard High-Pressure Laminate Covering: Average test values shall be within the range of 1,000,000 ohms* (1.0×10^6) *to 20,000 megohms* (2.0×10^{10} *ohms*), *as determined by testing in accordance with the test method for conductive flooring specified in Chapter 3 of NFPA 99, but modified to place one electrode on the floor surface and to attach one electrode to the understructure. Resistance shall be tested at 500 volts.*

Static-conductive floor tiles are typically not used in computer rooms for two noteworthy reasons. When conductive tiles are used on a floor, occupants must wear conductive footwear or grounding foot straps in order for static charges to dissipate through their shoes into the tile. Rubber, crepe, or thick composition soles insulate the wearer from the floor and render conductive tile ineffective. Also, conductive floor tile can present electrical shock hazards since its electrical resistance is not high enough to safely limit current to the floor surface from an understructure system that becomes accidentally charged.

12.2.4.2 Conductive and Static-Dissipative Floor Tile for Sensitive Equipment Areas

Static-conductive floor tile is typically used in ultrasensitive environments such as electronics manufacturing facilities, assembly factories, clean rooms, and MRI rooms. Its surface-to-ground electrical resistance is very low; therefore it provides rapid static dissipation to ground.

The standard language for specification of electrical resistance of conductive tile is the following: *Surface-to-Ground Resistance of Conductive Floor Covering: Average test values shall be within the range of 25,000 ohms (2.5×10^4) to 1,000,000 ohms (1.0×10^6 ohms). Conductive tiles may be tested in accordance with either NFPA 99 or ESD S7.1.*

There is also a more resistant variant of static-dissipative tile that has a resistance range of 1,000,000 ohms (1.0×10^6 ohms) to 100,000,000 ohms (1.0×10^8 ohms). This tile is usually referred to as static-dissipative tile, rather than conductive tile. (Both tiles dissipate static electricity to ground, but conductive tile does it much faster.)

12.2.5 Airflow Requirements

12.2.5.1 Selecting the Appropriate Airflow Method

Airflow can be achieved through many ways including grilles and diffusers, but mission critical data centers typically require a large volume of air to cool the equipment, so perforated panels or grates are usually selected. Both are interchangeable with solid panels, so it is relatively easy to modify airflow as needs change.

Perforated Panels. Perforated panels are generally an all-steel design with a series of holes that create a 25% open area in the panel surface for air circulation. The airflow can be regulated by means of a damper located on the underside of the panel if desired. Perforated panels can be laminated with the same factory finish as solid panels. Shown is the Tate Access Floor's Perf 1000 with airflow chart (Figure 12.8).

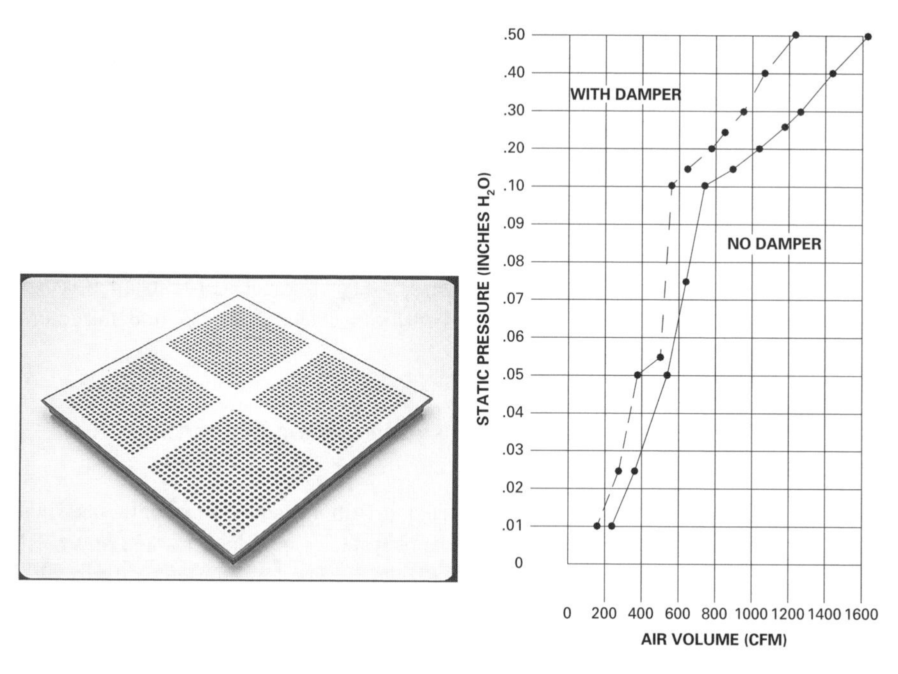

Figure 12.8 Tate Access Floor's Perf 1000 with airflow chart.

Grate Panels. Grate panels are generally made from aluminum, with a series of honeycomb slots that create a 56% open area for air circulation. The airflow can be regulated by means of a damper located on the underside of the panel if desired. Grate panels cannot be laminated with a tile, but they can be coated with conductive or nonconductive epoxy or metallic coatings or can simply be installed bare. Below is the Tate Access Floors GrateAire^{Tm} with airflow chart (Figure 12.9).

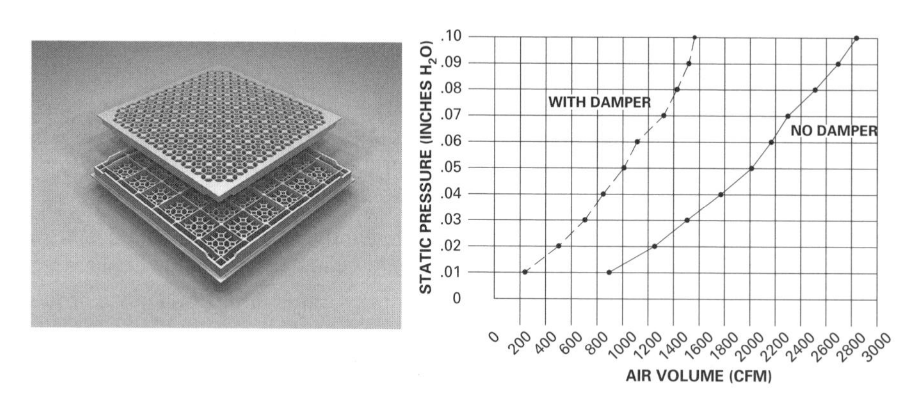

Figure 12.9 Tate Access Floor's GrateAire^{Tm} with airflow chart.

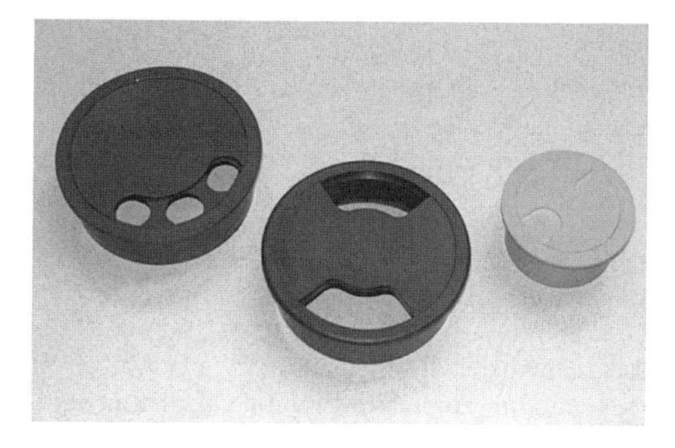

Figure 12.10 For areas where cables pass through to the plenum, grommets or cutout trim and seal restrict air leakage. (Courtesy of Tate Access Floors.)

12.2.5.2 Air Leakage

For areas without perforated or grate panels, the floor must remain properly sealed to control air leakage and proper air volume to the areas requiring airflow. Excessive leakage can reduce temperature control and cause poor air distribution performance. Be aware that air can leak through construction seams where walls, columns, and other obstructions extend through the access floor and through access floor, wall, and slab penetrations created for ducts, conduits, and pipes. Leakage between the panel seams is typically not an issue because of the stringer, but if necessary, gaskets can be applied to the stringer to reduce leakage.

New construction seams and utility penetrations created after the floor installation must be sealed. The access floor perimeter should remain sealed with snug-fitting perimeter panels and with the wall base firmly pressed against the floor. Perimeter seams where panels do not consistently fit tightly (within approximately 1/16 inch of wall), due to rough or irregular-shaped vertical surfaces, should be caulked, gasketed, taped, or otherwise sealed to maintain air tightness.

For areas where cable pass through to the plenum beneath the access floor is required, be sure to use grommets, or cutout trim and seal designed to restrict air leakage (Figure 12.10). These can be obtained from the access floor manufacturer or directly through a grommet manufacturer such as Koldlok®.

12.3 SAFETY CONCERNS

12.3.1 Removal and Reinstallation of Panels

Proper panel removal and reinstallation is important so that injuries to people and damage to the floor system are prevented. Stringer systems require particular care to maintain squareness of the grid.

12.3.1.1 Panel Lifters

Suction cup lifter (Figure 12.11): for lifting bare panels and panels laminated with vinyl, rubber, HPL, VCT or other hard surface coverings.

Perf panel lifter (Figure 12.12): for lifting perforated airflow panels (can also be used to lift grates).

12.3.2 Removing Panels

When underfloor access is required, remove only those panels in the area where you need access and reinstall them as your work progresses from one area to another. (The more panels you remove at one time, the more likely some pedestal height settings will be inadvertently changed.) The first panel must be removed with a panel lifter. Adjacent panels can be removed with a lifter, or by reaching under the floor and pushing them upward and grabbing them with your other hand.

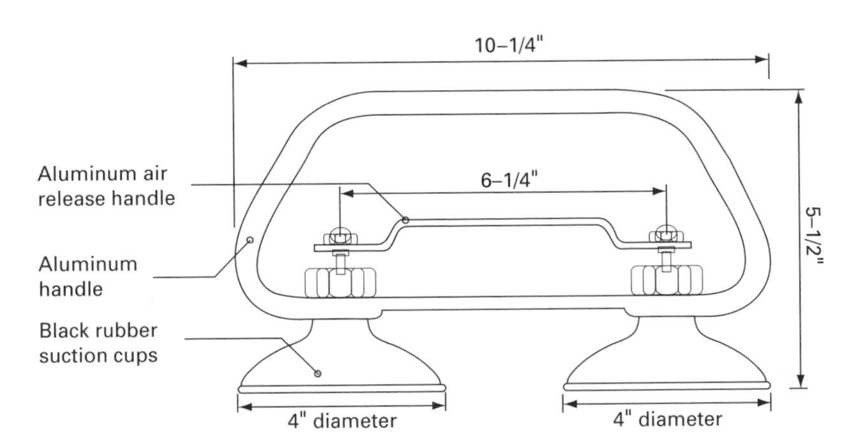

Figure 12.11 Suction cup lifter. (Courtesy of Tate Access Floors.)

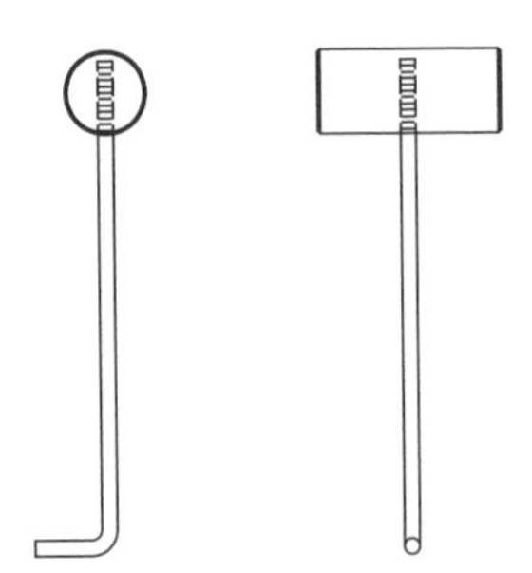

Figure 12.12 Perf panel lifter. (Courtesy of Tate Access Floors.)

12.3.2.1 Lifting a Solid Panel with a Suction Cup Lifter

Attach the lifter near one edge of the panel and lift up that side only (as if opening a hinged door). Once the panel's edge is far enough above the plane of the adjacent panels, remove it by hand. You may find that kneeling on the floor while lifting panels reduces back strain.

12.3.2.2 Precautions

- Never attempt to carry a panel by the lifter—the suction could break and allow the panel to fall from your hand.
- Do not use screwdrivers, pliers, or other tools to pry or lift panels.
- Do not change the height adjustment of the pedestals when removing panels.

12.3.2.3 Lifting a Perforated Panel

- A simple way to remove a perforated panel without a lifter is to remove an adjacent solid panel so that you can reach under it and push it out of the floor by hand. If you use a Perf panel lifter, hook the lifter through a perforation hole on one side of the panel and lift up until you can place your other hand under it; then remove it by hand.

12.3.3 Reinstalling Panels

All but the last panel to be reinstalled can be placed on the understructure *by hand* or with a lifter.

When using a lifter: Attach the lifter near one edge of the panel, sit the opposite edge of the panel on the understructure and lower the side with the lifter in your hand as if closing a door. *Kneeling on the floor while replacing panels may avoid back strain.*

12.3.3.1 Precautions

- Do not over-torque Posilock or stringer screws—your torque setting should be no more than 35 in.-lb.
- Do not attempt to reinstall panels by kicking them into place.
- Do not change the height adjustment of the pedestals when replacing panels.
- Make a final check to verify that panels are correctly in place and level.

12.3.4 Stringer Systems

- A stringer system can become out of square when people remove large blocks of panels from the floor and unintentionally twist the grid while performing underfloor work. Correcting an out-of-square grid condition is not an in-house, do-it-yourself task. It must be performed by a professional installer. To prevent grid-twisting from occurring in the first place, adhere to the instructions below. Be aware that two-foot stringers systems are more susceptible to grid-twisting than four-foot (basketweave) systems.

12.3.4.1 Precautions (Preventing Grid-Twisting)

- When you need to work under the floor, remove only as many floor panels as necessary to do the work. Keeping surrounding panels in the floor helps to maintain grid squareness: The more panels you remove at one time, the greater the possibility that the grid will be accidentally twisted. Replace panels as your work progresses from one area to another.
- Never pull electrical cables against the pedestals and stringers.
- When removing and reinstalling cut panels at the floor perimeter and around columns, reinstall each section exactly where it came from. Panels are custom-cut for each perimeter location and are not interchangeable. Walls are often wavy or out-of-square: Interchanging differently sized perimeter panels can cause floors to become out-of-square and can also cause panels to become too tight or too loose in the floor.
- Keep foot traffic and rolling loads on the access floor away from the area where the panels and stringers are to be removed. Failure to do this can easily cause panels to shift and stringers to twist on the heads.
- When removing stringers for complete access, remove only those stringers necessary for access and replace them as your work progresses from one area to another.
- Don't walk on stringers or put lateral loads on stringers when panels are removed from the floor.
- When replacing stringers, make sure that the adjacent pedestal heads attached to each stringer are square to one another before final tightening of the stringer screws.
- Don't over-torque the stingers with your driver when reattaching the stringers to the pedestal heads. Set the torque on your screw gun between 30 and 35 in.-lb.
- Don't try to re-square the stringer grid by hitting stringers with a hammer; this can result in the formation of bumps on top of the stringers that will cause panels to rock.

12.3.5 Protecting the Floor from Heavy Loads

The following types of heavy loading activity may merit the use of floor protection: construction activity, equipment move-in, facility reconfiguration, use of lift equipment or pallet jacks on the floor, and frequent transportation of heavy loads in areas where delivery routes exist.

A look at the allowable loads for your panel type will tell you whether the loads you intend to move are within the capacity of the floor. The allowable rolling load indicates the load allowed on a single wheel rolling across a panel.

12.3.5.1 Determining if a Rolling Load is Within the Floor's Capacity

First determine the type(s) of floor panels in the load path. Be mindful of the possibility that there may be different types of floor panels installed between delivery routes and other areas of the floor. For demonstration purposes, we will refer to the panel load chart in Table 12.1. For one-time movement of a heavy load, refer to the column headed Rolling Loads: 10 Passes. ("10 Passes" means that tests demonstrate that the panel can withstand the allowable load rolled across its surface 10 times without incurring excess deformation.) To determine if a load is within the panel's allowable capacity, ascertain the wheel spread of the vehicle. When there's at least 24 in. between wheels (front to back and side to side), there will be only one wheel on any floor panel at a time (provided that the load is not moved diagonally across individual panels). If the payload is evenly distributed, divide the combined weight of the vehicle and its payload by four to determine the load on each wheel and compare it to the allowable load in the chart. If the payload is unevenly distributed, estimate the heaviest wheel load and compare it to the chart. **Caution**: Avoid moving heavy loads diagonally across floor panels, because this could allow two wheels to simultaneously roll over opposite corners of each panel in the path. Failure to avoid this could result in excess panel deformation.

12.3.5.2 Floor Protection Methods for Heavy Rolling Loads

There are several methods of protecting the floor from heavy rolling loads:

- Place load-spreading plates on the floor along the load path.
- Reinforce the floor with additional support pedestals along the load path.

Table 12.1 Panel Load Chart

Panel Type	Allowable Loads			
	Concentrated Loads (lb)	Rolling Loads: 10 Passes (lb)	Rolling Loads: 10,000 Passes (lb)	Uniform Loads (lb./ft2)
ConCore 1000	1000	800	600	250
ConCore 1250	1250	1000	800	300
ConCore 1500	1500	1250	1000	375
ConCore 2000	2000	1500	1250	500
ConCore 2500	2500	1500	2000	625
All Steel 1000	1000	400	400	250
All Steel 1250	1250	500	500	300
All Steel 1500	1500	600	600	375

- Combine methods 1 and 2.
- Install heavy-grade floor panels along the load path.
- Use load moving equipment equipped with air bearings (also called air casters, air cushions).
- Use a moving vehicle with six or more wheels that are spread more than 24 in. apart.

12.3.5.3 Load-Spreading Plates

The load path can be covered with one of the following types of load-spreading plate:

- One layer of ¾-in.-thick plywood sheet.
- Two layers of 1/8-in.-thick steel plate.
- One layer of 1/4-in.-thick aluminum plate.

When moving very heavy loads, you should place either one layer of 1/4-in. aluminum plates or 1/8-in. steel plates on top of ¾-in. sheets of plywood. Overlap the ends of the metal plates to a position near the center of the wood sheets to keep the load from becoming too severe when the wheels reach the ends of wood sheets. Caution: When using ¾-in. spreader plates or any thicker combination of plates, you need to create a step-down at the end of the plate so that the load will not drop to the access floor. You can accomplish this by using any rigid material that's approximately half the thickness of the load spreader.

12.3.5.4 Floor Reinforcement

The floor panels in a heavy load path can be reinforced by having the access floor installer install additional support pedestals beneath them. Five-point supplemental support for each panel is achieved by positioning one additional pedestal under each panel center and one under the midpoint of each panel edge where two panels abut. Each pedestal will be glued to the subfloor with approved pedestal adhesive and will be height-adjusted so that each head is supporting the undersides of panels. (The number of additional pedestals required amounts to three per panel.) This method of floor reinforcement is often used in delivery corridors where heavy loads are moved on a daily basis and in areas where lift equipment is used to perform overhead maintenance work.

12.3.5.5 Load-Spreading Plates and Panel Reinforcement

On occasions when a one-time rolling load may slightly exceed floor capacity, reinforcement pedestals should be installed along the load path and load-spreading plates should be placed on top of the floor.

12.3.5.6 Heavy-Grade Floor Panels

Floor panels with different load ratings can be installed alongside one another to create a more supportive path for heavy loads. All panel types are typically interchangeable, but keep in mind that heavier-grade panels may require heavier duty understructure than standard-grade panels. Installation of floor panels and pedestal heads should be performed by the access floor installer.

12.3.5.7 Load-Moving Equipment Equipped with Air Bearings

Load-moving equipment equipped with air bearings has been successfully used to move equipment on access floors. The equipment uses compressed air to float loads across the floor on a low friction air film. The bearing area of the air casters is considerably larger than that of wheeled casters, so that use of load-spreading plates can typically be avoided. Perforated floors require only a thin polystyrene sheet to create an air seal.

12.3.5.8 Additional Equipment Movement Precautions

- Before rolling a load on the floor, make sure that all floor panels in and around the load path are in place to prevent the floor from shifting.
- When moving heavy loads on a floor *without* spreader plates, select a vehicle with the largest wheels available. The larger and wider the wheels, the less potential there is for localized panel indentation.
- When moving motorized vehicles on an access floor, avoid sudden stops and rapid accelerations. Hard starting and stopping can damage the understructure supporting the floor.
- Do not allow heavily loaded vehicles to drop to the floor when stepping down from a higher level.

12.3.5.9 Precautions for Raised Floor Ramps

Ramps and landings should be protected with $^3/_4$-in. plywood before moving very heavy loads on them. There should be plywood extending to the edge of the raised floor where it meets the top of the ramp. The plywood at the top of the ramp should be connected to the plywood at edge of the floor by a section of sheet metal attached to each piece of wood with sheet metal screws. A metal plate may be needed at the bottom of the ramp to step up to the plywood (the plate can be held in place by gravity).

12.3.5.10 Using Lift Equipment on the Floor

Whenever possible, overhead work that requires the use of heavy lift equipment should be performed prior to installation of the access floor. Where lifts will be occasionally used on an access floor for overhead maintenance, the system should be selected accordingly: Use heavy-grade panels installed on a four-foot bolted stringer understructure system.

A lift with a side-to-side wheel-spread greater than 24 in. is recommended so that the wheels will be on separate floor panels. To determine the compatibility of a particular lift with the floor, add the weight of the expected platform load to the weight of the unit and apportion the combined load to each wheel. Compare it to the allowable load in Table 12.1 in the column headed Rolling Loads: 10,000 Passes. Two protective measures can be used to fortify floor areas that will require overhead maintenance throughout the life of the building. The first is to reinforce the floor with five additional support pedestals per panel as previously mentioned. The second is to place load-spreading plates on the floor whenever maintenance with a lift is required. These two steps should be used simultaneously if the lift equipment has a wheel spread that is not fully 24 in. between wheels. Installation of additional support pedestals should be performed by the access floor installer.

12.3.5.11 Using Electronic Pallet Jacks on the Floor

The floor protection guidelines also apply to the use of electronic pallet jacks, which can be very heavy when loaded. A pallet jack's weight is unevenly distributed to its wheels; therefore the load per wheel needs to be calculated according to the Table 12.2. The wheel loads can then be compared to the allowable load in the load chart. An empty electronic jack typically bears 80% of its weight on the drive wheel and 10% on each fork wheel. To calculate the load per wheel of a loaded jack, enter the appropriate weights into Table 12.2.

Caution: Verify the distance between the fork wheels of your truck–many are built with less than 24 in. between them. If two fork wheels will simultaneously ride across a single column of panels, you need to combine the weight of the two wheels (or a portion of weight on the two wheels) to accurately determine the load on the panels. Using a pallet jack with both fork wheels simultaneously riding across a single column of panels over an extended period of time will cause panel deformation if the load exceeds the panel's rating. To accommodate a jack with this condition, you may need to have the floor reinforced with additional pedestals and/or a heavier grade of floor panel. Installation of floor panels and additional pedestal heads should be performed by the access floor installer. You should also avoid moving the jack diagonally across floor panels, because this could allow two wheels to simultaneously roll over opposite corners of each panel in the path.

Table 12.2 Load calculation table for electonic pallet jack

Jack and Payload	Drive Wheel Load	Fork Wheel Load
Empty jack weight:	80% of jack weight:	10% of jack weight:
Payload weight:	40% of payload:	30% of payload:
Total weight:	Total load on drive wheel:	Total load on fork wheel:

Caution: Verify the distance between the fork wheels of your truck (see text).

12.3.5.12 Using Manual Pallet Jacks on the Floor

To calculate the load on each wheel of a manual pallet jack, add the weight of the payload to the weight of the jack, then apportion 20% of the total weight to the steering wheel and 40% of the total weight to each fork wheel. The wheel loads can then be compared to the allowable load in the load chart.

12.3.5.13 Protecting the Floor's Finish from Heavy Move-in Loads

Laminated floor panels and painted finish floor panels are susceptible to scratching, minor indentations, and marring by the wheels of heavily loaded vehicles. Vinyl tiles are especially susceptible due to the ability of wheels to leave permanent impressions. To avoid causing this type of damage, the floor can be covered sheets of fiberboard (such as masonite), plywood, heavy cardboard, or rigid plastic (such as plexiglass). The sheets should be taped end to end with duct tape to ensure that there's a continuous path.

12.3.5.14 Floor Protection Methods for Heavy Stationary Equipment

Heavy stationary loads are most often concentrated loads (see definition on page 235) consisting of equipment with legs or feet. For these loads, refer to the Concentrated Loads column in the Table 12.1. A few types of concentrated loads may slightly exceed panel capacity and prescribe the use of additional panel support. The most common items are uninterruptible power supply units (UPS) and safes.

There are generally three ways to support these types of heavy concentrated loads:

- Reinforce the floor by installing additional support pedestals under the equipment legs.
- Permanently place a load-spreading plate (such as an aluminum sheet) on the access floor under the equipment.
- Substitute heavy-grade panels in the areas supporting the heavy equipment.

12.3.5.15 Determining if a Static Load Is Within the Floor's Capacity

To determine whether a load is within the floor's capacity, determine the leg footprint of the equipment. If the legs are more than 24 in. apart (and the equipment is situated square with the floor panels), the load will automatically be distributed to separate panels. If this is the case, divide the load amongst the legs (disproportionately, if the load is uneven) and compare the load on each panel to the allowable load in the load chart. If a piece of equipment will have two legs on a single panel, combine the loads on the two legs. **Caution**: If there will be separate pieces of equipment within close proximity, determine whether the two legs will reside on a single panel. Combine the loads for comparison to the chart.

12.3.5.16 Reinforcing the Floor with Additional Pedestals

The floor is reinforced by installing an additional support pedestal under (or near) each leg of the equipment and adhering it to the subfloor with approved pedestal adhesive. (The additional pedestals consist of standard pedestal bases with flat pedestal heads.) Each pedestal's height is adjusted so that they are supporting the undersides of panels.

12.3.6 Grounding the Access Floor

Access floor understructure must be grounded to the building in some manner. The type of understructure used determines the quantity of ground wire connections. The final determination of the number, type, and actual installation method of building ground wires should be made by an electrician. Typical wire recommendation is to use #6 AWG copper wire. The ultimate design should be approved and designed by the project electrical engineer.

The floor can be bonded using components available at electrical supply companies. We recommend attaching the wire connector to the top portion of the pedestal assembly (to the stud or the top of the pedestal tube) or to the pedestal head itself (Figure 12.13). For stringer systems attach a ground connector to a minimum of one pedestal for every 3000 square feet of access flooring.

12.3.6.1 Using a Consolidation Point

One grounding design possibility is to have a copper bus bar located somewhere centrally under the access floor so that the facility has a common consolidation point to terminate all bonding conductors (Figure 12.14). A single bonding conductor can then be taken back to the electrical panel supplying power to the equipment in the room. This helps to eliminate any stray ground currents that could be floating in the understructure.

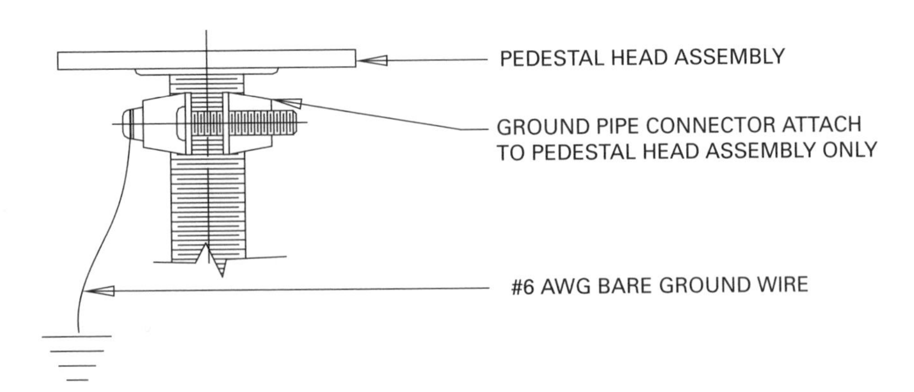

Figure 12.13 Typical grounding method. (Courtesy of Tate Access Floors.)

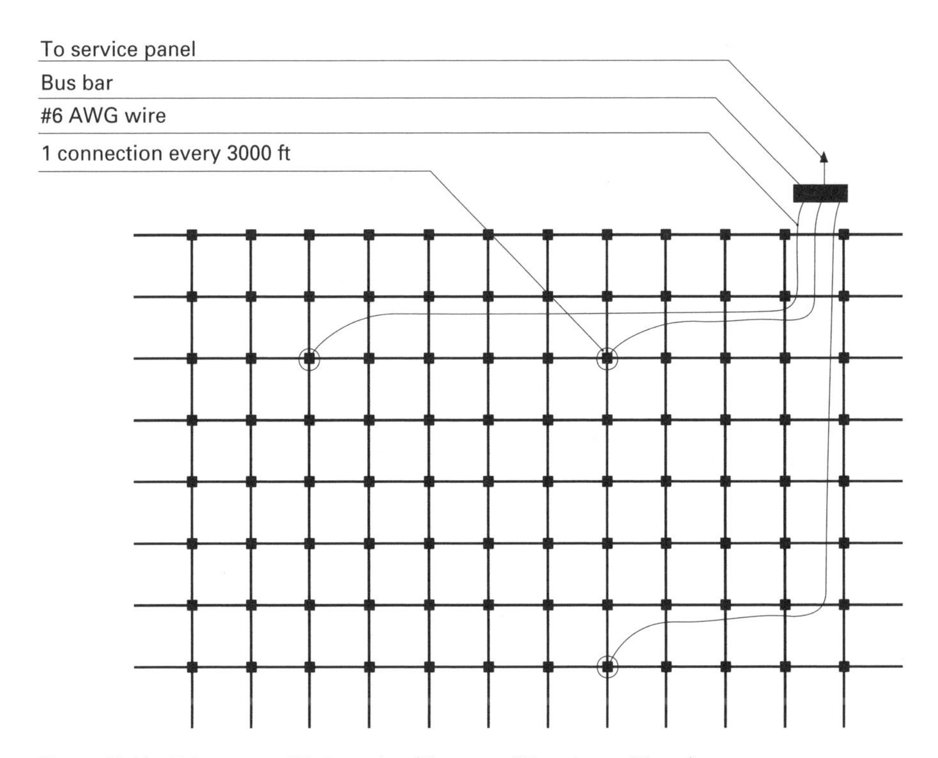

Figure 12.14 Using a consolidation point. (Courtesy of Tate Access Floors.)

12.3.7 Fire Protection

- From an access floor point of view, the only applicable code is NFPA 75 which simply states that access floors must be Class A for surface burning and smoke developed characteristics and that the understructure support must be noncombustible. The best way to guard against fire hazards is to select a system that is either of all steel-welded construction, or all steel-welded construction with a cementitious core. Avoid products that are made of wood or polymeric materials.

12.3.8 Zinc Whiskers

Some form of corrosion protection is necessary on steel access floor components, but the type of coating selected is important to avoid future problems.

Zinc whiskers (also called zinc needles) are tiny hair-like crystalline structures of zinc that have been found to sometimes grow on the surface of electroplated steel. Electroplating is a commonly used method of galvanizing steel and has been used on a variety of steel products now present in data centers and other computer-controlled environments. In recent years, whiskers have been found growing on electroplated components of computer hardware, cabinets, and racks, as well as on some woodcore access floor panels. In some cases these electrically conductive whiskers are dislodged from electroplated structures and infiltrated circuit boards and other electronic components via the room's pressurized air cooling system, causing short circuits, voltage variances, and other signal disturbances. These events of course can cause equipment service interruptions or failures.

An alternative to electroplating, which is applied through a high-energy electrical current, is hot dip galvanizing, which is applied through a molten bath. The hot dip process does not produce the "zinc whisker" effect and therefore should be specified.

12.4 PANEL CUTTING (FOR ALL STEEL PANELS OR CEMENT FILLED PANELS THAT DO NOT CONTAIN AN AGGREGATE)

Cutouts for accessories or cable passageways and panel cuts for system modifications can often times be provided by the access floor manufacturer, but if you opt to cut floor panels yourself, always use the proper equipment and follow the recommended safety requirements below.

12.4.1 Safety Requirements for Cutting Panels

When using a hand-held heavy duty (industrial) reciprocating saw, follow these guidelines:

- Use a bench or worktable with clamps to cut the panels.
- Work in a well-lighted area.
- Be sure tools are properly grounded and dry.

Use personal protection gear:

- Safety glasses (or full-face shield) and ear protection.
- Long-sleeve shirt or sleeve protectors.
- Lightweight work gloves to protect from sharp metal edges and hot saw dust.
- Steel toe safety shoes.

12.4.2 Guidelines for Cutting Panels

Cut the panels in an area where cement dust can easily be cleaned up. After cutting a cement-filled panel, wipe the cut edge(s) with a damp cloth before returning it to the floor. We do not recommend sealing cut panels, because the cement-fill mixture contains a binder that prevents cement dust emission after the cut edges have been wiped clean.

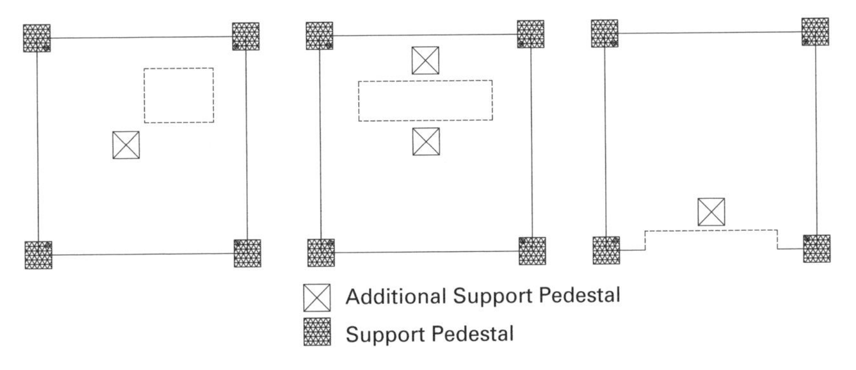

Figure 12.15 Additional support pedestal locations. (Courtesy of Tate Access Floors.)

12.4.3 Cutout Locations in Panels; Supplemental Support for Cut Panels

To maintain structural integrity in a panel with a cutout, locate the cutout at least 3 inches from the panel's edge. To maintain the original deflection condition in a panel with a cutout, or to have a cutout extending to the edge of a panel, you need to reinforce the panel near the cut with additional support pedestals (Figure 12.15). Each additional pedestal should be glued to the subfloor with approved pedestal adhesive and be height-adjusted so that the heads are supporting the underside of the panel. Reinforcement may not be required for all cutout sizes and locations. Consult with the manufacturer for specific requirements.

12.4.4 Saws and Blades for Panel Cutting

A cutout extending to the edge of a panel can be cut with a heavy-duty handheld reciprocating saw or a band saw. An interior cutout can only be cut with a heavy-duty handheld reciprocating saw. Use bi-metal saw blades with approximately 14 teeth per inch.

12.4.5 Interior Cutout Procedure

1. Lay out the cutout on the panel.
2. Drill entry holes in two opposite corners of the layout large enough for the saw blade to pass through without binding, as shown in (Figure 12.16).
3. Cut the hole.

Cutouts for floor accessories and air terminals: Debur the perimeter of cutouts made for air terminals, grills, or electrical boxes where no protective trim will be installed. Deburring can be accomplished with a metal file or coated abrasive sandpaper.

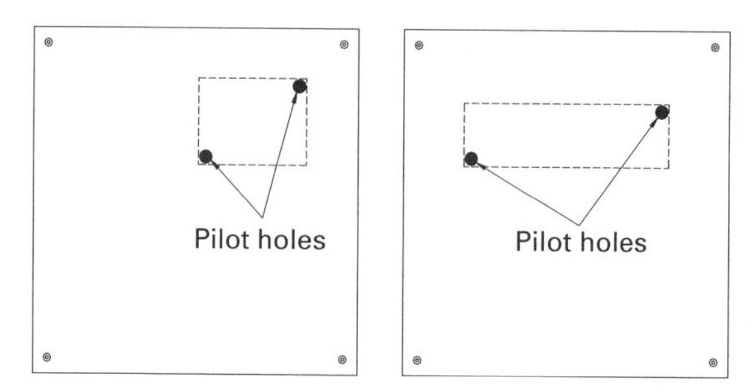

Figure 12.16 Interior cutout procedure. (Courtesy of Tate Access Floors.)

Cable cutouts: Install protective trim around exposed edges of cable cutouts, as shown in Figure 12.16.

12.4.6 Round Cutout Procedure

Round cutouts up to 6 in. in diameter can be made with a bi-metal hole saw. A drill press is best suited for this operation. Operate the drill press at a very slow speed. If a hand-held drill will be used, pre-drill a hole at the center of the cutout location to stabilize the hole saw.

For round cutouts larger than 6 in. in diameter, lay out the circle on the panel and cut the panel with a reciprocating saw. Drill an entry hole for the saw blade along the edge of the circle just inside of the layout line and make the cut. Debur the cut.

12.4.7 Installing Protective Trim Around Cut Edges

Rectangular cutouts used as a passageway for cables must have protective trim along the cut edges. Protective trim components typically include universal cutout trim, molded corners, and screws. A lay-in foam plenum seal is available to seal the opening (plenum seal is typically a sheet of Armaflex insulation).

12.4.7.1 Cutting and Installing the Trim

To determine how long to cut the straight pieces, take note of how the molded corners will hold them in place (Figure 12.17). Cut the sections straight at each end so that they'll fit under the corners. Place the sections along the cutout and secure each section with a molded corner by attaching the corners to the panel with the screws. You will need to pre-drill holes in the panel for the screws with a 9/64-in. metal bit. If the cutout extends to the edge of the panel, attach the trim near the edge (*without* using the corners) by fastening the trim directly to the panel with a screw or a pop rivet. If a screw is used, countersink the screw into the trim piece.

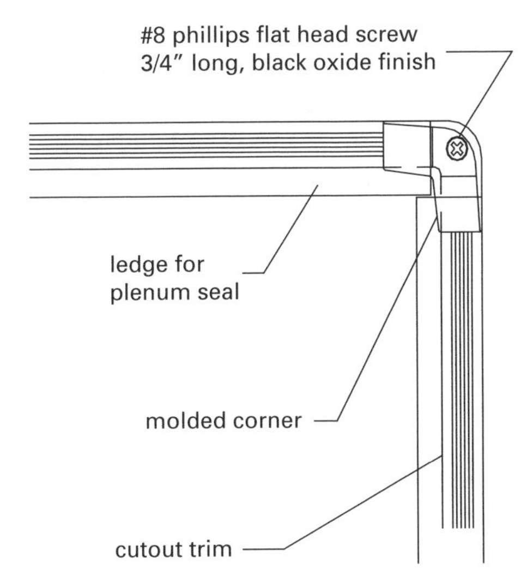

Figure 12.17 Cutout protection. (Courtesy of Tate Access Floors.)

12.5 ACCESS FLOOR MAINTENANCE

For preservation of all tiled floor panels, the most important thing to remember is that application of abundant and excessive amounts of water to the floor when mopping will degrade the glue bond of the tiles to the panels.

12.5.1 Standard High-Pressure Laminate Floor Tile (HPL)

Do

- Keep the floor clean by light damp-mopping with a mild multipurpose ammoniated floor cleaner or mild detergent.
- Use a nonflammable organic cleaner such as Rise Access Floor Cleaner on difficult to clean spots. (Rise can also be used for regular maintenance of high-pressure laminates.)
- Provide protection from sand and chemicals tracked in on shoes by providing "walk-off mats" at entrances.
- Rotate panels in high-use areas to areas of low traffic to spread years of wear over the entire system.

Don't

- Flood the floor or use anything other than a damp mop. Abundant quantities of water can attack the adhesive bond and cause delamination of the tiles from the floor panels. Do not permit the cleaner to get into the seams between the floor panels.

- Wax or seal the tile—it's never necessary. If a floor has been mistakenly waxed, the wax should be removed as soon as possible. Some commercial wax removers may cause damage to the floor, so select a product such as Owaxo, which has been specially designed to remove wax from high-pressure laminate surfaces.
- Use strong abrasives—steel wool, nylon pads, or scrapers—to remove stains.

Damp-Mopping Procedure for Standard High-Pressure Laminate Floor Tile

When light soiling is widespread and spot cleaning is impractical, use this damp-mopping procedure:

- Sweep or vacuum your floor thoroughly.
- Damp-mop with warm water and mild multipurpose ammoniated floor cleaner.
- Dip a sponge mop into a bucket of warm water, wring it out well, and push the sponge across the floor, pressing hard enough to loosen the surface dirt.
- Damp-mop a small area at a time, wringing the sponge out frequently to ensure the dirt is lifted and not just redistributed.
- When damp-mopping a large floor, change the water several times so the dirt doesn't get redeposited on the floor.
- Rinsing is the most important step. Detergent directions often state rinsing is not necessary. While this is true on some surfaces, any detergent film left on a floor will hold tracked-in dirt.
- Ideally, use one sponge mop and bucket to clean the floor and another mop and bucket solely for rinsing.

12.5.2 Vinyl Conductive and Static Dissipative Tile

Do

- Keep the floor clean by light damp-mopping with a neutral cleaner.
- Provide protection from sand and chemicals tracked in on shoes by providing "walk-off mats" at entrances.
- Rotate panels in high-use areas to areas of low traffic. This spreads years of wear over the entire system.
- Use a diluted commercial stripping agent on particularly bad spots.
- Use "Warning" or "Caution" placards if any traffic is possible while the floor is wet. Vinyl flooring will become slippery when wet.

Don't

- Flood the floor or use anything other than a damp mop. Abundant quantities of water can attack the adhesive bond and cause delamination of the tiles from the floor panels. Do not permit the cleaner to get into the seams between the floor panels.

- Conductive and static-dissipative tiles: Don't apply wax or any synthetic floor finish on these tiles. The use of any such material will build an insulating film on the floor and reduce its effectiveness.
- Use strong abrasives or scrapers to remove stains.

Damp-Mopping Procedure for Conductive and Static Dissipative Vinyl Tile

- When light soiling is widespread and spot cleaning is impractical, use this damp-mopping procedure:
- Sweep or vacuum your floor thoroughly.
- Damp-mop with warm water and commercial neutral floor cleaner.
- Dip a sponge mop into a bucket of warm water, wring it out well, and push the sponge across the floor, pressing hard enough to loosen the surface dirt.
- Damp-mop a small area at a time, wringing the sponge out frequently to ensure the dirt is lifted and not simply redistributed.
- When damp-mopping a large floor, change the water several times so the dirt doesn't get redeposited on the floor.
- A good sponge mop for cleaning is one with a nylon scrubbing pad attached to the front edge. This pad is similar to those recommended for use on Teflon pans.
- Rinsing is the most important step. Detergent directions often state rinsing is not necessary. While this is true on some surfaces, any detergent film left on a floor will hold tracked-in dirt.
- Ideally, use one sponge mop and bucket to clean the floor and another mop and bucket solely for rinsing.

12.5.3 Cleaning the Floor Cavity

The necessity of cleaning the floor cavity will depend on environmental conditions—the only way to tell if cavity cleaning is necessary is by occasional inspection. Normal accumulations of dust, dirt, and other debris can be removed by a shop vacuum. Perforated panels in computer rooms may need to be cleaned a couple of times a year by running a wide shop-vac nozzle across the underside of the panels.

12.5.4 Removing Liquid from the Floor Cavity

If a significant amount of liquid enters the access floor cavity by spillage or pipe leakage, vacuum extraction is the most effective way of removing it. The drying process can be accelerated with dehumidifiers, fans, and/or heaters. Extraction and drying should occur as soon as possible, and no more than 48 hours after the event.

For areas subject to excessive moisture or potential water leakage, water detection systems such as Dorlen's Water Alert® may be installed in the underfloor cavity. These systems monitor the subfloor in places where water leakage may otherwise go undetected. *Note*: We do not endorse any particular brand of water detection product, we only make access floor users aware of products available to fulfill special requirements.

12.6 TROUBLESHOOTING

From time to time you may be required to fix minor problems that are typically caused by unintentional mistreatment of floor systems. *Repairs that are not addressed here should be performed by a professional access floor installer.*

12.6.1 Making Pedestal Height Adjustments

You may need to make pedestal height adjustments to correct vertical misalignment conditions. Adjust the pedestal's height by lifting the head assembly out of the base tube enough so that you can turn the leveling nut on the stud. Turn the nut upward to decrease the pedestal height and turn it downward to increase the height. Once the head is reinserted in the tube, make sure that the anti-rotation mechanism on the bottom of the nut is aligned with the flat portion of the tube. **Caution**: A minimum of 1 in. of the stud must remain inside the tube to keep the head assembly safely secured to the base assembly. When making a height correction in a stringer system, it is not usually necessary to remove the stringers to make the adjustment. With the four panels surrounding the pedestal removed and the stringers attached, you can lift the head just enough to turn the nut on the head assembly.

12.6.2 Rocking Panel Condition

If a panel rocks of the understructure, it can be due to one of the following conditions:

- There is dirt or other debris on the stringers that needs to be cleaned.
- There is dirt or other debris stuck to the underside of the panel that be cleaned.
- One pedestal heads are vertically misaligned. Correct this by adjusting the height of the misaligned pedestals (see Section 12.6.1). Remove three panels surrounding the misaligned pedestal but keep a rocking panel in place to gauge the levelness. Raise or lower the pedestal head until the remaining panel appears level and no longer rocks. (Placing a level on the panel will help to make this adjustment.) Replace each of the panels and check for rocking before replacing the next one.

12.6.3 Panel Lipping Condition (Panel Sitting High)

If the lip of one panel is sitting higher than the lip of an adjacent panel, it can be due to one of the following conditions:

- There may be dirt or other debris on the stringers that needs to be cleaned.
- There may be dirt or other debris stuck to the underside of the higher panel that needs to be cleaned.
- A pedestal may have been knocked loose from the subfloor and may be tilting in one direction, as shown in Figure 12.18. Check this by removing two panels where the slipping exists. If any pedestal base plate is not flat on the subfloor and appears loose, you will need to remove the old adhesive and re-glue the pedestal to the subfloor. To do this, remove all stringers and panels surrounding the loose pedestal, cover the entire pedestal base plate with pedestal adhesive, and reposition it on the subfloor (pedestals are spaced 24 in. on center). While the adhesive is wet, reattach the stringers and replace just two panels to serve as a guide for precise pedestal positioning. Move the pedestal as necessary until it is perpendicular to the access floor and the stringers are level. Raise or lower the pedestal head until the panels are level and do not rock. (Placing a level on the panel will help to make this adjustment.) Replace all panels and verify that they do not rock.

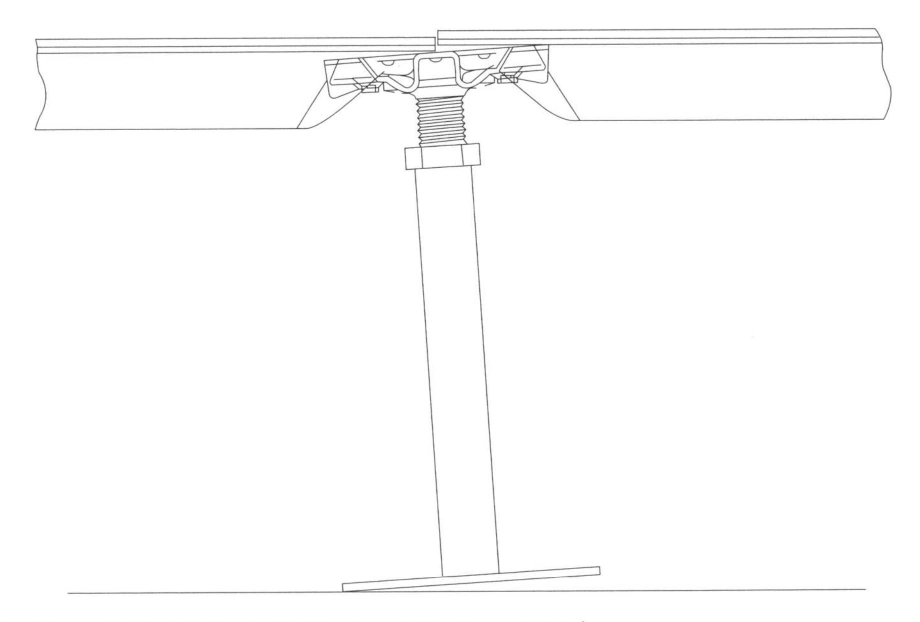

Figure 12.18 Tilted pedestal. (Courtesy of Tate Access Floors.)

12.6.4 Out-of-Square Stringer Grid (Twisted Grid)

This condition is created when people remove a large block of panels from the floor and then unintentionally twist the grid while performing underfloor work. Correcting an out-of-square grid condition is not an in-house, do-it-yourself task. It must be performed by a professional installer. To prevent grid-twisting from occurring in the first place, adhere to the instructions below. Be aware that 2-ft stringer systems are more susceptible to grid-twisting than 4-ft (basketweave) systems.

To Prevent Grid-Twisting

- When you need to work under the floor, remove only as many floor panels as necessary to do the work. Keeping surrounding panels in the floor helps to maintain grid squareness: The more panels you remove at one time, the greater the possibility that the grid will be accidentally twisted. Replace panels as your work progresses from one area to another.
- Never pull electrical cables against the pedestals and stringers.
- When removing and reinstalling cut panels at the floor perimeter and around columns, reinstall each section exactly where it came from. Panels are custom-cut for each perimeter location and are not interchangeable. Walls are often wavy or out of square: Interchanging differently sized perimeter panels can cause floors to become out-of-square and can also cause panels to become too tight or too loose in the floor.
- Keep foot traffic and rolling loads on the access floor away from the area where the panels and stringers are to be removed. Failure to do this can cause panels to shift and stringers to twist on the heads.
- When removing stringers for complete access, remove only those stringers necessary for access and replace them as your work progresses from one area to another.
- Don't walk on stringers or put lateral loads on stringers when panels are removed from the floor.
- When replacing stringers, make sure that the adjacent pedestal heads attached to each stringer are square to one another before final tightening of the stringer screws.
- Don't over-torque the stingers with your driver when reattaching the stringers to the pedestal heads. Set the torque on your screw gun to between 30 and 35 in.-lb.
- Don't try to re-square the stringer grid by hitting stringers with a hammer; this can result in the formation of bumps on top of the stringers that will cause panels to rock.

12.6.5 Tipping at Perimeter Panels

Tipping of perimeter panels toward a wall generally occurs when the perimeter panel is improperly supported at the wall. Adjust the leveling nut on the perimeter pedestal head, raising or lowering the head until the panel sits level.

12.6.6 Tight Floor or Loose Floor: Floor Systems Laminated with HPL Tile

If floor panels laminated with HPL have become too tight (or loose) after installation, it is typically due to one of the following causes.

- *Temperature/Humidity Change.* A significant increase of temperature or humidity causes the HPL on the panels to expand, making the panels difficult to remove and replace. (A decrease of temperature or humidity can cause the floor panels too be loose on the understructure.) To avoid these conditions, keep the room temperature and humidity levels constant. If an environmental change does occur and the floor seems to become tight or loose, it should return to its original condition after the room conditions return to the original settings.
- *Application of Excessive Amounts of Cleaning Water.* Putting excessive amounts of water on the floor wile mopping caused the water to migrate between the floor panels, which caused the tile to expand. This can be prevented by using the proper mopping technique: Wring excess water out of the mop before putting it on the floor; don't leave puddles of water on the floor. As stated in the maintenance section of this manual, abundant quantities of water can also attack the adhesive bond and cause delamination of the tiles from the floor panels.
- *Interchanging of Perimeter Panels.* Floor panels that have been cut to fit at the room perimeter and around columns were removed and were not returned to their original locations. During installation, panels are custom-cut for each perimeter location and are therefore not interchangeable. Walls are often wavy or out of square, and interchanging of perimeter panels can cause some areas of the floor to become loose or too tight. When removing and reinstalling cut panels at the floor perimeter and around columns, reinstall each cut panel section exactly where it came from.

Chapter 13

Fire Protection in Mission Critical Infrastructures

Brian K. Fabel

13.1 INTRODUCTION

Fire protection in facilities housing mission critical operations is a long-debated subject. Opinions, often based on misleading or false information, are as numerous as one can imagine. Lacking a firm understanding of the problems and potential solutions can lead to LCD (lowest common denominator) or "silver bullet" approaches. In reality, LCD solutions rarely ever meet the needs of operators, the end result being a waste of money, and silver bullets generally miss the target. The purpose of this chapter is to provide an understanding of how fires behave, an objective accounting of the various technologies available, the expected performance of these technologies, and the applicable codes.

Regardless of what anyone says, fires do occur in data centers, computer and server rooms, telecommunications sites, and other mission critical environments. However, there is enormous pressure to suppress and conceal this information because it represents a competitive disadvantage for the owner, engineer, and equipment makers. One manager reported that after two fires in an eight-year period, the manufacturer of the fire-prone hardware requested that the events be kept quiet. What customer can rely on an operator who has fires in their "mission critical" facility?

Although fires are less likely in these environments, they usually are a result of an overheated or failed component and occur in circuit boards, cable bundles, terminal strips, power supplies, transformers, electric motors (in HVAC units), and so on, basically anything that generates heat and is made in whole or part of combustible materials. Unless the fire has enormous consequences, such as the infamous Hinsdale Fire, it will never be reported to the general public.

Maintaining Mission Critical Systems in a 24/7 Environment By Peter M. Curtis

Makers of equipment will defend their products with statements such as "It's made of noncombustible (or fire rated) materials. It won't burn!" Nothing could be further from the truth. "Fire-rated" or limited-combustible materials are simply certified to standards for resistance to burning—they still burn. A common example is a "two-hour" fire-rated door. The fire rating only confirms that the door resists a complete breach by fire under a very controlled condition for a period of two hours. Actual results in a real fire may be very different. Mission critical facilities are filled with plastics in the form of circuit boards, data cable jackets, power cord jackets, enclosures, and so on, and all such plastics will burn.

Like power and cooling, fire protection in mission critical facilities is more of a performance-driven issue than a code-driven issue. Bear in mind that although fire codes pertaining to mission critical facilities exist, they require few fire protection measures beyond that required for general commercial facilities. Fire codes are, in fact, consensus standards and are developed for one purpose only: life safety. This concept is not well understood by many engineers and owners. The codes are written to ensure that in the event of a fire, a facility will not collapse until the last occupant is given an opportunity to safely escape. Fire codes are not meant to protect property, operational continuity, or recovery, except to the extent necessary for life safety.

Fire codes do govern how we implement fire protection technology. Even when elective, any installation of fire protection technology is required to comply with national, state, and local codes in effect in the jurisdiction at the time of the installation, modification, or upgrade. This means that a given set of fire protection technologies installed in a given facility in one locality may be subject to different required elements than the same technologies installed in an identical facility in a different locality. Add to that the idea that jurisdictions must adopt each edition of codes through the legislative process and discover that different editions of the same code with differing requirements may be in effect in any given city. It isn't safe to assume that the most recent edition of a particular code is in effect, and it may be the case that a jurisdiction is working from a code that is several editions obsolete.

13.2 PHILOSOPHY

In order to determine what technology is appropriate for an application, an extensive review of the facility, its mission, building, and fire codes in effect, and risk parameters should be conducted. Location, construction, purpose, contents, operational parameters, occupational parameters, emergency response procedures, and so on, should all be included in the review. Where is the facility located? Is it close to a firehouse? Is the facility occupied 24/7, or is it dark? What is the cost of lost assets? Lost business? Who is going to respond to a fire alarm?

Once a review has been conducted, then the exercise of determining how you intend to protect your facility can be confidently executed. The basic pieces can be broken into control, detection, alarm, and extinguishment, or suppression.

13.2.1 Alarm and Notification

There are two basic types of fire alarm systems: conventional and addressable. Conventional control systems (Figure 13.1) are microprocessor-based and process inputs on an on-or-off basis. Signaling and control functions for peripheral equipment such as shutdowns of HVAC or EPO triggers are accomplished with relays.

Addressable systems are programmable-computer based and communicate with devices through a digital bus, typically called an SLC (signal loop circuit). The bus can carry information and commands between the main processor (control panel) and programmable devices in the field such as smoke and heat detectors, input and output modules, relays, and remote displays and annunciators.

With the exception of small, simple applications, the proper choice for most mission critical applications is the addressable-type fire alarm control system. The main reason is cost: the initial cost of installation, and the cost of owning, operating, and maintaining a system. Conventional systems are normally modular, less capable, and of lower cost for a small system (a few thousand square feet). However, when multiple zones, hazards, alarm signaling requirements, and so on, are added to the operational characteristics, the added capabilities have to be bolted-on at incremental cost. A simple system with a few smoke detectors and pull stations on one detection circuit and two or three horn/strobes on one notification circuit is a perfect application for a conventional system. Twenty to thirty input and output devices are probably approaching the maximum cost-effectiveness with a conventional,

Figure 13.1 Conventional fire alarm control panel. (Photo courtesy of Fenwal Protection Systems.)

hardwired system. But for facilities requiring several hundred detectors, controls, and alarm devices, a conventional control panel would then be supporting perhaps dozens of wired circuits and a large number of special function modules. Meanwhile, addressable panels are capable of supporting hundreds of devices in multiple control schemes right out of the box. Most can accommodate 100 to 250 programmable detection, input, or output devices on a single pair of wires.

The operation and maintenance of an addressable system can be lower if it is properly set up. Many of the addressable systems have diagnostics that can provide information in significant detail (some even internet accessible) that can contribute to better decisions (thereby lowering costs) as well as help optimize the maintenance of the system. For example, a dirty smoke detector monitored by a conventional panel will show up as an alarm condition. To discover this, a technician must visit and conduct an inspection of the device and test it. This may be of little consequence if the alarm happens during the workday and a technician is available; however, if it happens at 3 AM on a Saturday night, there may be some significant expense involved. This is a realistic scenario because 75% of the total hours of each week are outside "normal" work hours. Now contrast that with the ability to dial-in or access an addressable panel over the Internet to discover a false alarm due to a dirty detector, the response to which can wait for a less costly solution.

With an addressable system, changing and expanding the system becomes less expensive as well. Many operational characteristics can be changed with a few minutes of programming rather than changing modules and pulling new wires as is frequently required with conventional systems.

Determining how a fire alarm is going to be communicated is another step in the process. It may be the most overlooked issue in mission critical fire protection. Alarms can be initiated upon smoke, heat or flame detection, gaseous agent or sprinkler water flow, or actuation of a manual pull station. Different alarms can be operated to indicate a variety of events such as an incipient fire detection or fire detection in different areas or rooms. A bell may be used to indicate fire in a power room while a high–low tone from a horn may be used to indicate fire in a server room or subfloor. Voice commands may be used in some scenarios.

Most gaseous suppression systems require a second, or confirming, fire detection before an automatic release will occur. This sequence is typically followed by three distinct alarms: 1, fire alarm; 2, pre-discharge alarm; and 3, discharge alarm. Interviews with employees working in facilities reveal that very few occupants understand the meaning of any given alarm and frequently react inappropriately in an emergency situation. Visitors, contractors, and new employees are typically clueless.

It is a good idea to develop alarm strategies that include audio–visual components, signage, and training. In grade school, everyone knew what to do when a particular alarm sounded because of training (fire drills and tornado drills for those of us in the Midwest, and atomic bomb drills for those old enough to remember). Fire and Building Codes and Listing services (Underwriters Laboratories, etc.) will govern what types of devices can be used, but there are more than enough options to accomplish just about any strategy. The key to making it work is to keep it simple enough to make sense, provide periodic training, and provide signage that can be read from a distance.

Changing the mode of operation of occupant notification and evacuation systems is also more easily accomplished with an addressable system. Changes in tones, frequencies, audio/visual toggling, and so on, can be accomplished through programming. Many manufacturers have addressable notification circuits that can be added into a signaling loop circuit as easily as a smoke detector.

13.2.2 Early Detection

The NFPA standard for the protection of Information Technology facilities (NFPA 75) requires smoke detection, and the standard for fire protection in Telecommunications Facilities (NFPA 76) requires early warning (EWFD) and/or very early warning fire detection (VEWFD) in sites carrying telecommunications. However, other detection technologies exist that can be used to eliminate or reduce loss. The technologies available for use in a mission critical facility include a variety of smoke and heat detectors, flame detectors, liquid leak detectors, and combustible/corrosive gas detectors. The wide variety of applications in a mission critical facility may warrant the use of any or all of the available technologies.

For example, a data center may include a server room with a raised floor, power room, battery room, emergency generator room, fuel storage, loading dock, staging/transfer area, offices and conference rooms, restrooms, break/kitchen area, transmission room, storage areas, and so on. Different types of fires with different characteristics occur naturally in each of these areas. Fires in server rooms are typically electrical/plastics fires that may fester for days undetected by personnel or conventional detection systems because of slow growth and high airflow and cooling conditions. Fires in generator rooms and fuel storage areas will ignite and grow rapidly, producing extreme heat and smoke production within seconds or minutes. Battery room fires can generate huge amounts of smoke rapidly with little increase in temperature. In each case, the optimum solution is different.

Beyond the specific applications, the data center may be located in a multi-tenant or high-rise building with other hazards that must be considered. Think about it: An emergency backup diesel generator installed in a suburban, single-story, one-tenant building is a different application than the same generator installed on the 20th floor of a downtown office tower.

13.2.3 Fire Suppression

Automatic fire suppression as a standard in commercial buildings didn't gain traction until the mid- to late 1980s. Fires in high-rises and densely occupied facilities in the 1970s and 1980s resulting in large losses of life, such as the Beverly Hills Supper Club Fire and the MGM Grand Hotel Fire, drove building agencies to require sprinklers throughout many types of facilities. This push was strictly life-safety related. Prior to this era, computer rooms and many telephone buildings were protected with Halon (Halon 1301 to be exact) or carbon dioxide (CO_2). They were the only suitable options, Halon being the much-preferred option due to the life-safety issues associated with CO_2. The primary reason for installing Halon systems was "business continuity protection." In many buildings, in occupancies that required sprinklers, owners were given exemptions under the codes if Halon systems were installed as an alternative to sprinklers.

Mission critical facilities were rarely protected with automatic sprinklers until the publication of the Montreal Protocol in 1987 and the ensuing Copenhagen Resolution calling for the cease-production of chlorinated fluorocarbons (CFCs) because of suspicions that these substances were contributing to the reduction of the ozone layer in our stratosphere. Halons belong to the family of chemicals affected by these treaties. Halon 1301 ceased production in the United States on December 31, 1993 (Halon 1301 is still produced in several countries in Asia, Africa, and South America in even greater quantities on a global basis today). The final rule on Halon 1301 was promulgated by the U.S. Environmental Protection Agency on April 6, 1998. Since no suitable alternatives to Halon existed at the time, many owners and engineers defaulted to sprinklers, particularly "pre-action" and "pre-action/dry-pipe double-interlock" systems, and this mind-set continues to this day. The emergence of MIC (microbial/corrosive attack of piping systems) is rapidly causing owners to rethink fire suppression strategies.

The determining factor in selecting the type of fire suppression to be used in a mission critical facility should be "loss avoidance." Life safety is already mandated by the building and fire codes in place. "Loss avoidance" includes loss of physical assets, collateral damage, lost business production, and lost future business. Future business? Yes, customers will leave if they feel let down. Think about what a customer would do if their mobile phone doesn't provide the coverage expected. Someone in every organization knows what these risks are, and access to that information is required in order to develop a proper fire protection strategy.

There are a number of water-based and gaseous suppression systems suitable for use in mission-critical facilities. Sprinklers are mandated in most cases, and although very effective in extinguishing fires in most "ordinary hazard" spaces, sprinklers do not provide effective extinguishing capabilities in most mission critical facilities due to the nature of the contents. Rather, sprinklers provide "suppression" in these types of areas. In order for a sprinkler system to extinguish a fire, the water must make contact with the burning material. Mission critical facilities are filled with rack enclosures and cabinets with solid perimeters that can shield burning material from the discharge of water. A sprinkler system will effectively control (suppress) such fires and prohibit the fire from spreading to other areas of the facility.

Sprinklers are primarily meant for the protection of the building for life-safety purposes, not business continuity, and collateral damage from a sprinkler release must be considered. Numerous case histories abound where the sprinkler did the job of extinguishing a fire and caused tens or hundreds of thousands of dollars in water damage. While only one sprinkler head discharges in most cases, thousands of gallons of water may be dispersed over a 10- to 15-ft radius, dousing not only the fire but also anything else within the spray and splash zone. Sprinkler water will fill a subfloor and then run off to other areas and lower levels. Typically in a sprinkler release, only the cost of smoke damage is greater than that of water damage.

Watermist systems, once only applicable for generator and fuel storage applications, have been developed for use in electrical equipment areas. Creating very small water droplets, Class 1 watermist systems in particular produce a discharge that behaves more like a gaseous system than a sprinkler system. In addition, the mist produced in some systems is electrically nonconductive, which reduces the likelihood of damaging electrical shorts that would happen in a sprinkler discharge. The relatively small amount of water compared to a sprinkler discharge means that collateral damage is low. A typical closed-head watermist system will deliver only 50–100 gallons of watermist in a 30-minute period. Class 2 and 3 watermists are suitable only for use in generator and fuel storage areas and should be avoided in electronic equipment areas.

Gaseous systems are designed to fill a protected space to a given concentration and not only extinguish a fire, but prohibit the re-ignition of a fire due to heat or electrical input. The available gaseous systems on the market today (referred to as Clean Agents) are clean (no residue), electrically nonconductive, noncorrosive, and safe for human exposure in the prescribed concentrations and limitations. Several types of both chemical agents and inert gas agents are available.

Chemical clean agents work in a fashion similar to that of Halon, first cooling and then chemically attacking (reacting with hydrogen) the combustion reaction. Inert gas agents work as CO_2 does, physically displacing air and lowering the resulting oxygen content of the area below that which will support combustion.

The "Clean Agents" currently available in the U.S. Market include the following:

- HFC 227ea, also known as FM200® and FE227®
- IG541, also known as Inergen®
- FK-5-1-12, also known as Novec1230® and Sapphire®
- HFC-125, also known as FE-25® and ECARO-25®
- HFC-23, also known as FE-13®
- IG-55, also known as Argonite®

It is frequently falsely reported that gaseous agents react with oxygen to extinguish fire. You may have heard it said that Halon "sucks the oxygen" out of air. This is completely untrue, and competing interests propagate the myth in a very irresponsible way. As a matter of fact, there are no fire suppression agents available that react with oxygen.

Other types of suppression systems that may be found in mission critical areas include foam–water systems and dry-chemical systems. These systems would be limited for application in diesel-generator, fuel storage, or kitchen areas.

13.3 SYSTEMS DESIGN

13.3.1 System Types

All fires progress through the same stages, regardless of the type of fire or material that is burning. The only difference in how fires grow is the speed at which the fire progresses through the four stages and the amount of combustibles available. Materials with greater flammability grow faster (e.g., gasoline burns faster than coal). The four stages are (1) incipient, (2) smoking, (3) flaming, (4) intense heat.

A fire that starts in a waste can full of paper will move from the incipient stage to intense heat in a matter of seconds to minutes. A fire that begins when a pressurized fuel line on a hot diesel generator bursts will move from ignition to the fourth stage (intense heat) in a few seconds. A fire involving an overheated circuit board in a server cabinet or in a cable bundle will grow slowly, remaining in the incipient stage for several hours or possibly several days, producing so little heat and smoke that it may be imperceptible to human senses, especially in high-airflow environments. Fires that occur in battery rooms, HVAC units, and power supplies typically stay in the incipient stage for minutes, moving to the flaming and intense heat stages in minutes or hours.

Fires produce smoke, heat, and flames, all of which may be detected by automated systems. Smoke may consist of a number of toxic and corrosive compounds including hydrochloric acid vapors. In still air the smoke rises to the ceiling with the heated air from the fire. However, mission critical facilities are often served by massive amounts of cooling accompanied by enormous volumes of air movement, which change the dynamics of smoke and heat. Cooling systems designed to keep processing equipment running efficiently also absorb the heat from a fire and move smoke around in mostly unpredictable ways. The movement of smoke is unpredictable because environmental conditions are constantly in flux.

13.3.2 Fire and Building Codes

Compliance with building and fire codes is essential. Even when the technology to be used is not required by such codes, most codes mandate that the code be followed. Cities, counties, and states each have their own variety of building and fire code and adopt standard codes, such as the International Fire Code (IFC) or the National Fire Protection Association (NFPA) Standards. Other standards that may be required include Underwriters Laboratory (UL), Factory Mutual (FM), American National Standards Institute (ANSI), and others. For the purpose of this discussion, reference will be primarily to the NFPA codes and standards.

During the conceptual design phase of a project, it is a wise idea to meet with local authorities having jurisdiction (building department, fire department, inspection bureau, etc.) to discuss any fire protection design issues that are beyond code requirements. What may be acceptable to the Authority Having Jurisdiction (AHJ) in Washington, DC may not be acceptable to those in New York, NY.

In addition, features and functions marketed by manufacturers of listed products may be unacceptable to the local authorities. For example, a maker of smoke detection may be able to coordinate several smoke detectors for confirmation before sounding an alarm, thereby eliminating false alarms due to a single detector. It's a great idea, but the local AHJ may, and can, require that each detector operate without such coordination.

Successive revisions of the various codes, which occur every 3 to 4 years, may have dramatically different requirements. For example, the 1996 version of the Fire Alarm Code (NFPA 72) does not allow partial or selective coverage installations of smoke detectors, even for nonrequired systems. So for a $100,000\text{-ft}^2$ building that is not required to have smoke detection and smoke detection is desired in the 5000 square foot computer room, smoke detection must be installed throughout the entire $100,000\,\text{ft}^2$. The current (2007) edition allows partial and selective coverage, so smoke detection may be installed only in the computer room.

In facilities that are strictly IT or telecommunications facilities, the requirement of NFPA 75 and 76, respectively, apply throughout the facility. In IT and telecommunications operations within multi-use facilities such as hospitals, manufacturing plants, and high-rise buildings, the requirements of NFPA 75 and 76 only apply to the specific area of the mission critical function.

The "NFPA 75, Standard for the Protection of Information Technology Equipment" requires smoke detection in equipment areas. Fire extinguishing systems are also required based on a few rules: (1) If the equipment area is in a sprinklered building, the equipment area must be sprinkled. (2) If the equipment area is in an unsprinkled building, the area must be protected with sprinkler system, a clean agent system, or both. (3) The area under a raised floor is to be protected with sprinklers, carbon dioxide, or inert gas extinguishing system (chemical clean agents protecting

Table 13.1 NFPA Standards that Govern the Application of Fire Protection Technologies in Mission Critical Facilities

Standard	Number	Current Edition
Uniform Fire Code	1	2006
Standard for Portable Fire Extinguishers	10	2007
Standard for the Installation of Sprinkler Systems	13	2007
National Fire Alarm Code	72	2007
Standard for the Protection of Information Technology Equipment	75	2003
Standard for the Fire Protection of Telecommunications Facilities	76	2005
Life Safety Code	101	2006
Standard for Watermist Fire Protection Systems	750	2006
Standard on Clean Agent Fire Extinguishing Systems	2001	2004

both above and below the raised floor are suitable). Also, any combustible storage media device in excess of 27 ft^3 in capacity (3 ft × 3 ft × 3 ft) is required to be fitted with an automatic sprinkler or a clean agent system with extended discharge.

The "NFPA 76, Standard for the Fire Protection of Telecommunications Facilities" requires the installation of early-warning smoke detection in all equipment areas, and very-early-warning smoke detection in all signal processing areas 2500 ft^2 or larger in floor space.

13.4 FIRE DETECTION

A variety of fire detection devices is available in the marketplace, and each are meant for specific types of applications, although manufacturers and installers may have reasons other than applicability for recommending a given technology. Types of fire detectors include smoke, heat, and flame. Other detectors that can indicate the increased potential for fire include corrosive and combustible gas detection and liquid leak detection.

Smoke detection is the most common and applicable type of fire detection in mission critical facilities. Smoke detectors measure smoke concentrations in terms of percent of visual obscuration per foot. In other words, how much visible light is obscured. Smoke detectors can measure obscuration rates from the invisible range (0–0.2%/ft), which corresponds to the incipient fire stage, to the visible range (0.2–6.0%/ft), which corresponds to the smoking stage.

Smoke detectors are available as passive detectors (or "spot" and "beam" types) and active ("air-sampling" and "aspirating" types) detectors. Spot-type detectors are mounted on ceilings or in cabinet and under raised floors. Beam detectors are mounted on walls and project a beam across an open space. Air-sampling smoke detectors draw continuous flows of air from multiple points anywhere in a three-dimensional volume to an analyzer. Aspirating detectors use an air movement device (a fan) to draw air into its sensor. Active detectors bring smoke to the detector, and passive detectors wait for smoke to find it.

Smoke detectors are to be installed according to guidelines found in the codes. For area coverage, spot detectors are to be located in a matrix that ranges from 11 ft to 30 ft (125–900 ft^2 per detector), depending on the amount of airflow in the volume of the protected area.

In normally ventilated spaces, spot smoke detectors are spaced at 30 ft on center. So for a 900-ft^2 room that is 30 ft wide by 30 ft long, one smoke detector located in the center of the room is sufficient. However, in a 900-ft^2 hallway that is 10 ft wide and 90 ft long, three smoke detectors are required at 15, 45, and 75 ft along the centerline of the hallway.

In mission critical spaces, detector density is increased (spacing is decreased) due to high airflow movement rates beginning at the equivalent of 8 air changes per hour (Figure 13.2). The density increases to 250 ft^2 per detector at 30 air changes per hour, and the maximum is 125 ft^2 per detector at 60 air changes per hour. Because of high power densities in equipment rooms today, the maximum density (one detector per 125 ft^2) is typical. For the purpose of location, the ports of

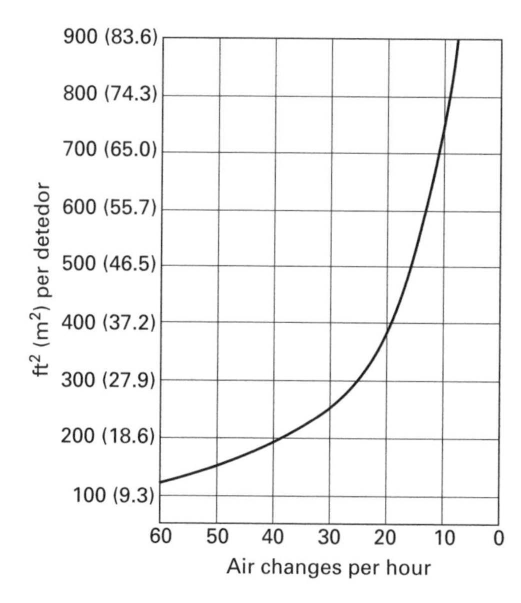

Figure 13.2 Detector density increases (spacing is decreased) along with higher airflow movement rates. (Source: National Fire Protection Association, NFPA 72, National Fire Alarm Code, chapter 5.7.5.3 (2007 ed.))

air-sampling smoke detectors are considered the equivalent of spot-type detectors and are subject to the same spacing rules.

Many mission critical facilities are being constructed with open ceilings, and the spacing of detectors must comply with "beam" or "joist" requirements. "Solid joists" are considered any solid member that protrudes 4 in. or more below a ceiling and is spaced at 3 ft or less, center-to-center. "Beams" are considered as members protruding more than 4 in. on centers of more than 3 ft center-to-center.

For smoke detectors, on ceilings less than 12 ft in height and beams less than 12 in. in depth, the detector spacing shall be reduced 50% in the dimension perpendicular to the beams. On ceilings greater than 12 ft or beams deeper than 12 in., each pocket shall have at least one smoke detector. Smoke detectors must be mounted on the ceiling, unless large (greater than 100 kW) fires are expected. For most mission critical facilities, this size fire is outside the expected range, except perhaps in standby generator or fuel storage areas.

Stratification is also a problem in areas with high ceilings and substantial cooling. In slow-growing fires without sufficient heat energy to propel smoke to the ceiling where smoke detectors are located, a fire can burn for an extended period of time without detection. The hot gases that entrain smoke are cooled and the smoke finds a thermal equilibrium and pools at a level that is well below the ceiling. The fire alarm code allows "high–low" layouts of smoke detectors. The layout should be an alternating pattern with the "high" level at the ceiling and the "low" level a minimum of 3 ft below the ceiling.

Smoke detectors in subfloors are also subject to the same spacing rules as those located at the ceiling level, but spacing is not subject to the high-airflow rules. It

Figure 13.3 High–low layouts of smoke detectors include some detectors 3 ft below the ceiling to overcome stratification. A spot detector (high) and air-sampling port (low) shown here. (Courtesy Orr Protection Systems, Inc.)

has been traditional practice, however, that smoke detectors below a raised floor are matched to the locations of the ceiling smoke detectors.

Historically, two passive technologies, photoelectric and ionization, have been widely used. Photoelectric detectors measure light scattering (ostensibly from the presence of smoke) and compare it to a threshold to establish an alarm condition. This threshold is also known as a detector's "sensitivity." Conventional photoelectric detectors are nominally set at 3%/ft obscuration, and ionization detectors are nominally 1%/ft. This corresponds to complete visual obscuration at 33-ft and 100-ft distances, respectively.

Prior to the advent of addressable/programmable fire detection systems, smoke and heat detectors were built to comply with published standards so much so that the performance characteristics were essentially the same regardless of brand. Only minor differences could be noted. Today, manufacturers develop technologies that are unique, and the performance characteristics and detection capabilities can be quite different from brand to brand, although they are listed to the same standard. To confuse matters, many manufacturers of detectors private label their technologies for a number of system makers, which may use different model names for the same device.

Modern smoke detectors are capable of being programmed to different thresholds down to and including levels in the invisible range, and many have early-warning thresholds that can provide advance notification of a potential problem. Air-sampling smoke detectors may have three or four programmable alarm thresholds that can be used for sequential interlocks such as early-warning alarm, equipment shutdown, fire alarm evacuation notice, gaseous agent release, sprinkler release, and so on.

NFPA 72, 75, and 76, dictate the layout and spacing of spot-type smoke detectors. While the Fire Alarm Code (NFPA 72) does not require smoke detection in mission critical areas, NFPA 75 and 76 do.

NFPA 75 requires smoke detection on the ceiling of electronic equipment areas, below raised floors where cables exist, and above ceilings where that space is used to recirculate air for other parts of a building. The spacing guidelines for IT areas are per the guidelines of NFPA 72.

NFPA 76 requires very-early-warning fire detection (VEWFD) for signal processing areas over 2500 ft^2 in size and early-warning fire detection (EWFD) in other smaller and all other equipment areas. In telecommunications facilities, a maximum of 200 ft^2 per detector VEWFD is permitted in signal processing areas, and 400 ft^2 for EWFD detectors in support areas such as battery, AC power, and DC power rooms. VEWFD is defined as detection with a maximum sensitivity of 0.2%/ft, and EWFD is defined as a maximum of 1.6%/ft. NFPA also requires VEWFD at the air returns of the signal processing areas at the rate of one detector (or sample port) for each 4 ft^2.

The bottom line in high airflow/cooling load environments for electronic equipment rooms, the best technology for discovering a fire before significant damage occurs, is air-sampling smoke detection (Figure 13.4). The systems draw a continuous sample from throughout an area, enabling the system to detect minute amounts

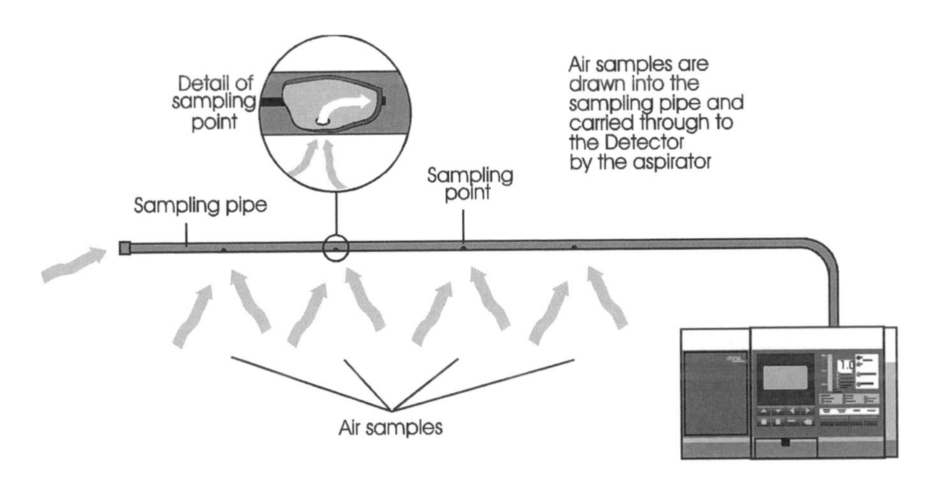

Figure 13.4 Air-sampling Smoke detection. (Image courtesy of Vision Systems.)

of heavily diluted and cooled smoke. Air-sampling systems can accurately detect slow-growing fires anywhere from a few hours to a few days in advance of spot-type smoke detectors. By actively exhausting the air samples, the systems can easily ignore spurious events and false alarm conditions. However, air systems must be designed, set up, programmed, and commissioned under watchful, experienced eyes or they can be an abortive nuisance.

The preferred construction material for air-sampling systems is plenum-rated CPVC pipe made from Blazemaster® plastic. The pipe is orange in color and is easily identifiable. It is manufactured by several different companies and is available from sprinkler distributors and air-sampling smoke detection manufacturers.

When the overheating of PVC-based plastics begins, the heated material off-gases hydrochloric acid (HCl) vapors. Detectors are made that will detect these small amounts of HCl vapors. There are no guidelines for spacing of these detectors, they can be placed selectively anywhere an overheat situation is probable or expected. Locations might include a high-power density rack or inside an enclosure.

In areas such as battery rooms and power rooms, lesser airflow reduces the requirements of smoke detection. Air change rates of 20 to 30 per hour are common, and 400 to 250 ft^2 per detector is typical. Passive spot detectors and air-sampling smoke detectors provide an excellent level of protection for these areas.

Although not prohibited, beam-type detectors and aspirating spot smoke detectors are designed for other types of environments and typically do not perform well in mission-critical facilities. Beam detectors are designed to work in large lobbies, atriums, and arenas. Aspirating spot detectors are designed for application in dirty environments and must be monitored and maintained frequently. Aspirating spot-type smoke detectors have a screen filter that becomes easily clogged with dust, even in "clean" areas, and can render a detector ineffective if not cleaned regularly.

In generator and fuel storage rooms, heat detectors are the primary choice for fire detection. Heat detectors are spaced in matrix format like smoke detectors, but the spacing is dictated by the particular model's listed spacing (usually 25–50 ft). In the way that smoke detector spacing is reduced for increased airflow, heat detector spacing is reduced by ceiling height (Table 13.2). The reduction begins at 10 ft and reaches a maximum at 30 ft above the floor. It is common to see ceiling heights in generator rooms anywhere from 12 to 25 ft.

Figure 13.5 Typical heat detectors. (Photo courtesy of Fenwal Protection Systems.)

Table 13.2 Heat Detector Spacing Reduction Based on Ceiling Height

Ceiling Height Above		Up to and Including		Multiply Listed Spacing by
m	ft	m	ft	
0	0	3.05	10	1
3.05	10	3.66	12	0.91
3.66	12	4.27	14	0.84
4.27	14	4.88	16	0.77
4.88	16	5.49	18	0.71
5.49	18	6.1	20	0.64
6.1	20	6.71	22	0.58
6.71	22	7.32	24	0.52
7.32	24	7.93	26	0.46
7.93	26	8.54	28	0.4
8.54	28	9.14	30	0.34

In areas with solid "beam" or "joist" ceilings, the spacing of heat detectors must be reduced. The rules are similar to those for smoke detectors, but different, and one should take time to understand the differences.

In solid-joist ceilings, spacing of a heat detector must be reduced by 50%. On ceilings with beams that are greater than 4 in. in depth, the spacing perpendicular to the beam must be reduced by 2/3. So for an area with 14 ft-high ceilings and 12-in. beam construction, a heat detector with a listed spacing of 50 ft would be reduced to 42 ft (50 × 0.84) running parallel to the beam, and then by 1/3 (42 × 2/3) to 28 ft perpendicular to the beam. On ceilings where beams are greater than 18 in. in depth and more than 8 ft on center, each pocket created by the beams is considered a separate area requiring at least one heat detector in each pocket.

Heat detectors must be mounted on the bottom of solid joists. They may be mounted on the bottom of beams that are less than 12 in. in depth and less than 8 ft on center. Otherwise, they must be mounted at the ceiling.

The types of heat detectors available are many and varied: fixed-temperature, rate-of-rise, rate-compensated, programmable, averaging, and so on. There are spot-type and linear- or cable-type. Technologies employed included fusible-elements, bi-metallic springs, thermistor sensors, and so on.

In most indoor applications, spot-type heat detectors are suitable. The temperature threshold should be set at 80–100 °F above the maximum ambient temperature for the area. In electronic equipment areas, a temperature threshold of 50–60 °F above ambient is sufficient. It must be noted that heat detection in electronic equipment spaces is typically a design to safeguard against an accidental pre-action sprinkler release. Heat detection is not a substitute, and smoke detection is still required in mission critical areas.

For conventional systems, rate-compensated detectors will provide the fastest response to a given temperature setting. They are not prone to false alarms due to temperature swings and are restorable and can be tested. Fixed-temperature and rate of rise detectors are less responsive and less expensive than rate-compensated detectors, which are 25–30% more expensive on an installed cost basis. Fusible element heat detectors are the least expensive, but also the slowest reacting and cannot be tested without requiring replacement.

Linear heat detectors (or cable-type) are meant for industrial equipment protection and therefore are suitable for fire detection in diesel generator and fuel storage applications. Linear detectors are subject to spacing rules like spot detectors, but can be distributed and located anywhere within the three-dimensional space and can be mounted directly to the equipment. This is extremely effective in areas with very high ceilings or excessive ventilation. Linear detection can also be mounted directly to switchgear, battery strings, cable bundles, or located in cable vaults or utility tunnels.

Linear heat detection is available in fixed-temperature and averaging-type detectors. The averaging-type detector can be programmed to respond to fast-growing fires that cause high temperatures in specific locations, or slow-growing fires that cause a general increase of the temperature in an area.

Flame detectors are also useful in generator and fuel storage areas. If the quickest response to a diesel fuel fire is desired, a flame detector will respond almost instantaneously to flames whereas heat detectors may take several minutes, depending on the location of the detector relative to the fire. Flame detectors can be placed anywhere, but must be pointed in the general direction of the fire in order detect it.

There are two primary types of flame detectors: ultraviolet (UV) and infrared (IR). Both look for light spectrum and both are subject to false alarms due to normal light sources including sunlight. Combination detectors such as "UV/IR" and "triple IR" are available and are much more reliable and stable than single-spectrum detectors.

Flame detectors are expensive, however, costing as much as 30 or 40 times the cost of a single heat detector. However, in large bays, flame detectors may not only be superior detection, but also cost effective compared with heat detectors. If the quickest detection to a fast fire is required, flame detectors become the best choice.

Combustible gas and liquid leak detection is also a consideration in mission critical facilities. Although not "fire" detection, these types of detectors can discover

Figure 13.6 Typical flame detectors. (Left: Triple IR. Right: UV/IR combination. (Photos courtesy of Det-tronics Corporation.)

the presence of substances that indicate a breach that heightens the probability of fire. Guidelines for required spacing or locations do not exist, but common sense dictates where and when the technologies are used.

In wet-cell battery rooms, hydrogen off-gas can be a problem. If not properly vented, it can pool and be ignited by static electricity or arcing contactors. Combustible gas detection can alert personnel to the presence of hydrogen due to a failed exhaust fan, or cracked battery case. They can also sense natural gas leaks or fuel vapors in generator and fuel storage areas.

Liquid leak detectors can be used to detect fuel tank and fuel line leaks. Linear- or cable-type detectors can be located within containment areas or attached directly to fuel lines. Some detection systems are designed so they are sensitive only to hydrocarbons such as diesel fuel and will not false alarm when made wet by mopping or water spray.

13.5 FIRE SUPPRESSION SYSTEMS

Today there are a number of types of automatic fire suppression systems available to an owner. Making a choice on what technology or combination of choices to utilize can be difficult, if taken seriously. The suppression technologies include sprinklers, watermist, carbon dioxide, Halon 1301, and the "clean" agents.

For mission critical facilities, the 2003 edition of "NFPA 75 Standard for the Protection of Information Technology Equipment" requires the following:

8.1 Automatic Sprinkler Systems.

8.1.1 Information technology equipment rooms and information technology equipment areas located in a sprinklered building shall be provided with an automatic sprinkler system.

8.1.1.1 Information technology equipment rooms and information technology equipment areas located in a non-sprinklered building shall be provided with an automatic sprinkler system or a gaseous clean agent extinguishing system or both (see Section 8.4).

8.1.1.2 Either an automatic sprinkler system, carbon dioxide extinguishing system, or inert agent fire extinguishing system for the protection of the area below the raised floor in a information technology equipment room or information technology equipment area shall be provided.*

8.1.2 Automatic sprinkler systems protecting information technology equipment rooms or information technology equipment areas shall be installed in accordance with NFPA 13, Standard for the Installation of Sprinkler Systems.*

8.1.3 Sprinkler systems protecting information technology equipment areas shall be valved separately from other sprinkler systems.

8.1.4 Automated Information Storage System (AISS) units containing combustible media with an aggregate storage capacity of more than 0.76* m^3 *(27* ft^3*) shall be protected within each unit by an automatic sprinkler system or a gaseous agent extinguishing system with extended discharge.*

Most data centers, server rooms, and so on, that fall under the mandate of NFPA 75 are found in facilities that are required by "NFPA 1 the Uniform Fire Code" and "NFPA 101 Life Safety Code" to be sprinkled. These types of facilities include hospitals, airports, high-rise buildings, industrial occupancies, and large business occupancies. Gaseous extinguishing systems are neither required nor prohibited in these occupancies.

Many large data centers are in stand-alone buildings or low-rise business occupancies that do not require sprinklers under NFPA 1 and 101. However, in these facilities, NFPA 75 requires either sprinklers or clean agent fire suppression for the "information technology equipment rooms and areas."

NFPA 75 also requires sprinklers, CO_2, or inert gas clean agent (IG-541, IG-55) suppression below a raised floor (this requirement is lifted if the equipment and underfloor spaces are protected with a total-flood clean agent system). In addition, the standard requires fire suppression in automated media/information storage units.

For facilities processing telecommunications, fire suppression systems are required by "NFPA 76 Standard for Fire Protection in Telecommunications Facilities" only if the telecommunications facility is located in a multi-tenant building that is not constructed to specific noncombustibility ratings.

Sprinklers save lives and property and are designed in accordance "NFPA 13 Standard for the Installation of Sprinkler Systems." For mission critical facilities, the primary application is considered "light hazard," and this would include all electrical and electronic equipment areas. Areas containing standby diesel generators are considered "ordinary hazard" applications.

The basic types of sprinkler systems are wet, dry, or pre-action types. Combination dry and pre-action systems are commonly known as "double-interlock" systems. Most building sprinkler systems are wet-type, meaning the pipes are normally filled with water. The sprinkler heads are closed and opened one at a time by heat, either bursting a glycerin-filled glass bulb or melting an element. Most activations of wet sprinklers involve only one sprinkler head.

Dry systems utilize a valve held shut by compressed air within the piping. When a sprinkler head is activated, the pressure within the piping is relieved and the valve opens, allowing water to flow to the open head. Dry systems are typically found in applications where freezing is a problem, such as cold storage facilities and loading docks.

Pre-action systems can be closed-head or deluge (open head) distribution systems. Pre-action systems utilize an independent detection system to provide a releasing signal that opens a valve, allowing water to flow to an activated head or, in the case of deluge systems, to all the heads served by the valve. Pre-action closed-head systems are typically found in mission critical applications. Pre-action deluge systems are found mostly in industrial fire protection applications. Pre-action closed head systems require that a releasing signal and a head opens before delivering water to the protected space.

Pre-action/dry-pipe or double-interlock systems are the best technology choice for conventional sprinklers in mission critical areas (Figure 13.7). The pipes are dry and supervised with compressed air. Any leaks in the piping are identified with

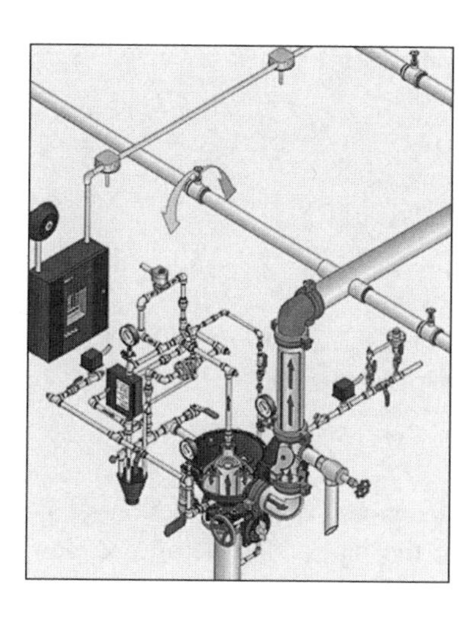

Figure 13.7 Pre-action/dry-pipe, or double-interlock systems are the best technology choice for conventional sprinklers in mission critical areas. (Image courtesy of Viking Sprinkler Corporation.)

Table 13.3 Advantages and Disadvantages of Sprinkler Systems

Advantages	Disadvantages
Most effective Life Safety	May not extinguish a fire in mission critical environment
Inexpensive suppression system	Collateral damage to building, equipment, business operations
Inexpensive restoration of system after suppression	Extensive clean up required
Low tech lifecycle	

low-pressure switches and excessive operation of the air compressor. The valve opens only when it has received a releasing signal, and the pressure within the piping has been relieved. Unlike a standard pre-action system, the valve does not open and water does not enter the piping until a sprinkler head opens.

There are both advantages and disadvantages to sprinklers systems (Table 13.3). Sprinkler systems were anathema to mission critical facilities infrastructure until the late 1980s and early 1990s, when it was well understood that Halon 1301 was not going to be produced for much longer, and many feared it would be banned at some point in the future. In the early 1990s, pre-action and double-interlock systems became commonplace. However, by the mid-1990s a problem with corrosion began to show up. The sprinkler code was changed and pre-action and double-interlock systems were required to use galvanized pipe, the thinking being that galvanized pipe would better resist the corrosion. It has now been discovered that the problem is even worse with galvanized pipe (Figure 13.8).

Once thought to be rusting, microbiologically induced corrosion, (MIC) is a corrosive attack on the piping by the wastes generated by microbes living in the dark,

Figure 13.8 Galvanized pipe damaged by corrosion. (Photo courtesy of Orr Protection Systems, Inc.)

wet, oxygen-rich atmosphere inside "dry" sprinkler pipes. When pre-action sprinklers are tested, some of the water is trapped in the pipe either because of low points, seams, or internal "dams" on grooved pipe. Microbes are then set to work gobbling up oxygen and water and leaving highly acidic wastes that attach to the rough surface of the pipe. Galvanized piping has an even greater profile than standard steel pipe; hence the corrosive action is more aggressive. The speed at which this action occurs depends on the amount of water left standing in the piping. It is only a matter of time until this problem surfaces in every pre-action or double interlock system installed. Treating the piping with microbicides, along with thoroughly draining and mechanically drying, is the only way to prevent this problem from occurring. Still, several other corrosion processes are at work in sprinkler pipes.

In mission critical areas, sprinkler systems must be separately valved from other areas of the building in which they are located. Which means that if a server room is located in the middle of an office high-rise, it would require tapping into the main fire water riser or branch piping for the water supply and piping to the double-interlock valve located close to the room. Each valve is limited to serve $52,000\,\text{ft}^2$ for light and ordinary hazard spacing which is sufficient for most information technology equipment rooms and telecommunications signal processing rooms, with the exception of extremely large facilities.

Each sprinkler double-interlock system is outfitted with several devices: releasing solenoid valve, tamper switch, pressure switch, flow switch, and air compressor. The releasing solenoid valve is operated by a control system and opens to bleed pressure from the double-interlock valve, allowing it to open. The tamper switch is a supervisory device that monitors the shutoff valve (which can be integral to the double-interlock valve). If the shutoff valve is closed, the fire alarm system will indicate that the system is disabled. The flow switch monitors the outgoing piping for water flow. In turn, it is monitored by the fire alarm system that will sound an alarm if water is flowing through the pipe. The pressure switch monitors air pressure in the pipes (branch piping) going out above the protected space to the sprinkler heads. A low air pressure indication means that either a sprinkler head has opened or there is a leak in the piping that needs to be repaired. The fire alarm panel will indicate this situation as a trouble signal. The air compressor supplies compressed air to the branch piping. Each system typically has its own compressor that may be attached directly to the piping; however, several systems can be served with one common compressor.

Sprinkler heads are spaced in a matrix at the ceiling according to design calculations, roughly on 10- to 15-ft centers, depending on the specific sprinkler head flow characteristics and water pressure in the system. Sprinkler heads are designed to open at standardized temperature settings. For mission critical facilities, these temperatures are 135 °F, 155 °F, 175 °F, and 200 °F. Typically, heads with 135 °F threshold would be used in electronic equipment areas, 155 °F or 175 °F in power and storage areas, and 200 °F in unconditioned standby generator areas. Keep in mind that because of thermal lag and the speed of growth of a fire, the temperatures surrounding the sprinkler head may reach several hundred degrees higher than the threshold before the head opens.

Sprinkler heads must also be located below "obstructions." In data centers with overhead cooling units or ductwork that are more than 3 feet wide, sprinkler heads must be located below the equipment.

A common strategy for releasing a double-interlock system from a fire alarm control system is to require "cross-zoning" or "confirming" detection, meaning that two devices or thresholds must be met before sending a releasing signal to the double-interlock valve. This is a strategy that goes back many years to Halon systems and is meant to reduce the likelihood of a premature or accidental release. "Any two" strategies are easy to program but may not provide the level of false alarm suppression required to eliminate accidental releases. Many cases exist where two smoke detectors in a subfloor close to a CRAC unit have gone into alarm causing a release of the suppression system because the reheat coils in the CRAC kicked on for the first time in a few months, burning off accumulated dust on the coils. If the confirming strategy in those cases had been "one in the subfloor, one on the ceiling," the pain and cost of the accidental release would have been avoided. Normal operating characteristics of the environment, plus the capabilities of the detection and control systems, should be taken into account when determining what confirming strategies should be. In facilities with gaseous total flood suppression systems, the sprinkler release signal should only be initiated only after discharge of the gaseous agent.

The sprinkler system should have means to manually release the valve. A manual pull station, properly identified with signage and located at each exit from a protected space, should be provided. A backup manual mechanical release should be located at the double-interlock valve.

The codes do not require that your mission critical equipment be shut down, but encourage an automatic emergency power off upon a sprinkler release. Remember, sprinklers are meant and designed for life safety, not business or asset protection. It is obvious that any water that makes contact with energized electronics will cause permanent damage. Manufacturers of computers have said that their equipment can be "thrown into a swimming pool, then dried out, and it will run." Maybe so, but if the piece is plugged in and powered up when thrown in, it will be destroyed.

In addition, sprinkler water is not clean: It is dirty and full of microbes, rust, and lubricants and other debris that are incidental to the construction of sprinkler systems. Old rags and cigarette butts have been found among other things in the discharge of sprinkler systems. Each sprinkler head in a light hazard area will discharge approximately 20 gallons per minute (a garden hose delivers about 3–5 gallons per minute) until the fire department shuts off the valve. The fire department will arrive within a few minutes of receipt of the alarm, but will not shut off the sprinkler valve until an investigation is complete, which typically takes an hour. Expect the cleanup after a sprinkler discharge to take several days. Ceiling tiles will be soaked and may collapse onto the equipment. Equipment that has been affected by the discharge should be disassembled and inspected, dried with hair dryers or heat guns, and cleaned with compressed air.

13.5.1 Watermist Systems

Watermist systems atomize water to form a very fine mist that becomes airborne like a gas. This airborne mist extinguishes fires with four mechanisms: cooling, radiant heat absorption, oxygen deprivation, and vapor suppression. The primary fire extinguishing mechanism of water is cooling, and the rate of cooling that water provides is proportional to the surface area of the water droplets that are distributed to the fire. The smaller the water droplet, the greater the surface area for a given quantity of water, increasing the rate of cooling. Cooling rates for watermist systems can be several hundred times that of conventional sprinklers. The airborne mist creates a screen that absorbs radiant heat (when entering burning structures, firefighters use "water fog" nozzles that produce a fog screen to absorb the radiant heat); and because the water droplets are so small, they turn to steam quickly when they impinge on the hot surface of the burning material. The conversion to steam expands the water by a factor of 1760 (one cup of water becomes 1760 cups of steam), and this expansion pushes air away from the fire, depriving it of oxygen where it needs it. In watermist environments, any airborne combustibles such as fuel or solvents are rendered inert.

Watermist systems were developed originally for applications aboard ships in an attempt to use less water and use it more effectively. In the early 1990s when researchers in the United States were scrambling for alternatives to Halon, much effort was put into the development of watermist systems for mission critical applications. Most of that effort ended up with the development of Class II mists that are only applicable to diesel standby generator fires. However, Class I watermist systems have been developed and have been listed for use as fire suppression and smoke scrubbing applications in electronic equipment areas. The key to using watermist in electronics areas has been the extremely small water droplet size and the use of deionized water.

Watermist systems (Figure 13.9) operate much like gaseous systems in that they are stand-alone and require no outside power to operate. The primary components of a basic watermist system include a water storage vessel, compressed gas agent and vessel (usually nitrogen or compressed air), a releasing valve, piping, and nozzles. They operate very simply: After receiving a releasing signal, the compressed gas pushes the water out through the pipes at elevated pressure where it is atomized at

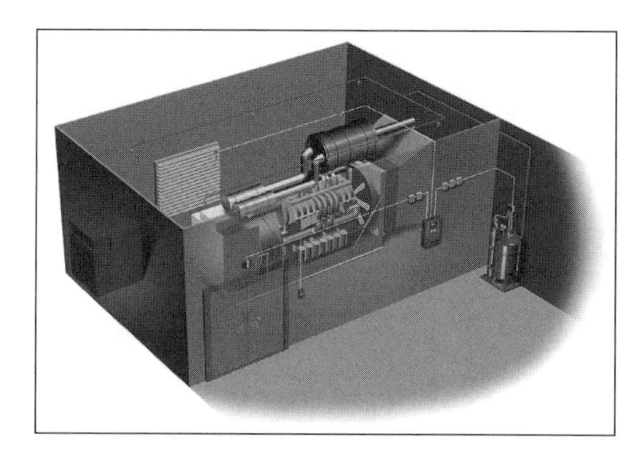

Figure 13.9 Typical watermist system. (Image courtesy of Fire Corporation.)

the nozzles. The supply is designed to run for 10 minutes in the case of a typical total flood system. The systems are pre-engineered and typically utilize four to eight nozzles.

Most systems available in the United States are total flooding type in compliance with standards developed by Factory Mutual for use in a maximum volume of 9160 ft^3 (think 916 ft^2 with a 10-ft ceiling). However, Class I systems are available in larger total flooding systems, local application systems, and even double-interlock systems that comply with NFPA 13.

Total flood watermist systems are applicable in standby generator areas. Most systems will deliver 70–100 gallons of watermist over a 10-minute period. The water is delivered as a mist and will create an environment very similar to a foggy morning and will reduce visibility to 5–10 ft. The systems standby at atmospheric pressure, and when released create 300–400 psig in the piping.

There are a number of benefits to a total flood watermist system in generator areas including:

- Little or no collateral damage
- Fast suppression of class B (fuels) and class A (solids) fires
- Will not thermally shock and damage a hot diesel engine
- Safe for human exposure
- No runoff
- Inexpensive recharge costs

Class I watermist (Figure 13.10) systems are also available as "local application" systems that can be targeted to a piece of equipment, such as a transformer or diesel engine, rather than as a total flooding system which requires distribution throughout an entire volume. So individual engines may be protected in a generator bay that may contain several.

Class I watermist system may also be designed for use in data centers, primarily as an alternative to double-interlock sprinklers. Watermist systems are available from only one manufacturer at this time for these applications. The systems use deionized water and stainless steel piping systems to deliver approximately 2 gallons

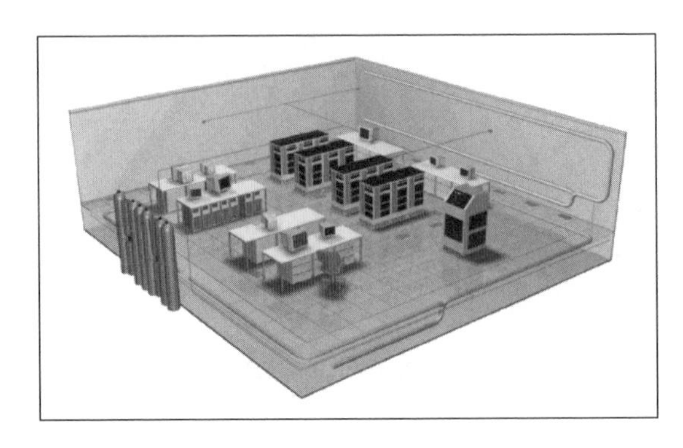

Figure 13.10 Class I watermist system. (Image courtesy of Marioff Corporation.)

per minute of watermist, which is electrically nonconductive. Like double-interlock sprinklers, the piping is supervised with compressed air, and the sprinkler heads are closed until heat-activated. The nozzles are located at the ceiling like standard sprinkler heads and have the same temperature ratings. The spacing of the nozzles depends on the size of the nozzle, but a 2-gallon-per-minute model is at approximately 12-ft centers. A single water supply and compressor can serve several different areas in several modes (double-interlock, total flood, local application), so a single system can protect all the spaces of an entire mission critical facility.

Watermist double-interlock systems have both advantages and disadvantages (Table 13.4). Unlike double-interlock sprinklers, MIC has not been a problem for double-interlock watermist systems (the nature of the stainless steel piping is not conducive to the problem). These systems are very expandable and can serve applications of several hundred thousand square feet and high-rise buildings.

A double-interlock watermist system will have a water supply, compressed gas supply and/or compressor, solenoid releasing valve, supervisory devices, zone valves, piping, and nozzles. The systems operate much like a standard double-interlock sprinkler, a releasing signal, and one activated nozzle before delivering watermist to the protected space. The system supervises the piping at 100 psig and

Table 13.4 Advantages and Disadvantages of Watermist Double-Interlock Systems

Advantages	Disadvantages
Safe for human exposure	Expensive to install
Low water usage	Requires additional maintenance skill set
Electrically non-conductive agent	
No collateral damage	
No thermal shock	
Low restoration costs	

the valve/compressed gas manifold at 360 psig. When released, the system will generate approximately 1400 psig in the discharge piping. The system is designed to operate for 30 minutes.

The nozzles of the double-interlock watermist system deliver the atomized water in a high-speed, cone-shaped discharge. Liquid water will plate out on any surface that is directly in the path of the discharge, wetting it. However, outside the cone of the discharge, the watermist forms a nonconductive cloud that will act like a gas and penetrate cabinets and enclosures enabling the mist to extinguish a fire. The mist will also "scrub" toxins from any smoke that is present.

Class I watermist systems are also available and listed for use as smoke scrubbing systems (only in subfloors in the United States) and can be used with sprinklers and gaseous agent systems. The smoke scrubbing systems are similar to the fire suppression systems except in the delivery of mist. The mist is distributed through a 4-in.-diameter plastic pipe and vacuumed through the same pipe on the opposite side of a space. The mist, carrying smoke and its constituents, is collected in a small reservoir. These systems are modular, based on a 30-ft by 15-ft area, and utilize about 4 gallons of water. Approval testing has demonstrated dramatic reductions of HCl vapors in fires involving jacketed cables and circuit boards.

13.5.2 Carbon Dioxide Systems

Currently, the code for carbon dioxide fire suppression systems (NFPA 12) prohibits the use of total flood systems in normally occupied spaces. Under the definitions of the NFPA code, mission-critical facilities such as data centers, telecommunications sites, and so on are considered to be normally occupied. This prohibition was made in response to demands by the U.S. Environmental Protection Agency to make changes as a life safety issue—there are one or two deaths in the United States each year as a result of exposure to carbon dioxide in total flooding applications. At this time, it is not known whether the EPA will accept or continue to push for more changes. The prohibition does not affect applications in subfloors that are not occupiable, those that are 12 in. or less in depth.

Carbon dioxide (CO_2) extinguishes fire by pushing air away from a fire. Systems are designed to either envelop a piece of equipment (local application) or fill a volume (total flood). Total flood applications in dry electrical hazards require a volume of 50% CO_2 in the protected space, pushing air out of the space and lowering the oxygen content to 11% from the normal 21% within 1 minute. Exposed to this environment, humans will pass out and cease breathing, hence the life-safety problem. Human brain function is impaired at oxygen levels of 16%, and serious permanent brain damage will occur at 12%.

CO_2 systems in mission critical applications are released like any other gaseous system, from a fire alarm and control system automatically after the confirmed detection of a fire or upon activation of a pull station. They distribute the CO_2 from high-pressure gas cylinders through valves into a manifold and out through piping to open nozzles at the ceiling of the protected space or in the subfloor. Pipe sizes for CO_2 systems in mission critical applications are typically 2 in. and less.

New and existing systems must be outfitted with a pneumatic time delay of at least 20 seconds (in addition to any pre-release time delays), a supervised lock-out valve to prohibit accidental release to avoid exposure to humans, a pressure switch to be supervised by the alarm system, and a pressure-operated siren (in addition to any electrically operated alarm devices in the space).

Subfloors protected with CO_2 will require (a) 100 lb of CO_2 per 1000 ft^3 of volume for protected space less than 2000 ft^3, and (b) 83 lb of CO_2 per 1000 ft^3 of volume for protected spaces of more than 2000 ft^3 (200-lb minimum). So, for a subfloor that is 50 ft $\times$ 50 ft and 12 in. deep, 208 lb ($50 \times 50 \times 1 \times 0.083$) of CO_2 is required. CO_2 cylinders are available in a variety of sizes up to 100 lb. This application requires a minimum of three cylinders; so three 75-lb cylinders would be used for a total of 225 lb. The cylinders are about 10 in. in diameter and 5 ft tall. One-hundred-pound cylinders are about 12 in. in diameter and 5 ft tall.

If the subfloor is used as a plenum for air conditioning, upon activation of a CO_2 system the air-conditioning units must be shut down in order to keep the gas in the subfloor. Also, any ductwork or cable chases below the floor leading to other areas will require consideration to avoid leakage of the gas. Ductwork should be fitted with motorized dampers to seal off and contain the gas. Cable chases can be filled with fire-stopping compounds or "pillows."

If a CO_2 system is in place, a series of warning signs must be installed throughout the space, at each entrance and in adjoining areas. The signs must comply with ANSI standards, which call for specific, standardized wording and symbols. In addition, self-contained breathing apparatus should be located at entrances to spaces protected by CO_2 so that emergency response personnel can enter and remove any incapacitated personnel or perform any necessary work required during a CO_2 discharge.

The advantage of a CO_2 system is that the agent is inexpensive and plentiful. Food-grade CO_2 is used and can be obtained from many fire protection vendors as well as commercial beverage vendors. The primary disadvantage of CO_2 systems is the life-safety issue.

13.5.3 Clean Agent Systems

Clean agents are used in total flood applications in mission critical facilities. They are stand-alone suppression systems and require no external power supply to operate. Like other systems, they are released by an electrical signal from a fire alarm and control system. The two basic types of clean agents extinguish fires in very different ways. Inert gas systems, like CO_2 systems, extinguish fires by displacing air and reducing available oxygen content to a level that will not support combustion. Chemical clean agents extinguish fires by cooling, and some also combine a chemical reaction.

Clean agents are safe for limited exposure; however, under the current clean agent code all agents have been assigned a maximum exposure limit of 5 minutes. Prior to the current issue, only inert gas agents were subject to a limit for these types of applications. The limit was established because exposure testing and computer modeling is based on a 5-minute duration.

There is no required hold-time for clean agents specified in the NFPA codes. Most designers and AHJ's have fallen back on the old 10-minute retention time specified in the Halon 1301 code (NFPA 12A). Some local jurisdictions may have specified retention times, but the NFPA 2001 code states that the acceptable retention time should be determined, then tested for verification using an approved room integrity test procedure.

13.5.4 Inert Gas Agents

The inert gas agents include IG-541 and IG-55. They are introduced to a typical mission critical application within 1 minute in concentrations of 35% and 38% by volume, respectively. IG-541 is a blend of 52% nitrogen, 40% argon, and 8% CO_2. It is so blended to create a low oxygen environment with a slightly elevated CO_2 concentration. This blend not only will extinguish a fire but also will cause human metabolism to utilize oxygen more efficiently. For this reason, humans can remain in IG-541 discharges for extended periods without suffering the effects of oxygen deprivation.

Pressure relief venting must be provided for all spaces protected with inert gas agents. The systems are designed to displace approximately 40% of the volume inside the room within 1 minute. This is like adding 40% more to a balloon that is already full and can't expand. The tightly sealed nature of modern mission critical facilities must be vented during discharge, or severe structural damage to walls and doors may occur. The venting must open during discharge and close immediately thereafter to provide a seal. Venting can be accomplished using motorized dampers, or in some cases nonmotorized back-draft dampers, but in all cases should utilize low-leakage-type fire/smoke rated dampers.

That inert gas agents create no environmental impact when compared with chemical clean agents is often pointed to as a selling point by manufactures and vendors. While it is true that nitrogen, argon, and carbon dioxide are three of the four largest naturally occurring gases on the planet, the manufacture of the heavy steel cylinder and high-pressure brass valve and fittings have significant environmental impact when compared with that for chemical agents. With all things considered and compared with environmental effects of more common machines such as automobiles, both inert gas and chemical clean agents have extremely benign environmental impact characteristics.

13.5.5 IG-541

IG-541 systems (Figure 13.11) are comprised of high-pressure gas cylinders, releasing valve and piping to open nozzles. IG-541 supplies are calculated in cubic feet, not pounds. For a mission critical space at 70 °F, the total flooding factor is roughly 0.44 ft^3 of IG-541 per cubic foot of protected volume. So for a computer room that

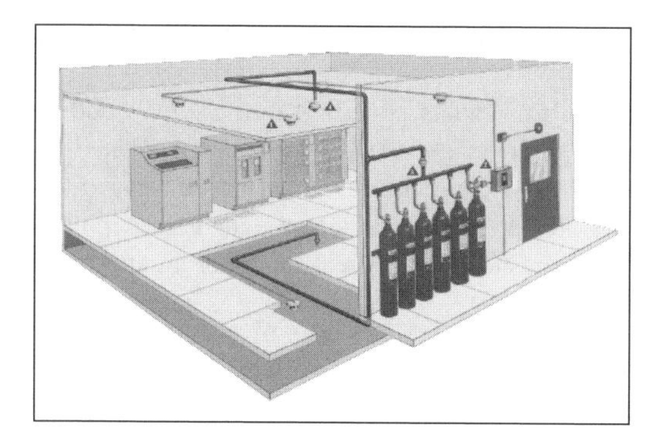

Figure 13.11 Typical IG-541 system. (Image courtesy of Tyco Fire & Security, Inc.)

is 30 ft long by 25 ft wide with a 10-ft ceiling height and a 24-in. raised floor, the amount of IG-541 required would be approximately 3960 ft^3. Cylinders are available in 200-, 250-, 350-, 425-, and 435-ft^3 capacities; therefore, this application would require 10 425-ft^3 cylinders. One nozzle on the ceiling and one under the raised floor should be sufficient.

IG-541 has three distinct advantages among clean agents:

- The recharge costs are less than chemical clean agents.
- Retention time is better than the chemical clean agents.
- Extended exposure at the specified concentration is not hazardous.

It also has some distinct disadvantages:

- There is only one manufacturer of the hardware.
- Few certified refill stations exist, and many major cities do not have one.

13.5.6 IG-55

IG-55 is 50% nitrogen and 50% argon and is a little different from IG-541 in that it has no provisions to help humans survive in the low oxygen atmosphere, and although humans are safe when exposed to the agent, they must leave an area where a discharge has occurred or risk suffering from oxygen deprivation.

IG-55 systems are comprised of high-pressure gas cylinders, releasing valve, and piping to open nozzles. IG-55 supplies are calculated in kilograms. For a mission critical space at 70 °F, the total flooding factor is roughly 0.02 kgs of IG-55 per cubic foot of protected volume. So for a computer room that is 30 ft long by 25 ft wide with a 10-ft ceiling height and a 24-in. raised floor, the amount of IG-55 required would be approximately 180 ft^3. Cylinders are available in effective capacities of 4.3, 18.2, and 21.8 kg; therefore, this application would require nine 21.8-kg cylinders or 10 18.2-kg cylinders. One nozzle on the ceiling and one under the raised floor should be sufficient.

IG-55 has three distinct advantages among clean agents:

- Recharge costs are less than chemical clean agents.
- Any compressed gas vendor can refill the cylinders.
- It has better retention than chemical clean agents.

It also has a distinct disadvantage:

- Extended exposure at the specified concentrations is hazardous to humans.

13.5.7 Chemical Clean Agents

The chemical clean agents available for use today include HFC-227ea, HFC-125, FK-5-1-12, and HFC-23. All of these agents extinguish fire primarily by cooling; they are fluorinated compounds that are very similar to refrigerants used in air-conditioning and refrigeration. For HFC-227ea and HFC-125, only two-thirds of the extinguishing process is cooling, and about one-third is a chemical reaction where the gas molecule breaks up in the presence of high temperatures (above 1200 °F) and the fluorine radical combines with free hydrogen, forming HF vapor and shutting down the combustion process. This production of HF is minor and contributes very little to the toxic nature of smoke from fires.

13.5.7.1 HFC-227ea

Since its introduction, HFC-227ea has been the primary successor to Halon 1301, making up 90–95% of clean agent system over the past decade. The agent is proven to be extremely safe and is actually used (under a different name) as a propellant in metered-dose inhalers for asthmatics. The agent is produced by two manufacturers, and more than a half-dozen manufacturers produce systems hardware. Systems are designed to provide a 7% by volume concentration in protected spaces within 10 seconds of activation. (Some manufacturers have been approved for concentration less than 7% in electrically de-energized applications.)

A typical HFC-227ea system (Figure 13.12) is made up of an agent container (or tank), releasing device, piping, and nozzles. HFC-227ea supplies are calculated in pounds. For a mission critical space at 70 °F, the total flooding factor is roughly 0.0341 lb of HFC-227ea per cubic foot of protected volume. So for a computer room that is 30 ft long by 25 ft wide with a 10-ft ceiling height and a 24-in. raised floor, the amount of HFC-227ea required would be approximately 307 lb. Agent containers are available in capacities from 10 lb to 1200 lb, depending on the manufacturer. Agent containers range in physical size from 6 in. in diameter and 2 ft tall to 2 ft in diameter and 6 ft tall. This application would require one 350-lb container. One nozzle on the ceiling and one under the raised floor should be sufficient.

Advantages of HFC-227ea include the following:

- Extremely safe for human exposure
- Less expensive than inert gas agents to install

Figure 13.12 Typical HFC-227ea system. (Image courtesy of Fire Corporation.)

- Relatively small amount storage compared to inert gas agents
- No room venting required
- Lowest environmental impact of all the chemical clean agents

Disadvantages include:

- More expensive recharge than inert gas agents

13.5.7.2 HFC-125

HFC-125 has only recently been approved for use as a total flooding agent in normally occupied spaces, although it was identified as the most likely successor to Halon 1301 in the early 1990s because of its physical characteristics. Under the old UL standard for extinguishing concentration testing, HFC-125 could not be used in safe concentrations. The new test standard, which better approximates the types of fires found in mission critical environments, along with new computer modeling, has paved the way for this agent to be used. Systems are designed to provide an 8% by volume concentration in protected spaces within 10 seconds of activation.

A typical HFC-125 system is made up of an agent container (or tank), releasing device, piping, and nozzles. HFC-125 supplies are calculated in pounds. For a mission critical space at 70 °F, the total flooding factor is 0.0274 lb of HFC-125 per cubic foot of protected volume. So for a computer room that is 30 ft long by 25 ft wide with a 10-ft ceiling height and a 24-in. raised floor, the amount of HFC-125 required would be approximately 247 lb. Agent containers are available in capacities from 10 lb to 1000 lb and are of the same construction as the containers for HFC-227ea. This application would require one 350-lb container. One nozzle on the ceiling and one under the raised floor should be sufficient.

Advantages of HFC-125 include the following:

- Least expensive to install of all the clean agents
- Relatively small amount storage compared to inert gas agents
- No room venting required

Disadvantages include:

- More expensive recharge than inert gas agents
- One manufacturer of agent and hardware

13.5.7.3 FK-5-12

FK-5-12 has only recently been developed for use as a total flooding agent in normally occupied spaces. FK-5-12 is actually a fluid at atmospheric pressure. It has an extremely low heat of vaporization, an extremely high vapor pressure, and a boiling point of 120 °F. Superpressurized with nitrogen, it is pushed through nozzles and is vaporized quickly. Systems are designed to provide a 4.2% by volume concentration in protected spaces within 10 seconds of activation.

A typical FK-5-12 system is made up of an agent container (or tank), releasing device, piping, and nozzles. FK-5-12 supplies are calculated in pounds. For a mission critical space at 70 °F, the total flooding factor is 0.0379 lb of FK-5-12 per cubic foot of protected volume. So for a computer room that is 30 ft long by 25 ft wide with a 10-ft ceiling height and a 24-in. raised floor, the amount of FK-5-12 required would be approximately 341 lb. Agent containers are available in capacities from 10 lb to 800 lb and are of the same construction as the containers for HFC-227ea. This application would require one 350-lb container. One nozzle on the ceiling and one under the raised floor should be sufficient.

Advantages of FK-5-12include the following:

- Very high safety margin (140%)
- Relatively small amount storage compared to inert gas agents
- No room venting required
- Lowest environmental impact of all the chemical clean agents

Disadvantages include:

- Most expensive clean agent system on the market
- More expensive recharge than inert gas agents
- One manufacturer of agent

13.5.7.4 HFC-23

HFC-23 has only recently been available on the market for several years but has not been used much in mission critical applications. It is mostly used in low-temperature industrial applications. Systems are designed to provide an 18% by volume concentration in protected spaces within 10 seconds of activation. Because of the high volumetric concentration and quick discharge time, rooms protected with HFC-23 must be vented in a fashion similar to that of the inert gas agents.

A typical HFC-23 system is made up of high-pressure cylinders, releasing device, piping, and nozzles. HFC-23 supplies are calculated in pounds. For a mission critical space at 70 °F, the total flooding factor is 0.04 lb of HFC-23 per cubic foot of protected volume. So for a computer room that is 30 ft long by 25 ft wide with a 10-ft ceiling height and a 24-in. raised floor, the amount of FK-5-12 required would be approximately 360 lbs. Agent cylinders are available in 74- and 115-lb capacities. This application would require four 115-lb containers. One nozzle on the ceiling and one under the raised floor should be sufficient.

Advantages of HFC-23 include the following:

- Very high safety margin (67%)
- Relatively small amount storage compared to inert gas agents

Disadvantages include:

- Room venting required
- One manufacturer of agent

13.5.7.5 Halon 1301

Although not manufactured in the United States and North America since 1993, Halon 1301 is still available in abundant quantities. Banking and conservation practices have created a situation where Halon 1301 is as inexpensive as it was in the 1980s.

The biggest problem facing owners of the Halon 1301 system is the availability of spare parts and technical support. Most manufacturers of Halon 1301 equipment quit making it years ago, and whatever is on the shelf is all that is left. So while Halon hasn't gone away, the equipment has. If parts can be found for repair or modification of a system, there is little likelihood that the hydraulics programs can be obtained from the manufacturer. Also, most technical support personnel today weren't even out of tech school in 1993.

The prevailing wisdom with regard to Halon 1301 systems is to take steps to conserve any Halon 1301 that is still in service and to prevent accidental release. This translates to upgrading control systems and devices and improving fire detection systems that might malfunction or false alarm, causing an accidental discharge. Beyond that, it is time to plan for the eventual replacement and decommissioning of Halon 1301 systems in an orderly fashion. If the replacement isn't budgeted, and parts can't be found to restore the system, it may be difficult to obtain the needed funds to install a well-designed replacement.

To put a few myths to rest:

- Halon 1301 is not banned in the United States. There is no federal or state legislation at this time that requires removal of Halon systems.
- You can only buy Halon 1301 in the United States if it was manufactured before January 1, 1994. Cheap Halon 1301 from overseas is illegal for purchase in the United States.
- Halon 1301 does not react with oxygen to put out a fire. It cools the fire and chemically attacks the combustion reaction.
- Halon 1301 is safe for human exposure. There has never been a case of a human death due to exposure to a properly designed Halon 1301 application.

13.5.8 Fire Extinguishers

Fire extinguishers for Class A fires (solid combustibles such as paper and plastics) are required in all commercial facilities. However, the extinguishers rated for Class A fires are typically dry-chemical (powder) type and, while perfect for offices and warehouses, are not appropriate for use in electronic equipment areas. The NFPA 75 Standard for the Protection of Information Technology codes requires CO_2 or halogenated hand portable extinguishers with at least a 2-A rating and specifically prohibits dry-chemical extinguishers in the information technology equipment areas. The "NFPA 76 Standard for Fire Protection in Telecommunications Facilities" also prohibits dry-chemical extinguishers from signal processing, frame, and power areas.

Bibliography

The U.S. Patriot Act

Ruszat, R., II, The USA Patriot Act, A Primer to Compliance Requirements, Trade Vendor Quarterly, Blakeley & Blakeley LLP Spring 03, http://www.nacmint.com/articles/art324.html

Croy, M., Forsythe Solutions Group Inc., What Business Continuity Means for Compliance, June 25, 2004, http://www.compliancepipeline.com/shared/article/printablePipelineArticle.jhtml;jsessionid=HRRSS1S45IVYWQSNDBCSKHSCJUMEKJVN?articleId=22102157

Federal Legislative and Regulatory Business Continuity Requirements for the IRS, February 28, 2003.

National Strategy for the Physical Protection of Critical Infrastructures and Key Assets, http://www.whitehouse.gov/pcipb/physical/securing_critical_infrastructures.pdf

Sarbanes–Oxley (SOX)

SEC rulemaking and reports issued under the Sarbanes–Oxley Act: http://www.sec.gov/spotlight/sarbanes-oxley.htm

http://www.sec.gov/about/laws/soa2002.pdf

http://www.liebert.com/support/whitepapers/documents/reg_compli.pdf

Knubel, J., The Relevance of the Sarbanes–Oxley Act to Non-Profit Organizations: Changes in Corporate Governance, Organization, Financial Reporting, Management Responsibility and Liability (2004), The Balanced Scorecard Institute, MD.

Russo, C., and Tucci, L., Will Cox cure SOX pain,? News Writers 05 Jun 2005, http://searchcio.techtarget.com/originalContent/0,289142,sid19_gci1095138,00.html

Sound Practices to Strengthen the Resilience of the U.S. Financial System, http://www.sec.gov/news/studies/34-47638.htm

Federal Real Property Council (FRPC)

Summary of major provisions, http://www.gsa.gov/Portal/gsa/ep/channelView.do?pf=y&channelId=-16603&pageTypeId=8203&channelPage=/ep/channel/gsaOverview.jsp&com.broadvision.session.new=Yes

Summerell, R., Today's agency must be responsible for real property, FederalTimes.com, July 11, 2005, http://federaltimes.com/index2.php?S=963195

NFPA 1600

Schmidt, D., NFPA 1600: The National Preparedness Standard, NFPA 1600 gives businesses and other organizations a foundation document to protect their employees and customers in the event of a terrorist attack, NFPA Journal®, January/February 2005, http://www.nfpa.org/itemDetail.asp?categoryID=914 &itemID = 22420&URL = Publications/NFPA%20Journal®/January%20/%20February%202005/ Features&cookie%5Ftest=1

American National Standards Institute, Inc., ANSI, *American National Standard Code for Pressure Piping, ANSI B31*, American National Standards Institute, Inc., New York, NY 10018.

International Fire Code Institute, *Uniform Fire Code Article 79*, International Fire Code Institute Whittier, CA 90601.

National Fire Protection Association, *Flammable and Combustible Liquids Code*, NFPA 30, National Fire Protection Association, Batterymarch Park, Quincy, MA 02269.

National Fire Protection Association, *Standard for the Installation and Use of Stationary Combustion Engines and Gas Turbines*, NFPA 37, National Fire Protection Association, Batterymarch Park, Quincy, MA 02269.

Underwriters Laboratories, Inc., *Standard for Steel Underground Tanks for Flammable and Combustible Liquids*, UL58, Underwriters Laboratories, Inc., Northbrook, IL 60062.

Underwriters Laboratories, Inc., *Standard for Steel Aboveground Tanks for Flammable and Combustible Liquids*, UL142, Underwriters Laboratories, Inc., Northbrook, IL 60062.

Underwriters Laboratories, Inc., *Standard for Nonmetalic Underground Piping for Flammable Liquids*, UL971, Underwriters Laboratories, Inc., Northbrook, IL 60062.

Underwriters Laboratories, Inc., *Standard for Glass-Fiber Reinforced Plastic Underground Storage Tanks for Petroleum Products*, UL1316, Underwriters Laboratories, Inc., Northbrook, IL 60062.

Underwriters Laboratories, Inc., *Standard for Protected Aboveground Tanks for Flammable and Combustible Liquids*, UL2085, Underwriters Laboratories, Inc., Northbrook, IL 60062.

United States Environmental Protection Agency, *Oil Pollution Prevention and Response; Non-Transportation-Related Onshore and Offshore Facilities*, 40CFR112, United States Environmental Protection Agency Headquarters, 1200 Pennsylvania Ave., Washington, DC 20460.

United States Environmental Protection Agency, *Technical Standards and Corrective Action Requirements for Owners and Operators of Underground Storage Tanks*, 40CFR280, United States Environmental Protection Agency Headquarters, 1200 Pennsylvania Ave., Washington, DC 20460.

UL 1008 *Standard for Transfer Switch Equipment.*

IEEE 602-1996 *IEEE Recommended Practice for Electric Systems in Health Care Facilities* (IEEE White Book).

IEEE 241-1990 *IEEE Recommended Practice for Electric Power Systems in Commercial Buildings* (IEEE Gray Book).

IEEE 446-1995 *IEEE Recommended Practice for Emergency and Standby Power Systems for Industrial and Commercial Applications* (IEEE Orange Book).

NFPA 70 2005 *National Electrical Code* (NEC).

NFPA 99 2005 *Standard for Health Care Facilities.*

NFPA 110 2005 *Standard for Emergency and Standby Power Systems.*

NEMA ICS 1.3-1986 *Industrial Control and Systems: Preventive Maintenance of Industrial Control and Systems Equipment.*

NEMA ICS 10-1999, Part 2 *Industrial Control and Systems: AC Transfer Equipment—Part 2, Static AC Transfer Equipment.*

Appendix A

Critical Power

EXECUTIVE SUMMARY

Electricity occupies a uniquely important role in the infrastructure of modern society. A complete loss of power shuts down telephone switches, wireless cell towers, bank computers, E911 operator centers, police communication networks, hospital emergency rooms, air traffic control, street lights, and the electrically actuated valves and pumps that move water, oil, and gas, along with the dedicated, highly specialized communications networks that control those physical networks. As familiar and pedestrian as electric power may seem, it is the first domino of critical infrastructure.

The public electric grid, however, is inherently vulnerable. Relatively small numbers of huge power plants are linked to millions of locations by hundreds of thousands of miles of exposed wires. Nearly all of the high-voltage lines run aboveground and traverse open country, and a handful of high-voltage lines serve entire metropolitan regions. Serious problems tend to propagate rapidly through the grid itself.

Thus, while the public grid must certainly be hardened and protected, most of the responsibility for guaranteeing supplies of critical power at large numbers of discrete, private grids and critical nodes ultimately falls on the private sector and on the lower tiers of the public sector—the counties, municipalities, and towns.

A New Profile for Grid-Outage Risks

Many essential services and businesses have critical power needs that have not been properly addressed, often because they have never been systematically assessed.

*Appendix A is reprinted with permission from © Digital Power Group.

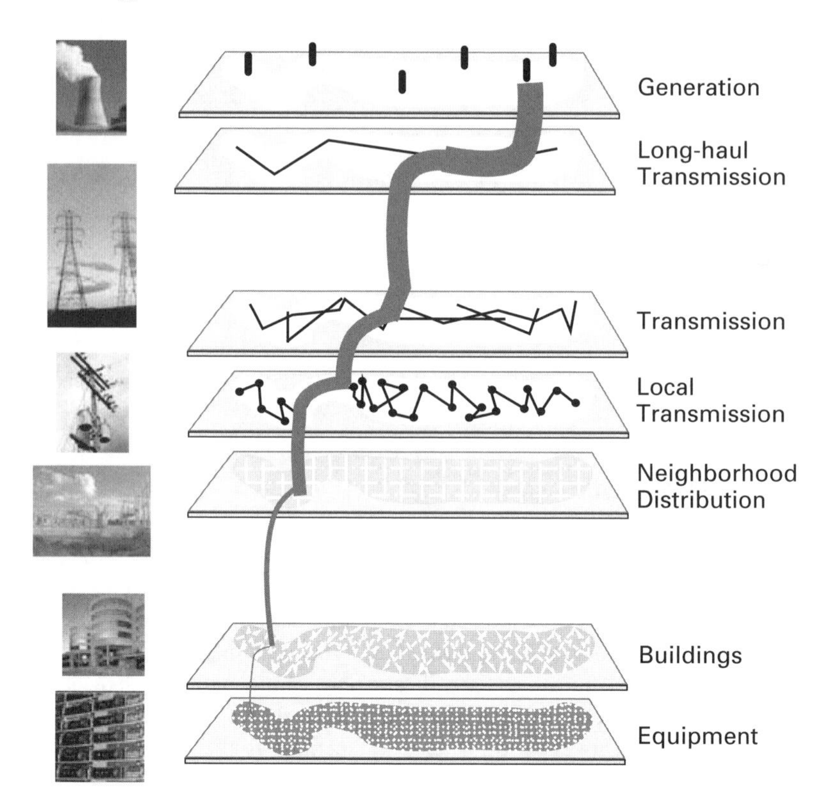

The tiers of the electric grid. Derived from "Distributed Energy Resources Interconnection Systems," U.S. DOE NREL (September 2002).

Even enterprises that have prepared properly for *yesterday's* power-interruption risk profiles may well be unprepared for *today's*. The risk-of-failure profiles of the past reflect the relatively benign threats of routine equipment failures, lightning strikes on power lines, and such small-scale hazards as squirrels chewing through insulators or cars colliding with utility poles. The possibility of a deliberate attack on the grid, however, changes the risk profile fundamentally—that possibility sharply raises the risk of outages that last a long time and extend over wide areas.

Even before 9/11, it had become clear that the digital economy requires a level of power reliability that the grid alone simply cannot deliver. Utilities have traditionally defined an "outage" as an interruption of 5 minutes or more. Digital hardware, by contrast, cannot tolerate power interruptions that last more than milliseconds. The challenge now is to deploy critical-power hardware that supplies exceptionally clean and reliable power to the critical nodes of the digital economy and that guarantees operational continuity for the duration of the extended grid outages that deliberate assaults on the infrastructure might cause.

Backup generators, uninterruptible power supplies (UPS), and standby batteries are already widely deployed. About 80 GW of off-grid backup generating capacity already exists—an installed base equal to about 10% of the grid's capacity. Roughly 3–5% of the public grid's capacity is currently complemented and conditioned by UPS's—about 25 GW of large UPS capacity in business and government buildings and another 10–15 GW of capacity in smaller desktop-sized units located in both businesses and residences. And end users have, as well, installed over 30 million large standby batteries.

Until recently, the deployment of much of this hardware has been directed at power *quality*—smoothing out spikes and dips that last for only fractions of a second—or short-duration issues of power *reliability*—dealing with grid outages lasting from minutes up to an hour or so. In the new geopolitical environment, however, planners must address the possibility of more frequent grid outages that last for many hours, days, or even longer. Assuring *continuity* during extended outages requires a different approach, and a different level of investment in local power infrastructure.

Report Authors

Mark P. Mills
Partner, Digital Power Group
Partner, Digital Power Capital

Peter Huber
Partner, Digital Power Group
Partner, Digital Power Capital
Senior Fellow, Manhattan Institute for Policy Research

Research Associates

Mary Catherine Martin
Heidi Beauregard

Digital Power Group
www.digitalpowergroup.com
1615 M Street NW, Suite 400
Washington, DC 20036
mmills@digitalpowergroup.com
phuber@digitalpowergroup.com

Sponsors & Supporters

Sponsorship of this White Paper was provided by EYP Mission Critical Facilities (Chair, Technical Advisory Committee), Cummins Power Generation, Danaher Power Solutions, EnerSys Reserve Power, Powerware, and Schneider Electric Square D.

We would also like to thank, for their ongoing assistance and support in our pursuit of this and related subjects, both Gilder Publishing and Forbes Publishing who (sequentially) published our investment newsletter, The Digital Power Report.

Tiers of Power

Much of the critical-infrastructure literature refers to the grid as a single structure, and thus implicitly treats it as "critical" from end to end. But the first essential step in restoring power after a major outage is to isolate faults and carve up the grid into much smaller, autonomous islands. From the perspective of the most critical loads, the restoration of power begins at the bottom, with on-site power instantly cutting in to maintain the functionality of the command and control systems that are essential in coordinating the step-by-step restoration of the larger whole.

The hardening of the grid does certainly begin at the top tier, in the generation and transmission facilities. Much of the modern grid's resilience is attributable to the simple fact that "interties" knit local or regional grids into a highly interconnected whole, so that any individual end user may receive power from many widely dispersed power plants. (This architecture also increases everyone's vulnerability to far away problems.)

Very large end-users rely on similar "intertie" strategies—one-tier lower down in the grid—to help secure their specific critical-power needs. Substations, deeper in the network and closer to critical loads, can also serve as sites for deployment of distributed generating equipment. With the addition of its own generating capacity, the substation is "sub" no longer—it becomes a full-fledged "mini-station." Opportunities for deploying new generation at this level of the grid—either permanently or when emergencies arise—are expanding, although still greatly under deployed. Utility-scale mobile "generators on wheels"—either diesels or turbines—offer an important additional option. Some substations already play host to small parking lots worth of tractor-trailers, each carrying 1–5 MW of generators. In the longer term, other sources of substation-level generation and storage may include fuel cells and massive arrays of advanced batteries.

For the most critical loads, however, none of these options is an adequate substitute for on-site backup power. On-site power begins with on-site supplies of stored electrical, mechanical, or chemical energy—typically mediated and controlled by the high-power electronics and controls of a UPS. Rechargeable batteries remain the overwhelmingly dominant second-source of power. But batteries store far less energy per unit of volume or weight than do liquid hydrocarbon fuels.

Thus, to cover the threat of longer grid outages, the backup system of choice is the stand-by diesel generator. Sized from 10s to 1000s of kilowatts, diesel gensets can provide days (or more) of backup run time—the limits are determined only by how much fuel is stored on-site, and whether supplies can be replenished. Diesel generators are strongly favored over other options because they strike the most attractive balance between cost, size, safety, emissions, and overall reliability. And the far-flung, highly distributed infrastructure of fuel oil storage tanks is effectively invulnerable to the kinds of catastrophic failures that could incapacitate power lines or gas pipelines across an entire region.

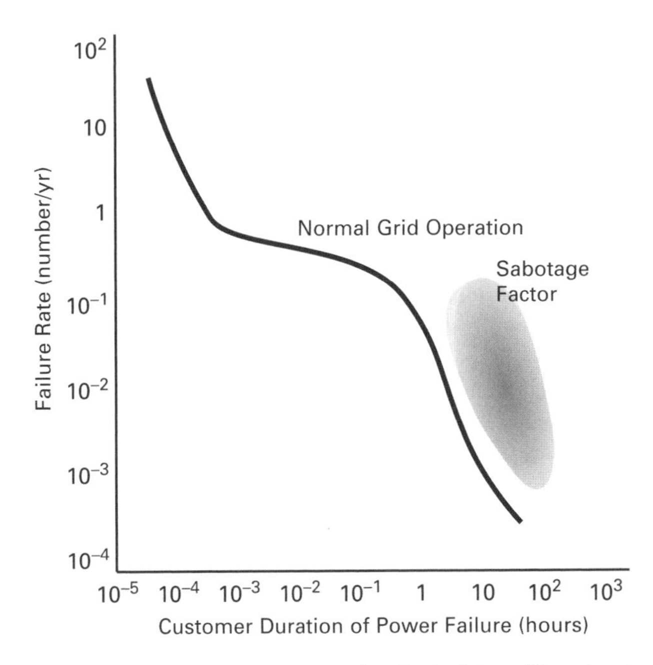

Electric failures: Customer perspective. Derived from "Powering the Internet-Datacom Equipment in Telecom Facilities," Advisory Committee of the IEEE International Telecommunications Energy Conference (1998).

To complement the hardware, monitoring and maintenance play a key role in maintaining power reliability, from the gigawatt-scale tiers at the top of the grid, down to the UPS and individual loads at the bottom. Real-time control plays an essential role in the stabilization of still-functioning resources, and the rapid restoration of power to critical loads after a major failure in any part of the grid. At the grid level, supervisory control and data acquisition systems (SCADA) are used by utilities and transmission authorities to monitor and manage distribution. At the user level, all the power hardware likewise depends increasingly on embedded sensors and software to monitor and coordinate—a nontrivial challenge as problems happen at the speed of electricity.

Reliability-centered maintenance—familiar in the aviation industry but still a relatively new concept for power—is becoming more important with the rising complexity of systems. Some of the most useful critical-power investments thus center on routine upgrades that replace older equipment with state-of-the-art hardware, which has built-in digital intelligence and monitoring capabilities. Changes as seemingly simple as speeding up the performance and automating of circuit breakers can greatly lower the likelihood of serious continuity interruptions precipitated by the power-protection hardware itself. Sensor- and software-driven predictive failure analysis is now emerging and will certainly become an essential component of next-generation reliability-centered maintenance.

Resilient Design

One of the most important—and least appreciated—challenges in the critical-power arena is to determine just how robust and resilient supplies of power actually are. It is easy to declare a power network "reliable," but difficult to ascertain the actual availability metrics. The aviation and nuclear industries have spent many decades developing systematic, quantitative tools for analyzing the overall resilience of alternative architectures and continuously improving the best ones.

But the tools of probabilistic risk analysis—essential for any rigorous assessment of reliability and availability—are still widely underused in critical-power planning. Employed systematically, they require power engineers, statisticians, and auditors to physically inspect premises, analyze multiple failure scenarios, draw on hardware failure-rate databases, and incorporate both human factors and external hazards. Proper critical design takes into account the key (though frequently overlooked) distinction between power reliability and the actual *availability* of the system thus powered. The analytical tools and the technologies required to engineer remarkably resilient, cost-effective power networks are now available. The challenge is to promote their intelligent use when and where they are needed.

Private Investment and the Public Interest

Significant niches of the private sector were making substantial investments in backup power long before 9/11, because electricity is essential for operating most everything else in the digital age, and because the grid cannot provide power that is sufficiently reliable for many very important operations. Backing up a building's power supplies can be far more expensive than screening its entrances, but improving power improves the bottom line, by keeping computers lit and the assembly lines running. Likewise, in the public sector: Secure power means better service.

Though undertaken for private or local purposes, such investments directly increase the reliability and resilience of the public grid as a whole. In the event of a major assault, the process of restoring power to all will be speeded up and facilitated by the fact that some of the largest and most critical loads will be able to take care of themselves for hours, days, or even weeks.

Even more important, the process of restoring power system-wide has to begin with secure supplies of power at the critical nodes. Coordinating the response to a major power outage requires functioning telephone switches, E911 centers, and police communications, and the grid itself can't be re-lit unless its supervisory control networks remain fully powered. The most essential step in restoring power is not to lose it—or at worst, to restore it almost immediately—at key nodes and subsidiary grids from which the step-by-step restoration of the larger whole can proceed.

Finally, in times of crisis, private generators can not only reduce demand for grid power, they can—with suitable engineering of the public-private interfaces—feed power back into limited parts of the public grid. Options for re-energizing the grid from the bottom up are increasing as the high-power switches and control systems improve.

In sum, the most effective way for government to secure the nation's critical power infrastructure is to encourage private sector investment in critical power facilities—not just by the relatively small numbers of quasi-public utilities and large federal agencies, but by private entities and state and local government agencies. Dispersed planning and investment is the key to building a highly resilient infrastructure of power.

Accordingly, as discussed in more detail at the end of this report, we identify eight major areas for coordinated action by policy makers, industry associations, and end users in the public and private sectors.

1. **Assess Vulnerabilities**. Policy makers should be leading and coordinating the efforts of user groups, critical power providers, and utilities to conduct systematic assessments of critical-power vulnerabilities, for specific industries, utility grids, and configurations of backup systems.
2. **Establish Critical-Power Standards for Facilities Used to Support Key Government Functions**. Federal and local organizations should work with the private sector to establish guidelines, procedures, and (in some cases) mandatory requirements for power continuity at private facilities critical to government functions.
3. **Share Safety- and Performance-Related Information, Best Practices, and Standards**. Utilities, private suppliers, and operators of backup power systems should develop procedures for the systematic sharing of safety- and performance-related information, best practices, and standards. Policy makers should take steps to facilitate and accelerate such initiatives.
4. **Interconnect Public and Private Supervisory Control and Data Acquisition Networks**. The supervisory control and data acquisition networks operated by utilities and the operators of backup power systems should be engineered for the secure exchange of information, in order to facilitate coordinated operation of public and private generators and grids. Policy makers should take steps to facilitate and accelerate that development.
5. **Secure Automated Control Systems**. The necessary integration of supervisory control and data acquisition networks operated by utilities and the owners of backup power systems requires high assurance of cyber-security of the networks in both tiers. Policy makers should take steps to advance and coordinate the development of complementary security protocols in the public and private tiers of the electric grid.
6. **Share Assets**. Policy makers and the private sector should take steps to promote sharing of "critical spares" for on-site generation and power-conditioning equipment, and to advance and coordinate the establishment of distributed reserves and priority distribution systems for the fuel required to operate backup generators.

7. **Enhance Interfaces Between On-Site Generating Capacity and The Public Grid**. Improved technical and economic integration of on-site generating capacity and the public grid can backup critical loads, lower costs, and improve the overall resilience of the grid as a whole, and should therefore rank as a top priority for policy makers and the private sector.
8. **Remove Obstacles**. Private investment in critical-power facilities creates public benefits, and policy makers should explore alternative means to remove obstacles that impede private investment in these facilities.

INTRODUCTION

Plans for the long-term protection and hardening of U.S. infrastructures are now underway, under the auspices of the Department of Homeland Security (DHS). Two of the leading public/private partnerships developing standards and plans for action include The Infrastructure Security Partnership (TISP) and the Partnership for Critical Infrastructure Security (PCIS). Through these and a web of other similar initiatives, a vigorous and broad effort is under way to identify, coordinate, and direct the reinforcement, protection, and security of key infrastructure components in both the public and the private sectors. Electric power is one of the six network-centered sectors that have been identified as key, along with information and communications, banking and finance, oil and gas, rail and air transport, and water. Four critical service sectors have also been identified: Government, law enforcement, emergency services, and health services.

All of these critical sectors are interdependent in varying degrees. But electricity occupies a uniquely important role. The loss of power shuts down telephone switches, wireless cell towers, bank computers, E911 operator centers, police communication networks, hospital emergency rooms, air traffic control, street lights, and the electrically actuated valves and pumps that move water, oil, and gas, along with the dedicated, highly specialized communications networks that control those physical networks.

The loss of power also takes out virtually all of the new systems and technologies being deployed for 24/7 security in both private and public facilities, not just communications and computing but everything from iris scanning to baggage x-raying, from security cameras (visual and infrared) to perimeter intrusion systems, from air quality monitors to air-scrubbers.

More broadly, the loss of power shuts down any factory, plant, office, or building that depends on computers, communications systems, pumps, motors, cooling systems, or any other electrically operated system. As the President of PCIS recently testified in Congress:

> *(T)he line between physical and cyber assets is becoming even more blurred by the widespread use of digital control systems–electronically controlled devices that report on kilowatt hours transmitted, gallons per hour of oil and water, cubic feet of natural gas, traffic on "smart roadways," and can actually control physical assets like flood gates; oil, gas, and water valves and flow controllers; ATM machines; and the list keeps growing.*[1]

The infrastructure of "critical power" is thus highly distributed. It is also multi-tiered—there are many different levels of "criticality" to address. Power is critical wherever it fuels a critical node, however large or small. Some nodes are as large as a military base, a bank, or a chip fab; some are as small as a single cell tower, a valve in a pipeline, or a crucial switch on the grid. At some critical nodes, the power only needs to be secure enough to permit an orderly shut down; others have to be robust enough to run autonomously for hours, days, weeks, or even longer. And the number of critical nodes continues to increase rapidly, as the entire nation grows increasingly electrical, increasingly digital, and increasingly automated.

The hardening of our electric power infrastructure thus requires actions that extend much deeper and more ubiquitously than is commonly recognized. The handful of gigawatt-scale power plants, along with the public grid, certainly must be protected. But there are hundreds of thousands of smaller nodes and private grids—ranging from tens of kilowatts to tens of megawatts in size—that must be protected. Reliability levels must be structured, tiered, and nested. However much is done to strengthen it, the three-million-mile grid will inevitably remain the least reliable (though also the most affordable) source of power, just as it is today; far higher levels of electrical hardening will be required in much smaller islands of reliability—in the nodes that provide communications, computing, key security and health services, and so forth. So, while the public grid itself must be hardened and protected, it is neither feasible nor economical to sufficiently harden the entire grid as much as it is possible and necessary to harden much larger numbers of discrete, private grids and nodes.

Securing the infrastructure of critical power will require, in other words, an approach similar to the one taken in addressing the Y2K bug in computer software and firmware several years ago. No top-down solution was possible; the points of vulnerability had to be identified, and the fixes implemented on a distributed, granular basis, with the private sector ultimately taking most of the initiative. The need to adopt that same approach for securing data and telecom facilities was reaffirmed in an Executive Order issued barely a month after the 9/11 attacks.

> *The information technology revolution has changed the way business is transacted, government operates, and national defense is conducted. Those three functions now depend on an interdependent network of critical information infrastructures. The protection program authorized by this order shall consist of continuous efforts to secure information systems for critical infrastructure, including emergency preparedness communications, and* ***the physical assets that support such systems****. Protection of these systems is essential to the telecommunications, energy, financial services, manufacturing, water, transportation, health care, and emergency services sectors.*[2] [emphasis added]

Electric power is certainly the most important of the "physical assets that support such systems." And most of the responsibility for securing the key nodes of the power infrastructure and guaranteeing supplies of critical power will ultimately fall not on utilities or the federal government, but on the private sector, and on the lower tiers—the countries, municipalities, and towns—of the public sector. Most of the critical nodes are owned and operated in these lower levels. Securing critical power must inevitably focus on numerous, smaller nodes and islands of power vulnerability.

The grid has always been vulnerable, and major segments of both the public and private sectors have long recognized the need for taking a distributed approach to securing their own particularized critical power requirements. The military certainly grasps the critical importance of power in all its systems, and deploys backup power in depth. So do federal, state, and local government agencies, hospitals, phone companies, wireless carriers, broadcasters, banks, insurance, financial trading companies, major providers of online and Web-hosting services, package distribution companies like UPS and Federal Express, major manufacturers, and pharmaceutical and biotech companies. Utilities themselves widely deploy backup-power systems to keep control centers, valves, switches, and other essential hardware running in power plants, when the power plant itself is unable to supply the grid that powers the plant's own, internal infrastructure.

But with that said, much of yesterday's planning for emergencies must now be reevaluated. In the past, many enterprises simply did not need to deploy backup power at all, because the risk-cost profile of expected power failures made it economically rational to simply do nothing, and simply shoulder the costs of the infrequent, and typically short, outages when they occurred. Natural disasters—mainly weather-related—presented the most significant threat of longer outages, and these risks too were reasonably well understood and bounded (Figure 1).

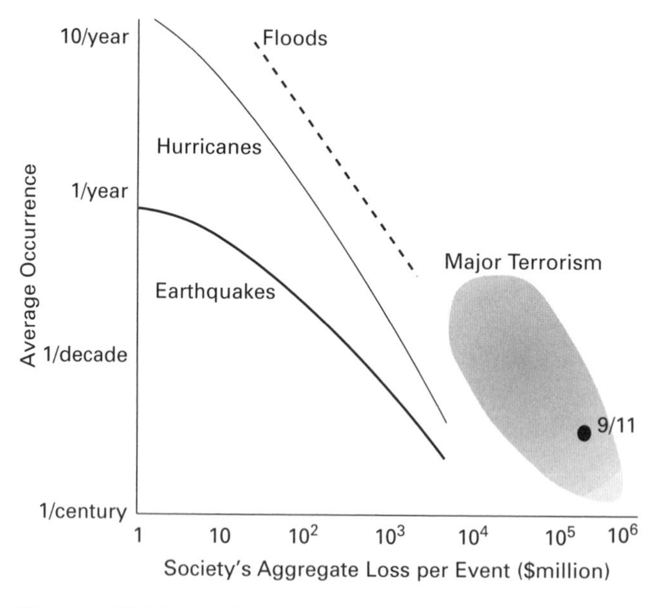

Figure 1 Wide-impact disasters. Businesses, emergency planners, insurance companies, and government now face the challenge of a new 'zone' of risk and consequence. Derived from Amin, Massoud, "Financial Impact of World Trade Center Attack," EPRI, DRI-WEFA (January 2002).

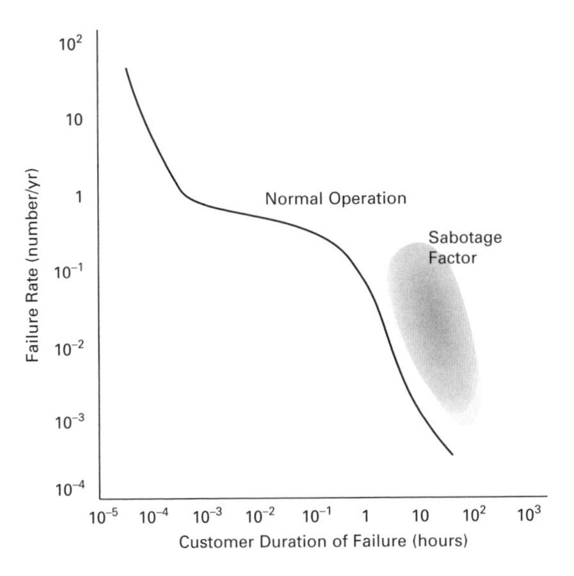

Figure 2 Electric grid failures: Customer perspective. The potential for sabotage creates an entirely new regime for customer outages, adding yet another dimension to protecting critical operations from the inherently unreliable grid. Derived from "Powering the Internet-Datacom Equipment in Telecom Facilities," Advisory Committee of the IEEE International Telecommunications Energy Conference (1998).

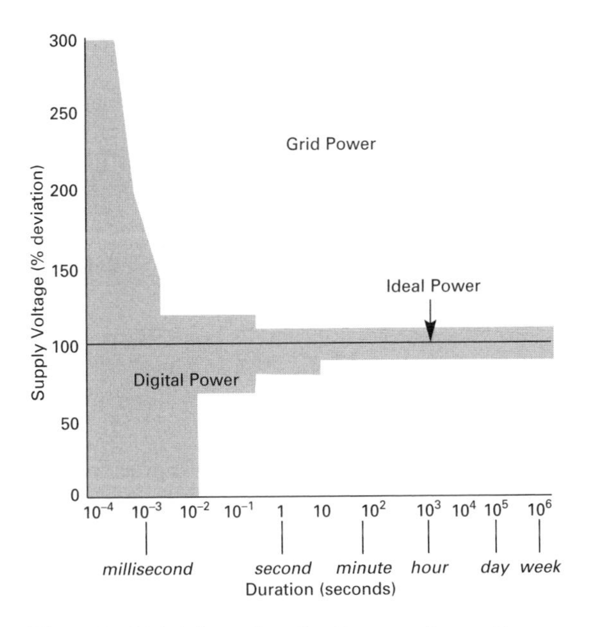

Figure 3 Digital demands and grid power. *Source*: Recommended Power Quality Envelope, IEEE 445–1987.

Much of the rest of the power-related analytic and engineering effort has been directed at power *quality*—smoothing out spikes and dips that last for only fractions of a second—or short-duration issues of power *reliability*—dealing with outages of minutes to an hour or so (Figures 2 and 3). Today's planning, by contrast, must address the threat of grid outages measured in many hours or days. This requires a different level of hardening of local power infrastructure, to assure *continuity* of operations. Heretofore, *that* challenge has been undertaken only by the likes of military bases and Federal Reserve banks.

The failure profiles of the past reflected the relatively benign and familiar threats of the past. Investments in power assurance were planned accordingly. The new environment requires a fundamental reassessment of where power requirements are critical and how supplies of critical power can be assured.

DEMAND

Congress has broadly defined the starting point to assess the scope and character of demand for critical power. The USA Patriot Act defines critical infrastructures as:

> *[S]ystems and assets, whether physical or virtual, so vital to the United States that the incapacity or destruction of such systems and assets would have a debilitating impact on security, national economic security, national public health or safety, or any combination of those matters.*

From this high-level starting point, most discussions of critical infrastructure jump quickly to lists of networked facilities and services, with electric power ranked somewhere down the list. If any attempt is made to rank levels of criticality, information and communications often come first, alongside government, law enforcement, emergency services, and health services. Banking and finance, electricity, oil, gas, water, and transportation are typically viewed as defining the second tier.

This gets things backward. However familiar and pedestrian electric power may seem, it is the first domino of critical infrastructure.

Powering Public Networks

It does not require a great deal of fine analysis to rank as "critical infrastructure" the principal information and material-moving networks of the modern economy—the telecom and financial networks that move bits, and the electricity, oil, gas, and water networks that move material and energy. Defining something as critical is, ultimately, a statement that many people depend in important ways on a system or service, the failure of which will set off a cascade of harmful consequences. Large networks are critical simply because so many people depend on them so much.

The security of all of these networks is the subject of urgent, ongoing assessment. Much of the analysis has been focused on physical and cyber security—protecting the physical structures themselves, or the computers that are used to control them. But their greatest vulnerability is the power on which every aspect of their control and operation ultimately depends. While the multiple layers of the nation's critical infrastructure are highly interdependent, *electric power* is, far more often than not, the prime mover—the key enabler of all the others.

The information and communications sectors are certainly all-electric—bits are electrons, tiny packages of energy moving through long wires, or oscillating in antennas to project radio waves, or exciting lasers to project information through fiber-optic glass. Much of the movement of water, gas, and oil depends on electric pumps and electrically controlled valves. Rail and air transportation are completely dependent on electronic communication and traffic controls; much of short-haul rail is also electrically powered. The agencies and enterprises that provide government, law enforcement, emergency services, and financial services are equally dependent on their communications and computers. And modern hospitals depend completely on all-electric technology.

Simulations of terrorist attacks on the public grid have thus demonstrated a process in which a very small number of well directed attacks precipitate multi-state power outages, which in turn disrupt telecommunications and natural gas distribution systems, and—soon thereafter—transportation, emergency services, and law enforcement. This was in fact how the disaster played out in the New York City area on 9/11. The collapse of the Twin Towers destroyed two Consolidated Edison substations that relayed electricity to a large area of Lower Manhattan. Successive layers of critical infrastructure then collapsed as a result. Power outages degraded landline telephone service and subway service. Communications failures then undermined or paralyzed evacuation and emergency response services. The Department of Homeland Security's Critical Infrastructure Assurance Office (CIAO—formerly in the Commerce Department) has analyzed "[t]he cascading fallout" from the 9/11 attack, much of it traceable directly to the loss of electricity for telecommunications. That same dynamic is analyzed in Critical Infrastructure Interdependencies, a study published by the U.S. Department of Energy (DOE) soon after the attack.[3]

Analyses of the vulnerabilities of other critical-infrastructure sectors have reached similar conclusions: The loss of electric power quickly brings down communications and financial networks, cripples the movement of oil, gas, water, and traffic, and it paralyzes emergency response services.

Telecommunications Networks (Table 1)

Broadcast, telephone, cable, data, and networks, other business networks are now completely dependent on electric power. A May 2002 report, prepared by major wireless and wireline telecom trade associations, of course emphasized the importance of communications as a critical infrastructure service in its own right, but also stressed that sector's dependence on electricity.[4] After the 9/11 attack, the report notes, "many customers in New York found that their communications problems

Table 1 Telecommunications Networks

Category	Number
Broadcast TV[5]	1,500
Broadcast radio[6]	10,000
Telephone (central offices)[7]	25,000
Cable (headends, etc.)[8]	10,000
Wireless Infrastructure	
Mobile Telephone Switching Offices[9]	2,800
Base stations[10]	140,000
Base station controllers[11]	1,900
Private satellite links[12]	700 (e)
Major internet data centers[13]	400
Internet Network Access Points (NAPs)[14]	11
ISPs[15]	7,000 (e)
ISP backbone connections[16]	12,000
Internet points of presence[17]	16,000
Commercial buildings with significant data centers and/or info networks[18]	>9,000
Critical financial networks (e.g., major banks)[19]	19,000

stemmed not from destroyed telecommunications hardware but from power failures and stalled diesel generators." To address the problem, however, the report mainly urged utilities to modify their "electric service priority systems" by "adding a limited number of specific telecommunications critical facilities that service National Security and Emergency Preparedness requirements." Little emphasis was placed on the industry's own ability and responsibility to plan for more prolonged grid outages, and take steps to secure its own power supplies.

Financial Services Networks (Table 2)

Government agencies and trade associations that regulate and represent this sector have devoted considerable effort to assuring "continuity" of operations in the event of another major attack. A January 2003 report by the General Accounting Office (GAO) on this sector's vulnerabilities notes that financial services are "highly dependent on other critical infrastructures," particularly the "telecommunications and power sectors."[20] This then leads to a discussion of the widespread power disruptions that could result from cyber attacks on the "supervisory control and data acquisition" (SCADA) systems used to control power and other energy distribution networks. A second GAO report issued in February 2003 concluded that while progress had been made, financial organizations remained directly or indirectly vulnerable to disruptions of their underlying supplies of power.[21] This GAO report recounts how a provider of telecom services to Wall Street had to shut down a key telecom switch in the late evening of September 11, 2001, because "commercial

Table 2 Financial Services Networks[22]

Institution type	Total	Large		Small and Medium	
		Number	Assets ($billions)	Number	Assets ($billions)
Federal Reserve	970	25	1,400	950	300
State banks	5,000	10	300	5,000	940
National banks	2,000	40	3,000	2,100	700
Federal and state thrifts	900	20	600	870	400
Federal credit unions	10,000	1	15	10,000	500
Total*	19,000	100	5,200	19,000	2,800

* Totals don't match due to rounding

power to that switch was lost, and backup power supplies (generator, then batteries) were eventually exhausted before . . . technicians could gain access to their facilities in order to restore power." The resumption of financial market services had to await the resumption of telecom services, which had to await the arrival of backup generators and the fuel to run them.

An April 2003 paper, by the Federal Reserve, likewise emphasized the financial sector's vulnerability to wide-scale disruptions of "transportation, telecommunications, power, and other critical infrastructure components across a metropolitan or other geographic area." The report also emphasized that backup sites "should not rely on the same infrastructure components" as those used by the primary site.[23]

Networks of Law Enforcement, Public Safety, and Emergency Services (Table 3)

A report by the U.S. Fire Administration (USFA)—part of the Federal Emergency Management Administration (FEMA), which is now part of DHS—lists as critical nodes E911 and other public safety communications centers and dispatch networks, fire, rescue and emergency medical service stations, pumping stations and water reservoirs for major urban areas, along with bridges, tunnels, and major roadways serving large population centers. Electric power is critical at all of these nodes. Government, police, and other emergency services all depend heavily on communications; hospitals and critical care facilities are also completely dependent on electricity to power the sensors, imaging systems, pumps, and other equipment used to form images and move materials through the modern hospital and its patients.

After ice storms crippled much of the Northeast coast in 1998, FEMA issued a number of power-related findings and recommendations.[30] Numerous government agencies involved in disaster response had lost their communications capabilities, the agency reported, because of a loss of electric power. Many broadcast stations likewise stopped transmitting news updates and emergency messages because of insufficient backup generator capacity. Power interruptions had caused the loss of

Table 3 Law Enforcement, Public Safety, and Emergency Services Networks

Category	Number
Hospitals[24]	5,800
Local health clinics[25]	23,000
Nursing homes[26]	17,000
E911 call centers[27]	6,000
Fire & rescue services[28]	45,000
Critical municipal buildings (incl. police)[29]	14,000

food supplies at 75% of Disaster Recovery Centers. FEMA specifically recommended the deployment of on-site auxiliary power capacity sufficient to keep key equipment operational "for the duration of a utility outage" at critical-care facilities in hospitals, nursing homes, broadcast stations, and at all National Weather Service (NWS) radio transmitters. Nevertheless, a comprehensive survey by the USFA in 2002 found that 57% of firehouses still had no backup power systems.[31]

Physical Networks: Electricity, Oil, Gas, Water, and Transportation (Table 4)

The largest and most important physical networks—the electric grid itself, water, oil, and gas pipelines, and transportation networks—are all highly dependent on electric power to drive pumps and activate valves and switches. A report from the USFA recounts how one major metropolitan area introduced "rolling brownouts" to curtail electrical power consumption, recognizing that this would interrupt domestic

Table 4 Physical Networks: Electricity, Oil, Gas, Water, and Transportation

Category	Number/Size
Electricity	
Transmission SCADA control points	
FERC grid monitor/control[32]	12
Network Reliability Coordinating Centers[33]	20
Regional Transmission Control Area Centers[34]	130
Utility control centers[35]	>300
Power plants[36]	10,500 (e)
Large (> 500 MW)	500 (e)
Small (< 500 MW)	10,000 (e)
Transmission Lines	680,000 miles
Transmission substations	7,000
Local distribution lines	2.5 million miles
Local distribution substations	100,000

Oil & Gas	
Oil & Gas SCADA systems	>300
Oil pumping stations[37]	3,000
Gas compressor/pumping stations[38]	4,000
Oil pipelines[39]	177,000 miles
Gas pipelines[40]	1.4 million miles
Oil wells[41]	520,000
Gas wells[42]	360,000
Off-shore wells[43]	4,000
Natural gas processing[44]	600
Oil refineries[45]	150
Oil product terminals[46]	1,400
Oil "bulk stations"[47]	7,500
Oil storage terminals	2,000 (e)
Gas storage facilities[48]	460
Gasoline service stations	180,000
Water & Wastewater	
Treatment facilities (1990)[49]	40,000
Community water systems[50]	56,000
Transportation	
FAA critical centers[51]	56
Airport control towers[52]	560
Rail control centers (99,000 miles)[53]	100
Major municipal traffic control centers[54]	>100

water consumption—but initially overlooking the fact that the intermittent shutdown of water pumps would interfere with firefighting as well.

The grid requires its own backup power to actuate valves and switches, to run pumps and lights in electric power plants when primary sources of power fail, and to assure the delivery of power to the communications and control networks that activate switches and breakers to stabilize the grid when transformers fail, lines go down, and large loads short out.

Supervisory Control and Data Acquisition (SCADA) for Physical Networks

The physical networks that move material, energy, and vehicles are all critically dependent on their control systems. The electric grid, oil and gas pipelines, and many industrial systems (and especially those managing hazardous chemicals), are monitored and controlled by means of complex, computer controlled, SCADA networks that collect and convey information about the state of the network and dispatch commands to actuate switches, circuit breakers, pumps, and valves. A critical-infrastructure report issued by the American Petroleum Institute in March 2002, for example, focuses primarily on physical and personnel issues relating to security, but notes the importance of networked computer systems that run refineries and pipelines, and the "electrical power lines (including backup power systems)" on which their operations depend.[55]

SCADA networks, which generate over $3 billion per year in global revenues from hardware and software sales, control annual flows of many hundreds of billions of dollars of energy, and also monitor and control all major transportation of water and wastewater. Utility operations, for example, typically center on SCADA master stations located in one, or sometimes two, key locations; in a few SCADA networks, regional control centers can take over local operations in the event of a major calamity that takes out the master stations. These master stations may monitor data from 30,000 or more collection points, as often as every two seconds. Much of the communication with the field equipment sensors occurs via analog and digital microwave technology; fiber-optic lines, satellite links, spread-spectrum radio, two-way radio, and other technologies, particularly in the backbones of these private communications networks.

And the importance of power to the continued operation of the networks that distribute materials and energy is growing. The recent push to make electric power markets more competitive is creating more points of interconnection and power hand-off, requiring more data transparency and much closer and more precise coordination of power flows. Gas and oil pipelines are becoming more dependent on electrically actuated and telecom-controlled switches. A significant number of natural gas pipeline shutoff valves, for example, are manually operated hand-wheels, that may, in emergency situations, take hours to locate and shut down. Remotely actuated valves allow rapid pipeline shutdown—but require secure power for the SCADA control network and the valves themselves.

In the past, these systems were generally designed and installed with minimal attention to security. Experts now view them as highly vulnerable to cyber attack.[56] Electronic intrusions could precipitate widespread power outages at regional and even national levels. SCADA systems simply *must* be kept running to prevent minor disruptions from turning into major ones. When major disruptions occur, the SCADA systems will be integral to the recovery process, because they provide the essential information and control capabilities to coordinate emergency responses. When SCADA networks go down, the networks they control go down too, and they can't be practically restarted until the SCADA networks themselves become operational once again.

Command and Control Systems for Transportation Networks

Much of the transportation system likewise depends on communications and data networks for coordination and control. And while very little of the transportation system itself is electrically powered, all of the control systems are. However much kerosene or diesel fuel they may have at hand to fuel their engines, planes don't fly without electrically powered air traffic control, railroads don't run without the communications and control networks that manage traffic flow and the configuration of the tracks, and ships and trucks don't move without the similar control/scheduling systems that synchronize movement through harbors and control the traffic lights.

In many respects, the four major families of networks—telecom, finance, government and public safety, and the physical (material, power, transportation)—all depend on each other. Electric power plants depend on railroads and pipelines for their raw fuel; railways and gas pipelines need electricity for supervision and control, and all of the networks depend on police, fire, and emergency services to maintain safety and public order. But electric power is, nevertheless, uniquely important. After major catastrophes, the process of restarting normal life begins with limited supplies of raw fuel—most often diesel fuel—that are used to fire up backup generators, which are used to restart everything else—the communications networks, computers, pumps, and valves that move information and financial data, and reactivate the government services and the material and energy-moving networks, which bring in still more fuel, and then still more power.

The Vulnerable Public Grid

An important 1997 report Critical Foundations by the President's Commission on Critical Infrastructure Protection provides a top-down analysis of the vulnerabilities of the public grid.[57] The report emphasizes how much other infrastructure networks (most notably telecommunications, finance, and transportation) have come to depend on electric power for their continued operation, and notes the "significant physical vulnerabilities" of "substations, generation facilities, and transmission lines." The report contains many useful recommendations for making the public grid more secure. Yet in the end, it simply fails to address the plain—and widely recognized—fact that the grid itself can never be made secure enough to guarantee power continuity at the most critical nodes.

In similar fashion, utilities themselves certainly recognize how much other sectors depend on power. All major utilities have established "electric service priority" (ESP) protocols to prioritize efforts made to restore power after a major outage. High on the list, of course, are life support, medical facilities, and police and fire stations. But what most utilities emphasize, understandably enough, is what they themselves can do to help restore grid power quickly to the customers that need it the most.

Thus, utilities consult with other sectors in setting these priorities, and in the post-9/11 environment there is, inevitably, a certain amount of lobbying under way to persuade utilities to reorder their power-restoration priorities. The telecommunications industry, for example, launched a "Telecommunications Electric Service Priority" (TESP) initiative to urge utilities to "modify their existing ESP systems by adding a limited number of specific telecommunications critical facilities that service National Security and Emergency Preparedness requirements."[58] The proposed list of such facilities is a long one: It includes all facilities engaged in "national security leadership, maintenance of law and order, maintenance of the national economy, and public health, safety, and welfare." TESP defines "critical facilities" as "those that perform functions critical to the monitoring, control, support, signaling, and switching of the voice telecommunications infrastructure."

Understandable and even necessary though it is, the lobbying for priority in the utility's power-restoration hierarchy is, ultimately, an acknowledgement that a power-dependent facility isn't "critical" enough to need power better than the grid can supply—or else it is an abdication of the responsibility to secure alternative, off-grid power supplies.

Utilities can and do establish service resumption priorities. They also continuously improve the robustness of the grid as a whole. Enormous amounts of investment have been made to improve the reliability of the public grid since a single faulty relay at the Sir Adam Beck Station no. 2 in Ontario, Canada, caused a key transmission line to disconnect ("open") on November 9, 1965, plunging the entire northeastern area of the United States and large parts of Canada into an eighteen-hour blackout. That seminal event led to the creation of the North American Electric Reliability Council (NERC) and a substantially more resilient grid. But the grid still has many points of vulnerability, and its inherent fragility will never be eliminated. While continuing to rely heavily on the grid in their normal operations, the operators of all truly "critical" nodes and networks also take steps to create their own, independent islands of secure power, to ensure continuity of operation in the case of a major grid outage.

What makes the grid essential is also what makes it vulnerable—it is a vast, sprawling, multi-tiered structure that reaches everywhere, and is used by everyone. Indeed, measured by route miles and physical footprint, the North American grid is by far the largest network on the planet (Figure 4).

The top tier of the grid is fueled (most typically) by coal, uranium or gas; each lower tier is typically "fueled," initially at least, by the electric power delivered from the tier above. "Generating stations" in the top tier dispatch electrical power through some 680,000 miles of high-voltage, long-haul transmission lines, which feed power into 100,000 "substations." The substations dispatch power, in turn, through 2.5 million miles of local distribution wires. The wires are extended and exposed, while the grid's power plants are huge (because big plants burn fuel more efficiently) and thus comparatively few and far between. Nearly all the high-voltage lines run above ground and traverse the open country, and a handful of high-voltage lines serve entire metropolitan regions. At the same time, a couple of large power plants can provide all the power required by a city of a half-million. Many communities are served by just a handful of smaller power plants, or fractional shares of a few bigger power plants.

Taken together, these attributes make the public grid inherently vulnerable to major disruptions. Demand overloads simultaneously stress all of the main power cables—typically three to five—serving a large city, and one failure can trigger others. On August 10, 1996, three lost transmission lines and some malfunctioning equipment triggered a series of cascading blackouts that paralyzed the West Coast, affecting 7.5 million customers in 11 states and two Canadian provinces. Sometimes purely physical stresses affect the length of the conduits that house power lines. In April 1992, for example, construction workers installing support pillars in the Chicago River punctured the roof of a freight tunnel beneath the river bottom; the ensuing flood shut down utility power for weeks in the heart of Chicago. In

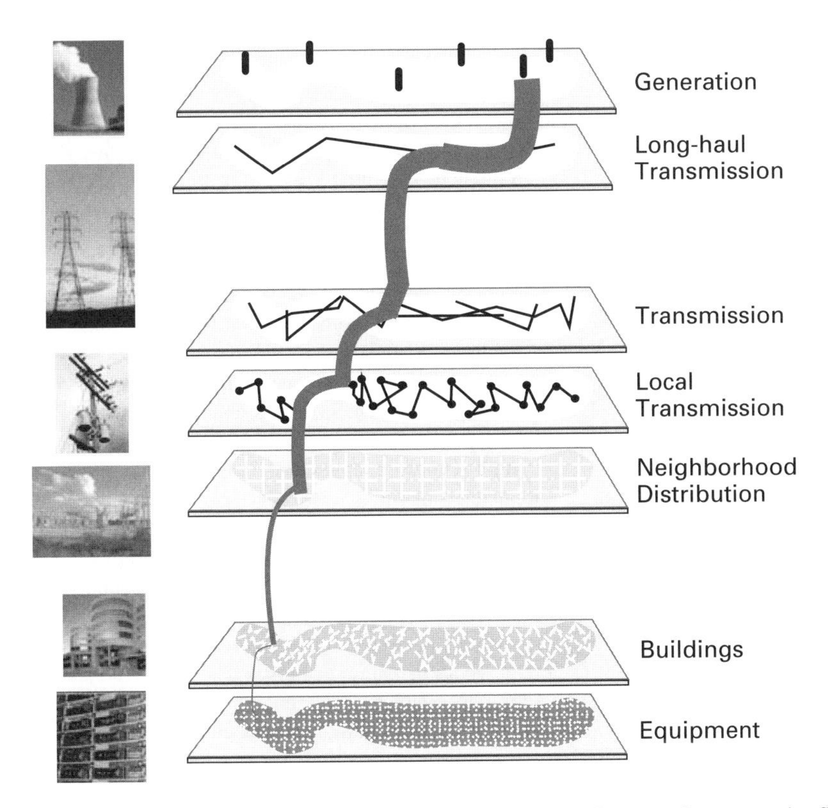

Figure 4 The multi-tiered grid. Derived from "Distributed Energy Resources Interconnection Systems," U.S. DOE NREL (September 2002).

response to that and similar events, the City of Chicago with the DOE undertook a uniquely comprehensive and prescient study of the consequences of power outages, which catalogues and categorizes critical-power customers (Table 5).

Serious problems tend to propagate through the grid, as fast as the power itself. Even a severe electrical failure on a single (large) customer's premises can travel upward through the grid to cause much more widespread disruptions. Failures at key points higher in the grid can black out far larger areas. The November 1965 Northeast blackout mentioned earlier was caused by the failure of one small relay in Ontario, which caused a key transmission line to disconnect; that triggered a sequence of escalating line overloads that cascaded instantaneously down the main trunk lines of the grid. Additional lines failed, separating additional plants from cities and towns that used their power. Generating plants in the New York City area then shut down automatically to prevent overloads to their turbines. While much has been improved since the 1960s, recent simulations confirm that deliberate attacks on a very limited number of key points on the public grid could still cause very widespread outages that would take, at the very least, many days to correct.

Table 5 Critical Municipal Facilities[59]

Type	Examples
Emergency services	Police stations, fire stations, paramedic stations, emergency communication transmitters
Water system	Water supply pumping stations, wastewater pumping stations and treatment plants
Transportation	Traffic intersections, aviation terminals and air traffic control, railroad crossings, electric rail systems
Medical	Hospitals, nursing homes, mental health treatment facilities, specialized treatment center (e.g., outpatient surgery, dialysis, cancer therapy), rehabilitation centers, blood donation centers
Schools	Nursery schools, kindergarten, elementary schools, high schools, colleges, business and trade schools
Day care	Registered facilities, sitter services, after-school centers
Senior	Senior citizen centers, retirement communities
Social service	Homeless/transient shelters, missions and soup kitchens, youth, family, and battered person shelters, heating/cooling shelters
Detention centers	Jails, youth detention centers
Community centers	Libraries, civic centers, recreational facilities
Public assembly	Stadiums, auditoriums, theaters, cinemas, religious facilities, malls, conference centers, museums
Hotels	Hotels, motels, boarding houses
High-rise buildings	Apartments, condos, commercial
Food services	Restaurants, supermarkets, food processing facilities
Industry	Hazardous material handling

Precisely because it is so critical, a late-2002 White House briefing involving the President's Critical Infrastructure Protection Board specifically noted that the electric power grid now stands "in the cross hairs of knowledgeable enemies who understand that all other critical-infrastructure components depend on energy for their day-to-day operations."[60]

A New Profile for Grid-Outage Risks

Installations of backup generators, uninterruptible power supplies (UPS), and stand-by batteries provide an initial—though backward-looking—measure of total demand for critical power as already determined by end-users themselves.

In many ways the presence of a UPS provides the most telegraphic and useful indicator of a critical electrical load on the premises. A UPS isn't cheap, it isn't deployed lightly, and—as its name reveals—its whole purpose is to assure the continuous provision of power, most often to digital loads. Thus, an array of power electronics, sensors, software, and batteries takes power from whatever source can supply it—the grid, the standby batteries, or a backup generator or fuel cell—and delivers power devoid of sags, spikes, and harmonics. Typically, the on-board battery backup is just sufficient to permit switching to a secondary source after the grid fails—a backup generator, or a larger bank of batteries. The smallest UPS units sit on desktops; the largest serve entire offices or small buildings. The UPS functions very much as a kind of silicon power plant—taking in "low grade" fuel (in this case, dirty and unreliable kilowatt-hours) to convert into a higher-grade fuel (sag- and spike-free, always-on kilowatt-hours).

Most of the large-unit capacity is in commercial buildings, with some significant share used in industrial environments to maintain critical non-computing digital loads. Fully half of non-desktop UPS capacity is supplied by 75-kilowatt (kW) or larger UPS units. One-third of that capacity is supplied by units larger than 200 kW.

Based on annual UPS shipments and the typical operating lifetimes of these units, we estimate that approximately 25 gigawatts (GW) of large UPS capacity is currently installed and running in businesses and government buildings in the United States, with another 10–15 GW of capacity in smaller desktop-sized units in both businesses and residences (Figure 5).

These remarkable figures provide a direct, quantitative estimate of how much U.S. power consumption is viewed as (in some sense) "critical" by end-users themselves. To put them in perspective, the public U.S. grid as a whole is powered by roughly 790 GW of large coal, nuclear, gas-fired, and hydroelectric plants—thus,

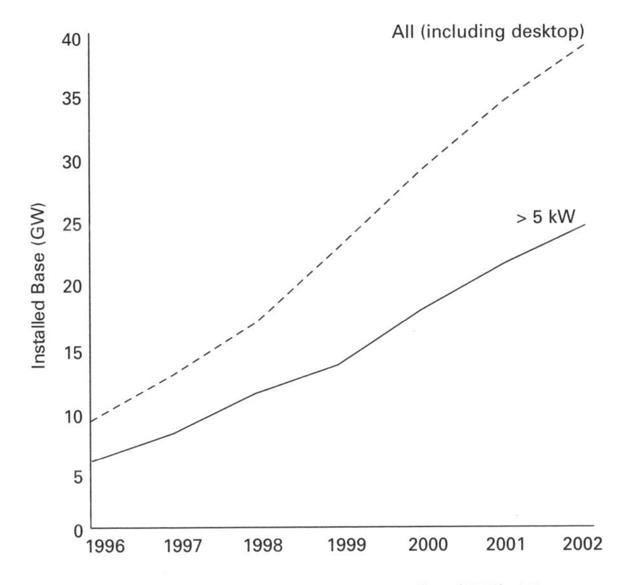

Figure 5 Total uninterruptible power supplies (UPS). The presence of a UPS provides the most telegraphic and useful indicator of what customers consider "critical" electrical loads. Derived from Powerware and Frost & Sullivan World UPS Market 2002.

about 3–5% of the *grid's* capacity is complemented and conditioned by UPS capacity in buildings and factories.[61]

Deployments of backup generators—the vast majority of them reciprocating engines that burn either diesel fuel or natural gas—provide a second rough indicator of end-users' assessments of their own "critical power" requirements. The indicator is imprecise, because generators are also widely used to supply power in off-grid locations. Nevertheless, trade estimates indicate that there is now 80 GW of backup generation capacity deployed in the United States—in the aggregate, about 10% of the grid's capacity. Over the past several years, no less than 1 megawatt (MW) of such distributed (grid-independent) capacity is now being purchased for every 6–10 MW of central-power-plant capacity brought on line (Figure 6).

One finds, for example, thirteen two-megawatt diesel generators installed outside AOL's two major centers in Prince William County and Herndon, Virginia. Real estate companies and data hotels like Equinix, Level3, and Qwest have likewise become major owner-operators of distributed generation (DG) power systems. And all the major engine makers assemble and lease power-plants-on-wheels—tractor-trailers with hundreds of kilowatts to several megawatts of generating capability. These units are positioned around the country for emergencies, parked in substations (to meet peak demand, or replace power from lost transmission feeds), or in parking lots near critical loads.

The installed base of standby batteries provides a third—again retrospective—measure of what end-users perceive to be their "critical power" needs. The sale

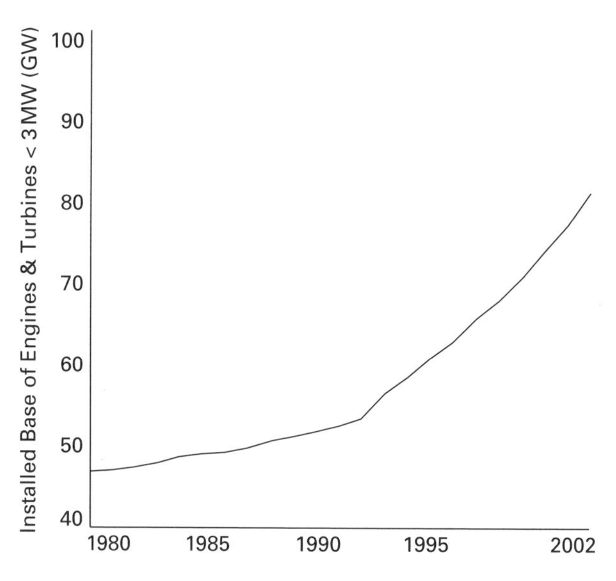

Figure 6 Off-grid backup power. Total purchases of "small," distributed generators for backup power (rarely connected to the grid) are a significant share of the 40 MW/yr of utility central station capacity added annually in the past two years. Derived from Diesel & Gas Engine Worldwide Engine Order Survey.

of long-life, high-performance backup lead–acid batteries soared through the year 2000—doubling over a few years to the point of creating delivery shortages. Even though the market is still working off that inventory, sales of new batteries remain 20% higher than 5 years ago—and the advent of data centers and wireless telecom has permanently moved such heavy-duty "industrial" batteries to the lead position in a business formerly dominated by batteries used in more traditional industrial motive power applications (e.g., lift trucks).[63] There is thus a collective total of about 40 million kilowatt-hours stored at any time in about 30 million lead–acid batteries distributed across the landscape (enough electricity to run one million homes for 24 hours) (Figure 7).

For every 500 kW of UPS, there are typically five tons of batteries nearby. Telephone company central offices and data centers, for example, still typically contain entire floors filled with lead–acid batteries. The floor space set aside for large UPSs—100 to 1000 kW AC and DC silicon power plants—is likewise dominated by lead and acid. There are batteries in every wireless base station and in every optical headend. Some 249 batteries on steel racks stand behind the planet's most precise atomic clock in Boulder, Colorado.

It would be a serious mistake, however, to infer from existing deployments of UPS systems, batteries, and generators that the full extent of the need for such facilities has already been recognized and addressed. To begin with, many enterprises simply fail to plan for rare-but-catastrophic events until after the first one hits. Thus, for example, the four large, broad-scale power outages of the last 10 years—Hurricane Andrew in Florida in 1992, Hurricane Fran in Virginia in 1996, and ice storms hitting the East Coast during the winters of 1998 and

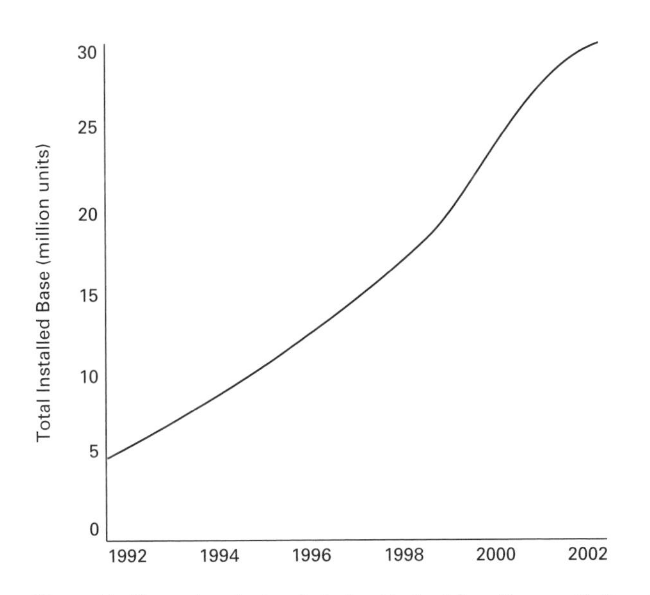

Figure 7 Heavy-duty backup batteries. Derived from Enersys, Battery Council International.

2002—precipitated a cascade of new orders for on-premises power supplies, and these orders continued to be placed for weeks after grid power had been restored.[64] Planning rationally for infrequent but grave contingencies is inherently difficult, and even risk-averse planners have a strong tendency to discount to zero hazards that are thought to be just "too unlikely" to worry about. Many essential services and businesses have critical power needs that have not been addressed only because they have never been systematically assessed.

It is impossible to estimate with any precision how many sites have critical power needs that simply remain unrecognized because they haven't yet been hit by disaster. We do know that there are hundreds of thousands of 10- to 100-kW sites nationwide—the electrical loads now created by tens of thousands of high-end wireless base stations, fiber repeater shacks, and digital offices that—unlike the phone company's central offices—have limited (and sometimes, no) backup. The national banks and financial exchanges have already deployed their backup power systems (although, as earlier noted in a recent GAO report, many are not adequate for the new challenges), but many smaller commercial, investment, regional banking, credit, and trading companies may not yet have done so. The federal government's buildings already have their backup power, but state and particularly local governments lag far behind them. The general unwillingness to confront hazards that are both very grave and very remote has always been a problem. But it is an especially serious one when risk profiles change—as they surely have in the post-9/11 era. Countless enterprises that were inadequately prepared for hurricanes and ice storms are at even greater risk now that sabotage and terrorism have changed the profile of credible threats.

Even enterprises that have prepared properly for yesterday's risk profiles may well be unprepared for tomorrow's. Many surveys, for example, have attempted to document the costs of outages for various industrial and commercial enterprises (Tables 6 and 7). These surveys typically address costs of equipment damage, loss of in-production materials and products, and lost employee productivity, among other factors, from outages that last *minutes* to (typically) *one hour*. But the choice of a one-hour out-age, rather than, say, a three-day outage, reflects the survey's assumptions about what kinds of outages are reasonably likely.

Table 6 Cost of Power Interruptions for Industrial Plants[65]

Interruption length	Cost/kW load interrupted	
	Minimum	Maximum
1 minute	$1	$17
1 hour	7	43
3 hours	14	130

Table 7 Cost of Power Interruptions for Office Buildings with Computers[66]

Interruption length	Cost/peak kWh not delivered	
	Minimum	Maximum
15 minutes	$5.68	$67.10
1 hour	5.68	75.29
>1 hour	0.48	204.33

The risk-of-failure profiles of the past reflect the relatively benign threats of the past–routine equipment failures, lightning strikes on power lines, and such small-scale hazards as squirrels chewing through insulators or cars colliding with utility poles. Most accidental grid interruptions last barely a second or two, and many "power quality" issues involve problems that persist for only tens of milliseconds (one or two cycles). In most areas of the country, grid outages of an hour or two occur, on average, no more than once or twice a year, and longer outages are much rarer than that. Accidental outages tend to be geographically confined as well; the most common ones involve blown circuits in a single building (and, most typically, caused by human error—ironically enough, much of it "maintenance" related), or interruptions confined to the area served by a single utility substation.

The possibility of deliberate attack on the grid, however, changes the risk profile fundamentally—that possibility sharply raises the risk of outages that last a long time and that extend over wide areas. The planning challenge now shifts from issues of power *quality* or *reliability* to issues of business *sustainability* (Figure 8). Planning must now take into account outages that last not for seconds, or for a single hour, but for days.

There is normally very little risk that several high-voltage lines feeding a metropolitan area from several different points on the compass will fail simultaneously, and when just one such line fails, all the resources at hand can be mobilized to repair it. Deliberate assaults, by contrast, are much more likely to disable multiple points on the network simultaneously. A National Academy of Sciences 2002 report drove this reality home with its stark observation: "[A] coordinated attack on a selected set of key points in the [electrical] system could result in a long-term, multistate blackout. While power might be restored in parts of the region within a matter of days or weeks, acute shortages could mandate rolling blackouts for as long as several years."[67] Users, who have perceived no need to plan systematically, even for those outages that fit the traditional statistical profiles, may now require systems to address the new risks. Enterprises that can afford simply to shut down and wait out short blackouts may not be able to take that approach in response to the mounting threats of longer outages.

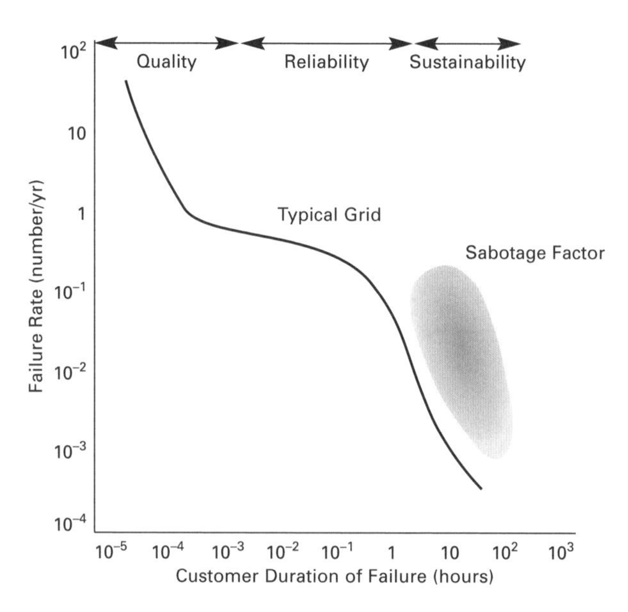

Figure 8 Electric grid failures: Customer perspective. The power engineering challenge now expands beyond the quality or reliability of electricity, to include **sustainability**—the ability to continue to operate in the event of outages that last not for seconds, minutes or an hour, but days. Derived from "Powering the Internet-Datacom Equipment in Telecom Facilities," Advisory Committee of the IEEE International Telecommunications Energy Conference (1998).

Powering Critical Nodes

Roughly one-half of U.S. electric power is consumed by mass-market users—residences and small businesses with peak power requirements under about 20 kW. Overwhelmingly, these consumers currently rely on grid power—power that is (roughly) 99.9% reliable, which retails for about 10¢/kWh, and that is generated at large, centralized power plants for about 3¢/kWh wholesale. The average residential customer experiences 90 minutes of power interruptions per year—with 70–80 minutes of that down time attributable to distribution system problems—but finds this quality of service acceptable, because the power is used mainly for noncritical purposes that operate on flexible schedules. The washing of clothes and dishes can be postponed; refrigerators and homes remain acceptably cool (or warm) even when the power goes out for several hours; for longer outages, people can temporarily relocate, if they must, to places such as shelters and other centralized facilities where power remains available. If these customers want greater reliability, they generally obtain it by buying power "appliances"—most typically desktop-UPS units for computers or small gasoline or diesel backup generators for other purposes.

At the other pole, aluminum smelters and large auto assembly lines, and enormous commercial complexes, among others, may deploy backup power facilities for certain nodes and control points within their facilities, but cannot backup their entire operations. These monster-load customers, with peak loads above about 10 MW, account for about 20% of U.S. electric consumption. Like it or not, they typically have no choice but to depend on the public grid. To make their power more reliable, they work directly with the utility; the most common approach is to engineer two independent links to the public grid, that connect, if possible, with independent high-voltage lines that are powered in turn by different power plants. Any other approach would be prohibitively expensive, because it would leave huge amounts of generating capacity standing idle most of the time (Tables 8 and 9).

(Even in this high-power arena there are, nonetheless, a relatively small subset of digital loads that easily exceed 1 MW of demand—large data and telecom centers—and that simply have no choice but to deploy adequate grid-free backup. The central new challenge for this class of customer is how to deal with the addition of the continuity metric.)

That leaves the middle tier of loads, from roughly 20 kW on up to a megawatt or so. This segment of demand accounts for about 30% of U.S. power consumption. A disproportionate share of private production and public services fall within this tier, however—air traffic control centers, supermarkets, city halls, factories, broadcast stations, frozen-food warehouses, office buildings, most data centers, and telephone exchanges, among countless others. These middle-tier nodes are also uniquely important in the process of recovering from major outages. Most users in the top and bottom tiers can ride out power outages for days, or even weeks, so long as power is restored reasonably quickly to tens of thousands of discrete nodes in this middle tier. Thus, in times of disaster, recovery always begins locally, in islands of self-help and resilience.

Large numbers of hospitals, government agencies, phone companies, and private enterprises have already taken steps to secure their power supplies from the

Table 8 Electric Demand: Commercial Buildings[68]

	Small	Midsize	Large
Average kW/building	10	65	740
Size ($\times 1000\,ft^2$)	1–10	10–100	>100
Number of buildings (thousands)	3000	1000	100
Percent of total commercial electricity	23	40	38

Table 9 Electric Demand: Manufacturing Establishments[69]

	Small	Midsize	Large
Average kW/establishment	100	2500	19,300
Size (employment)	1–99	100–499	500+
Number of establishments	328,000	30,000	4800
Percent of manufacturing electricity	15	38	47

bottom up. They do not count on the public grid alone to meet their critical power requirements—they identify critical-power loads and then add standby generators and switching systems to boost overall reliability. They define critical power in terms of specific loads and nodes, not the network as a whole. This bottom-up approach to securing critical power is an essential complement to top-down efforts to secure the grid. The Institute of Electrical and Electronics Engineers (IEEE) begins from that same premise in its comprehensive analysis and standards document on Emergency and Standby Power Systems.[70] First issued in 1987, and last updated in 1995, this document is still perhaps the most comprehensive analysis of the end-user initiatives that are required to ensure power continuity at critical nodes. A more recent IEEE publication in 2001 expands their standards and evaluations on power from the perspective of different classes of end users, focusing more generally on various nonemergency aspects of distributed generation for industrial and commercial markets.[71]

Like the critical public infrastructure that they mirror, most critical-power nodes are defined by the private networks they serve—the hubs that switch and route information or that control the movement and flow of key materials through physical networks, or that power essential safety systems and services.

Whether it serves a corporation or a city hall, a node that switches or dispatches more than a gigabyte per second of data for much of the working day—or a megabyte per second on a 24/7 basis, or more than 10 kW of analog or digital signal through the airwaves—likely plays a very important role in the day-to-day lives of many people. A typical telephone exchange requires about 100 kW to power it; a television station, about 100–250 kW; a radio station, 20–100 kW; a cell tower 10–20 kW. In some of these sectors—broadcasting, for example (see Table 1)—the number of critical nodes has not changed much in the past several years. In others—data centers and wireless telephone transceivers, for example—the number of critical nodes has been rising rapidly (Figure 9).

Powering the wireless communication infrastructure is particularly important because it is in many respects much less vulnerable than the wireline network. Emergency planners have long recognized that the broadcast networks are essential for mass announcements in times of crisis, as are the radio networks used to coordinate responses by police, fire, and emergency services. Assuring the power requirements of wireless telecommunications networks has emerged as a uniquely important priority in critical power planning, because wireless nodes can be replicated more cheaply, and secured much more effectively, than wired networks. Wireless networks also support mobility, which is often essential to providing critical services and when recovering from major disasters.

The public broadcast and wireless networks already rank as part of the "critical" infrastructure. What is still often overlooked, however, is that there are now tens of thousands of private nodes that route and dispatch comparable (or larger) volumes of data. However private or local it may appear, an information node that requires 10–50 kW or more of continuous power is likely to be one on which thousands of people depend in some very direct way. To put this number in perspective, a typical U.S. household is a 1.5-kW load (24/7 average), with 10–20% of that power

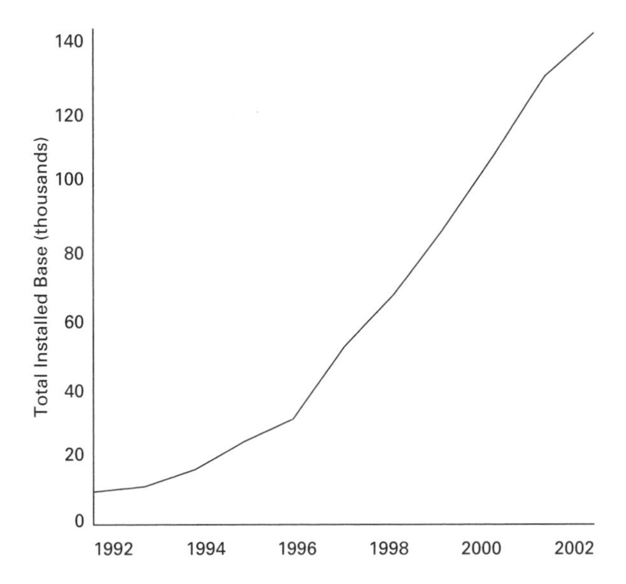

Figure 9 Cellular base stations. Cellular telephony is now a central part of emergency response for citizens, businesses, and emergency professionals, making the sustainability of these networks no longer solely an issue of customer satisfaction. *Source*: Cellular Telecommunications & Internet Association Semi-Annual Wireless Survey.

(150–300 continuous watts) consumed by televisions, telephones, computers, and related peripherals. Comparable computing and communications nodes in commercial and municipal centers define a new, critical class that have received far too little notice in planning to secure critical infrastructure.

Because these nodes are privately owned and operated, and so widely dispersed, it is extremely difficult to quantify their numbers or pin down with any precision the roles they play in maintaining communication, control and the continuity of critical operations. There are, however, some relevant high-level statistics. The total inventory of computing systems installed in all commercial buildings continues to rise rapidly (Figure 10).

Most of the 220 million microcomputers used by American business are located in approximately one million buildings in the mid-tier of the three-tiered hierarchy of power loads *(see Table 8)*. Further analysis of commercial building data suggests a profile of how the critical loads are distributed (Table 10).

It is even more difficult to define and catalogue critical-to-power *physical* networks in the private sector. A recent paper by the Electric Power Research Institute (EPRI) emphasizes the power vulnerability of enterprises engaged in *continuous process manufacturing* of such things as paper, chemicals, petroleum products, rubber and plastic, stone, clay, glass, and primary metals—all companies with manufacturing facilities that continuously feed raw materials, often at high temperatures, through an industrial process.[74] Many other large factories and corporations have their own internal water, water-treatment, and on-premises pipeline services.

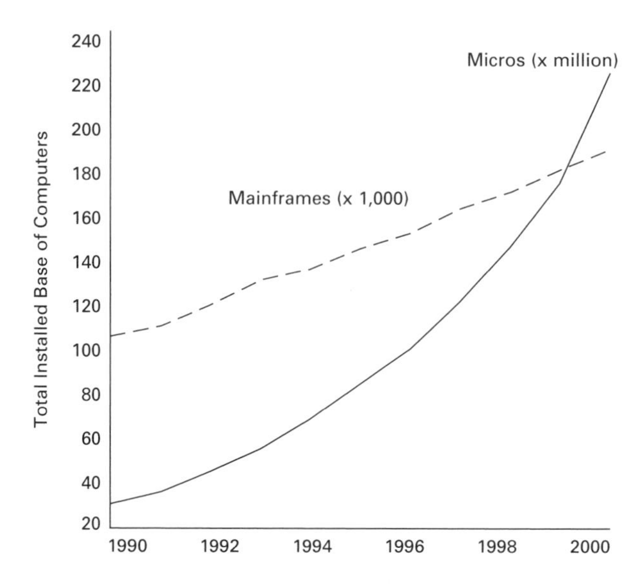

Figure 10 Computers in businesses (excludes PCs in residences). The commercial inventory (which excludes 58 million PCs in homes and excludes notebook PCs) provides an indication of the number of computers in data centers, or data rooms. The total commercial employment of fewer than 90 million people (i.e., well under 90 million "desktops") suggests substantially more than 120 million microcomputers clustered in critical nodes of varying sizes. Derived from "Information Technology Data Book 2000," "Home Computers and Internet Use in the United States: August 2000," and U.S. Census "Access Denied: Changes in Computer Ownership and Use: 1984–1997," U.S. Census.

Table 10 Electric Demand: Commercial Buildings[72]

Principal Building Activity*	Buildings (thousands)	kW/Building (avg)[73]
Critical Category #1		
Health Care	130	100
Public Order & Safety	60	40
Critical Category #2		
Office	750	60
Mercantile	670	40
Food (Sales+Service)	520	40
Service	470	20
Critical Category #3		
Lodging	150	70
Education	330	40
Public Assembly	300	40
Warehouse & Storage	500	25
Religious Worship	300	10

* There is of course not a one-to-one correlation in such data as many of the critical loads, for example in hospitals, are digital in nature, but would not be counted as "microcomputers" in the statistical sources used.

The total power requirements of industrial facilities are defined by individual industrial processes, and vary widely (Table 11). But digital technologies are rapidly taking control of almost *all* industrial processes, even the most familiar and mundane; as a result, there are now very few industries that can continue to function at any level without sufficient electricity to power their key command-and-control networks (Table 12). At the same time, there is steady growth in the number of processes that depend on electricity as their primary fuel—as, for example, when electrically-powered infrared ovens and lasers displace conventional ovens and torches for drying paint and welding metals. Not all of these applications are "critical," but

Table 11 Electric Demand: Manufacturing Sector[75]

Industry	Establishments (thousands)	kW/ Establishment (avg)
Primary Metals	4	8500
Petroleum and coal products	2	5950
Paper	5	5100
Chemicals	9	4650
Textile mills	3	1960
Transportation equipment	8	1350
Beverage and tobacco products	2	1030
Plastics and rubber products	12	870
Food	17	785
Computer and electronic products	10	780
Electrical equipment, appliances, components	5	690
Nonmetallic mineral products	11	680
Wood products	12	370
Machinery	20	280
Fabricated metal products	41	250
Textile product mills	4	240
Furniture and allied products	11	150
Leather and allied products	1	150
Printing and related support	26	110
Apparel	13	80
Miscellaneous	14	170

Table 12 Digital penetration in manufacturing[76]

Digital Technology	Share of Manufacturers (%)
Computer-aided design	85
Local area networks	70
Just-in-time systems	60
Computer-aided manufacturing	60
Robots	20

some significant fraction of them is, because they manufacture or service materials or components required for the continued operation of networks on the front lines.

In the end, "critical power" requirements must be defined application-by-application, and site-by-site. They depend on tolerance for interruptions and outages, however short or long, which depend in turn on how much outages cost the enterprise or agency itself, and those who depend on its goods or services. Secondary losses outside the enterprise may be far larger than those within it. A power failure at a cred-it-card-verification center, for example, will entail relatively modest internal losses, but may obstruct ordinary commerce for merchants and customers worldwide.

The costs are often very sensitive to the duration of the outage. Even the briefest interruptions can be enormously expensive when they take down the computers that run an airline's entire reservations system, or a financial institution's trading desk, or some manufacturing processes with long "reboot" procedures. Even a momentary loss of the power needed by an air traffic control center, or a process control system in a high-speed automated manufacturing plant, may cause catastrophic losses of life or capital. Refrigerated warehouses, by contrast, may be able to ride through outages lasting many hours without significant loss–but then at some point refrigerated foods begin to spoil, and losses become severe.

That critical power requirements are so highly variable and distributed presents both a problem and an opportunity in the formulation of public policy.

On the one hand, it is not possible to secure requirements for critical power by focusing on the grid alone. The grid is a shared resource; the whole thing simply cannot be hardened enough to meet the needs of the tens of thousands of truly critical nodes scattered throughout the country in both public and private sectors.

On the other hand, the efforts undertaken to harden these many individual nodes will have a direct, positive impact on the reliability of the public grid as a whole. As noted, large-area power outages are often the result of cascading failures. Aggressive load shedding is the one way to cut off chain reactions like these. The development of a broadly distributed base of secure capacity and well-engineered local grids on private premises will add a great deal of resilience to the public grid, simply by making a significant part of its normal load less dependent on it. Islands of especially robust and reliable power serve as the centers for relieving stresses on the network and for restoring power more broadly. In the aggregate, private initiatives to secure private power will have a very large beneficial impact on the stability and reliability of the public grid as well.

Fueling the Digital Economy

It is increasingly difficult to define just where the *digital economy* ends, and thus by extension, the needs for critical power. Microprocessors are now embedded everywhere, and are often in final control of such mundane activities as opening and closing cash registers and doors. EPRI defines the digital economy to encompass telecommunications, data storage and retrieval services, biotechnology, electronics manufacturing, the financial industry, and countless other activities that rely heavily on data storage and retrieval, data processing, or research and development operations. What is clear, in any event, is that all digital hardware is electrically powered; when the electrons stop moving, so do the bits.

However defined, the digital economy is by far the fastest growing segment of the overall economy. Largely as a result, more than 90% of the growth in U.S. energy demand since 1980 has been met by electricity (Figure 11).

Even during the most recent years of sluggish economic growth, demand for electricity has continued to rise by 2–3% annually—and by about 4% in 2002—growth rates that may appear modest but are quite substantial in absolute power (and hardware) terms considering the magnitude of annual U.S. electric use (3.6 trillion kilowatt-hours).

The nearly uninterrupted century-long growth in electric demand (driven, a priori by the preferential growth in technologies that use electricity over combustible fuels) has continued to increase the economy's dependence on kilowatt-hours. The nation is now at a point where electricity accounts for over 60% of all fuel used by the GDP-producing parts of the economy (industry, commerce, services)—in 1950, the figure was only 25% (Figure 12).

That the trend illustrated in Figure 11 will continue is strongly suggested by the nature of capital spending, which is skewed heavily toward electricity-consuming hardware. Some 60% of all new capital spending is on information-technology equipment, all of it powered by electricity, and the most recent data show that percentage rising (Figure 13). All the fastest growth sectors of the

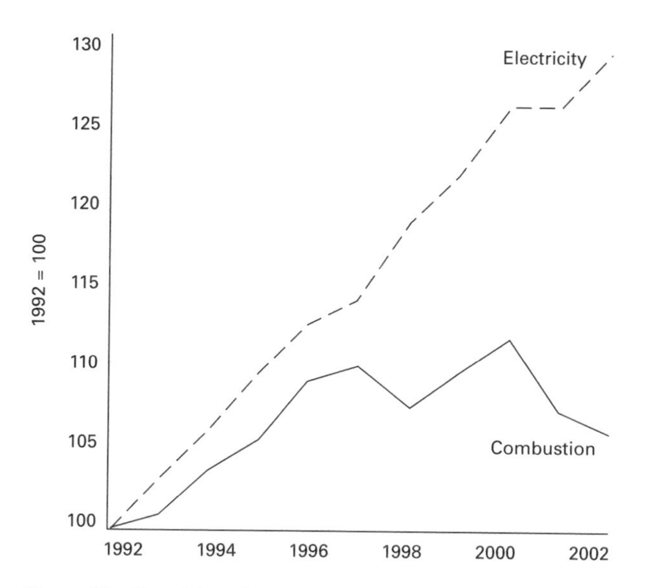

Figure 11 Growth in end-Use demand (commercial and industrial sector primary energy consumption). The technologies driving growth in the commercial and industrial sectors of the economy have been mainly fueled by electricity, thereby continually increasing the number and magnitude of critical uses of power. *Source*: EIA Monthly Energy Review (March 2003).

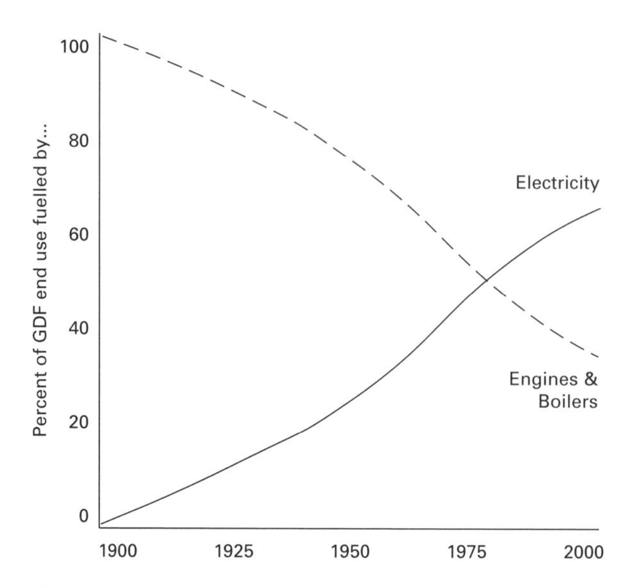

Figure 12 End-use dependence on fuel (Excludes residential energy: counts only fuels used by GDP-producing sectors; transportation, industry (incl. mining, agriculture) and services. The businesses, activities and technologies that comprise the GDP have grown continually more dependent on electricity. *Source*: EIA Annual Energy Review 2000, Bureau of Economic Analysis, and U.S. Census Bureau Historical Statistics of the United States Colonial Times to 1970.

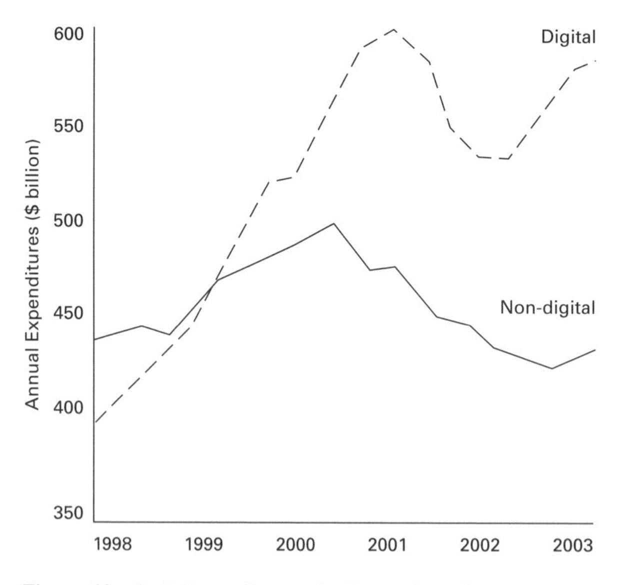

Figure 13 Capital spending on hardware. Spending on new hardware is heavily biased towards electricity-consuming and in particular highly power-sensitive and often "critical" digital equipment. *Source*: Commerce Department and Business Week.

economy—information technology and telecom most notably—depend entirely on electricity.

According to surveys conducted by the University of Delaware's Disaster Research Center, large and small businesses in five major business sectors see electricity as "the most critical lifeline service for business operations."[77]

EPRI estimates that approximately 9% of electricity used in the industrial sector is now used to power digital hardware, most of it in manufacturing electronic components and in automated process control (Table 13). And EPRI attributes 12% of all U.S. power consumption in 2001 to the operation and manufacture of digital devices (microprocessors, chips, and related systems), digital applications (e.g., home entertainment, digital office equipment, networks, data processing, and digital controls), and digitally enabled enterprises (businesses that are exceptionally dependent on digital technologies, including security, banking/finance, e-commerce, and data warehousing and management). Our own analysis of the same constellation of digital demands, from desktop to data center, and from factory to chip fab, led us to much the same conclusion as EPRI's, a couple of years earlier[78] (Table 14).

Table 13 Digital Demand (EPRI)[79]

Sector	TWh/year	Percent Share of Sector
Residential	150	13
Commercial	148	13
Industrial	93	9
Total U.S.	391	12

Table 14 Digital Electricity Demand[80]

Source	Percent of U.S. Electricity
Networks	~ **1–2**
Wired	~ 0.2–0.5
Wireless	~ 0.2–0.6
Data centers	~ 0.2–0.4
Factories	~ **2–6**
Chip fabs	~ 0.5
All other digital manufacturing	~ 2–6
Desktops	~ **6**
In offices	~ 2.6
UPS systems	> 0.5
In homes	~ 1.4
Cooling all of the above	~ 1.5
Total	**9–14**

Focusing only on the narrower universe of the digital desktop, the DOE's Energy Information Administration (EIA) periodically estimates that about 1.4% of national electric demand comes from PCs and ancillary hardware (e.g., printers, monitors) sitting on desks in homes, and 2.6% from machines in offices. (These approximations depend heavily on estimates about usage patterns—how many hours a day our computers, monitors, printers, and other desktop hardware are left running.) In its breakdown of residential consumption of electricity, the EIA also notes cryptically that the other "(e)lectronics, which include audio/video devices and PC add-ons such as scanners and printers, are estimated to account for 10% of all residential electricity use," or about 3% of total *national* electric use.

Cooling and backup power requirements add another, generally ignored, yet significant component to these totals. They add not only to the absolute magnitude of demand associated with digital loads, but frequently in the former case to yet another critical node. Indeed, a number of computing facilities within factories or offices have been forced to shut down during power failures despite the successful operation of a facility's UPS and backup generators—because the UPS and backup system was not installed to operate the air-conditioning system that cooled the computing room. Rapid overheating lead to the need for manual shutdown of the computing.

Several years ago, a British study was the first to note the related anomaly in the commercial sector—efficiency of conventional electric demand was rising rapidly (lighting, standard equipment, air conditioning)—but overall commercial building electric (and cooling) demand just didn't fall (it rose).[81] The study authors attributed it to the rising direct demand from digital equipment, including rising cooling demand. Recent data for Manhattan suggests a similar trend. Compared to the booming 1990s, despite the cooler economy and much higher office vacancies (15% vs. 5%), ConEd reports electric growth now running 25% higher in the past year or two—attributing demand to greater use of digital hardware, and to the greater use of air conditioning.[82]

Wireline and wireless communications infrastructures, together with large data centers, account for another 1–2% of total demand. The fastest-growth sector of demand in the EIA commercial building statistics, for example, is a miscellaneous grab-bag category called "other" loads that "(i)ncludes miscellaneous uses, such as service station pumps, automated teller machines, telecommunications equipment, and medical equipment"—this category now accounts for a remarkable 36% of all commercial building electric use (12% of national demand). The EIA doesn't parse these numbers further, but given the extensive list of equipment that is explicitly counted outside of "other," it seems likely that "telecommunications equipment" is a very important component here.

Finally, the manufacturing of digital equipment accounts for another 2–6% of total electric demand. In 1998, the U.S. Environmental Protection Agency (EPA) had estimated the nation's fabs (microprocessor fabrication factories) accounted for almost 0.5% of national electricity consumption. That number has surely grown since then. The 100 big semiconductor fabs in the United States operate at typical 10- to 20-MW loads per fab, with $1 million/month electric bills. And the manufacturing of information technology extends far beyond the fabs—in fact the fabs almost certainly rank last and least in the digital manufacturing sector's demand for power.

The macroeconomic statistics lend support to these estimates. The U.S. Department of Commerce estimates that the information portion of the economy accounts for at least 8% of the GDP.[83] The Federal Reserve estimates that information technology accounts for 20–60% of GDP growth.[84] And as a general rule, every 1% point of GDP growth drives a 0.7–1% point of kWh growth according to the EIA.[85] Thus, at the macroeconomic level, information technology—if it uses its proportionate share of energy—would appear to account for at least 8% of all *energy* use and a disproportionately higher share of *electric* use.

It is impossible to estimate with any precision how much economic leverage to attribute to these all-digital loads. What is intuitively clear, however, is that securing power supplies to digital loads generally is a high priority, because so much else cannot continue to function when the microprocessors and digital communications systems shut down.

Hard Power

The public grid's inherent vulnerabilities have been noted before. Its architecture—one of relatively small numbers of huge power plants linked to millions of locations by hundreds of thousands of miles of exposed wires—has been frequently criticized, though generally by those who do not understand either the technology or the economics of power production. What is equally clear, however, is that the public grid's architecture is exceptionally efficient. Power plants are thermal machines, and in the thermal world, bigger is almost always much more efficient than smaller. Today's 180-MW frame turbines now attain almost 60% thermal efficiency; a 30-kW microturbine attains 26%. Utilities running huge turbines produce 3¢/kWh electrons; in the best of circumstances, microturbines can perhaps generate 15¢/kWh power. Even as oil prices have gyrated, the average retail price of utility-generated power has fallen 10% since 1990, and wholesale prices are in virtual free-fall. This means that most of the demand will inevitably continue to be satisfied by grid power.

But even before 9/11, it had become clear that the digital economy requires much more than the grid alone can deliver. Utilities have traditionally defined an "outage" to be an interruption of 5 minutes or more. But the Information Technology Industry Council (ITIC) in the guideline known as the "ITIC curve" defines a power "failure" as any voltage that falls below 70% of nominal for more than 0.02 seconds, or below 80% of nominal for more than 0.5 seconds. Additional parameters address voltages below 90% of nominal for more than 10 seconds, and over voltage conditions that can (all too easily) fry sensitive electronics. The "brownout"—a grid-wide reduction in voltage—is the utility's first response to generating capacity shortages. But a brownout that merely dims bulbs can shut down digital equipment.

The challenge and the opportunity for both public and private planners are to address the new critical-power challenges in ways that solve both problems. Mounting threats from the outside give increasing reason to question the grid's reliability in any event. At the same time, every significant node in the digital economy defines a point of rising demand for power that is exceptionally clean, reliable, and sustainable—far more so than the grid can ever deliver, even in the absence of any threat of deliberate attack. In a recent survey, security directors at leading U.S. businesses ranked the threat of terrorism among their top five concerns, but few expected to see any increase in their budgets in the next few years.[86] The typical view is that security issues "won't generate revenue, they'll consume capital." The pressure to reduce costs exceeds the pressure to improve security.

With power, however, these two objectives can often be complementary. For many in the digital economy, grid power is inadequate in any event; this is why there has been so much investment already in backup generators, uninterruptible power supplies, and backup batteries. The challenge going forward is to extend investment in power *quality* and *reliability*—which many businesses need and are undertaking in any event—to assure *continuity* of operation in the event of larger and longer interruptions in grid-supplied power.

RESILIENT POWER

Airports have their individual towers, but the flow of commercial aircraft at altitude is controlled by "National Airspace System" (NAS) facilities—about two dozen regional control centers, together with another 19 air-traffic-control hubs co-located in airport control towers at the nation's busiest airports. For obvious reasons, all are deemed "critical." And a typical control tower requires about 200 kW of power to stay lit; the major centers need about 500 kW.

The Federal Aviation Administration (FAA) has long recognized that it "cannot rely solely on commercial power sources to support NAS facilities. In recent years, the number and duration of commercial power outages have increased steadily, and the trend is expected to continue into the future."[87] The FAA has thus deployed extensive critical-power backup facilities—in the aggregate, some 9000 batteries, 3000 generators, and 600 UPS systems. The Agency has deployed some 2000 kW of diesel generators, for example, in double-redundant architectures.

In a year-2000 report,[88] the Agency nevertheless recognized that the extensive backup facilities it does have are seriously insufficient and out of date; a major upgrade is now underway. All of its principal power systems are being equipped with high-resolution sensors, linked via Ethernet networks. A follow-up report in February 2003 reviews the capital investments still required, and stresses that further upgrades are urgently needed.[89] The Agency estimates that it would require approximately $100 million annually to replenish the entire inventory of its backup power systems every 15–20 years.

Tiers of Power

As noted above, the electrical grid is a multi-tiered structure (see Figure 4). Architecturally similar arrays of generators, wires, switches, and transformers appear within each of the grid's principal tiers. The generation and transmission tiers at the top have stadium-sized, gigawatt-scale power plants and commensurately high-voltage wires, building-sized transformers, truck-sized capacitors, and arrays of mechanical and electromechanical relays and switches (Figure 14). The distribution tiers in the middle have tennis-court-sized, megawatt-scale substations, van-sized transformers, and barrel-sized transformers mounted ubiquitously on poles and in underground vaults (Figure 15). The bottom tiers transform, condition, and distribute power within factories, commercial buildings, and homes, via power-distribution units, lower-voltage on-premise grids, and dispersed switches, batteries, and backup systems further downstream (Figure 16).

In the most primitive architecture, the grid is just power plant and wires, no more. Power is generated at the top tier, consumed at the bottom, and transported from end to end by a passive, unswitched, trunk-and-branch network. This was the structure of the very first grid, from Edison's Pearl Street Station in New York, in 1882. The higher up things fail, the more widely the failure is felt.

The modern grid is, of course, much more robust. Many different power plants operate in tandem to maintain power flows over regions spanning thousands of miles. In principal, segments of the grid can be cut off when transformers fail or lines go down, so that failures can be isolated before they propagate to disrupt power supplies over much larger regions. (The effectiveness depends on the level of spending on the public grid, which has been in decline for years.) Identical

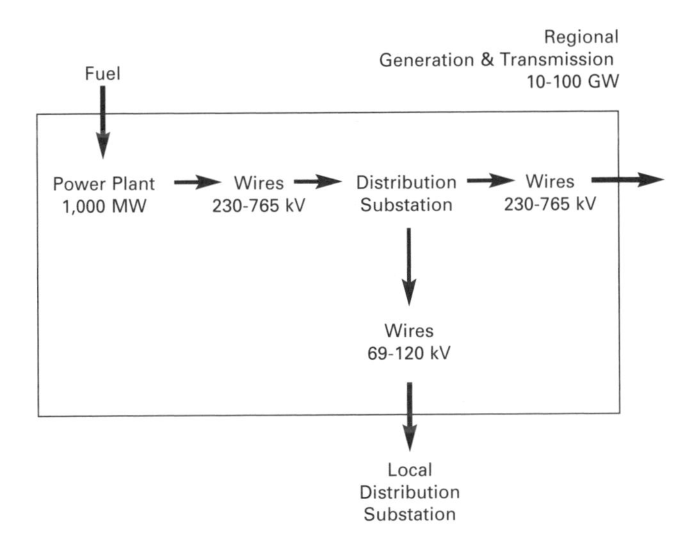

Figure 14

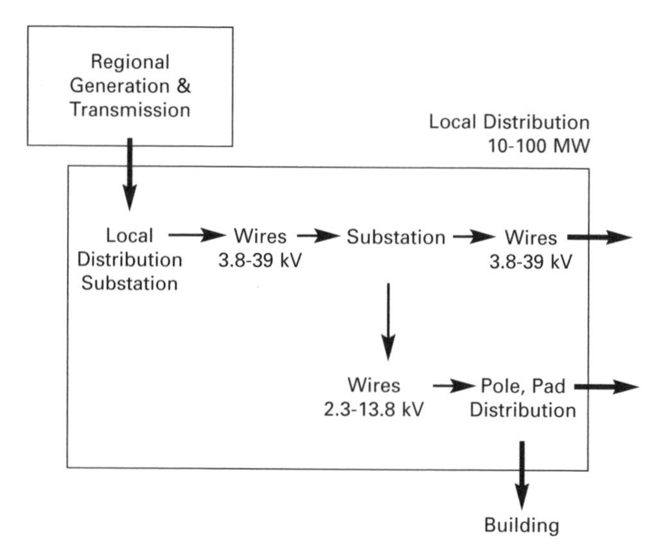

Figure 15

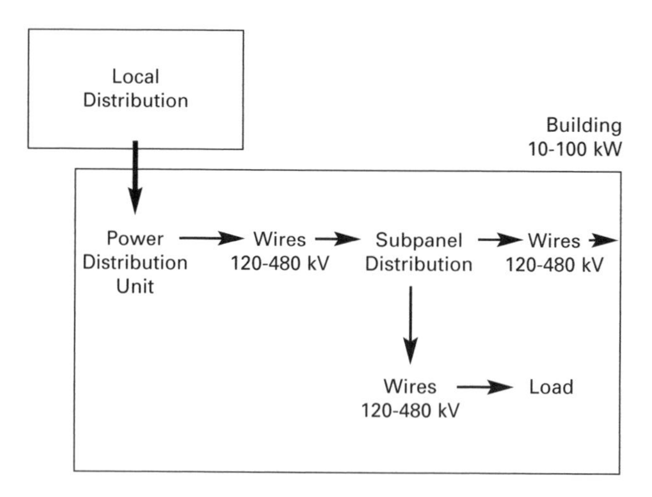

Figure 16

strategies of isolation and redundancy are used on private premises, to make the supplies of power to critical loads much more reliable than the grid alone can be counted on to deliver.

Switches control the flow of power throughout the grid, from power plant down to the final load. "Interties" between high-voltage transmission lines in the top tiers allow even the very largest plants to supplement and backup each other. Distributed generation facilities located in the middle tiers can power smaller segments of the grid, and keep them lit even when power is interrupted in the highest tiers. When power stops flowing through the bottom tiers of the public grid, critical-power

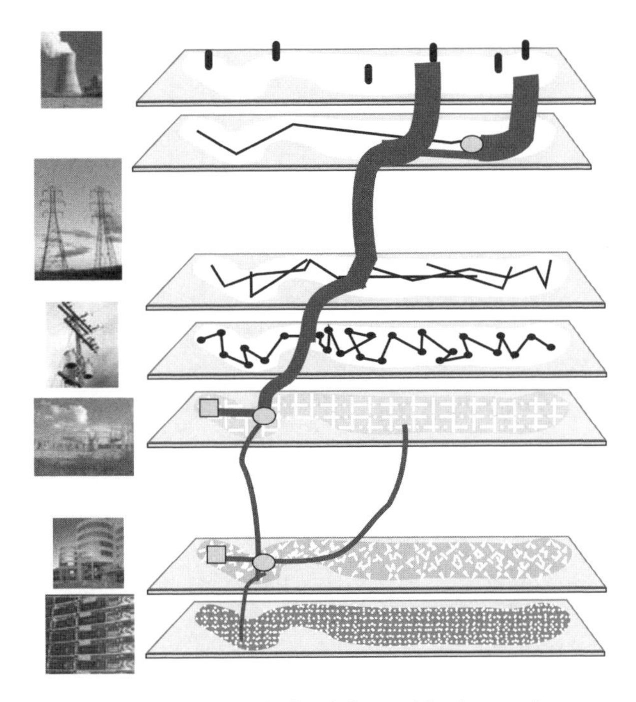

Figure 17 Adding "Interties," switches, and local generation.

circuits on private premises are isolated and private, on-premises generators kick in (Figure 17).

The first essential step in restoring power after a major outage is to isolate faults and carve up the grid into smaller, autonomous islands. From the perspective of the most critical loads, the restoration of power begins at the bottom, with on-site power instantly cutting in to maintain the functionality of command and control systems that are essential in coordinating the step-by-step restoration of the larger whole.

Adding Logic to the Grid: The Static Transfer Switch

The switches that perform these functions must operate very fast. As discussed earlier, a "blackout" for digital equipment is any power interruption that lasts more than a few tens of milliseconds. Such speeds are beyond the capabilities of an electromechanical switch, however (Figure 18). The most critical switching function must therefore be performed by a high-speed, solid-state device—a "digital" switch that can open and close circuits fast enough to maintain the flow of smooth, seamless power to digital loads. This device—the "static transfer switch" (STS)—thus plays an essential role in mediating between alternative sources of power. It opens and closes power circuits faster than sensitive loads can discern, and faster

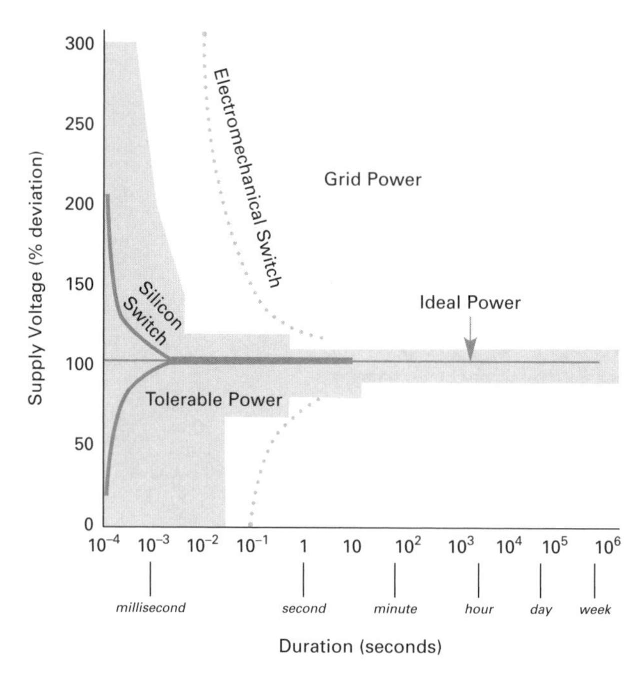

Figure 18 Adding fast switches.

than destructive harmonics can propagate, providing local isolation, selectively, and on command.

An STS is typically built around a silicon controlled rectifier (SCR)—a fat, 6-in., 1500-micron-thick silicon wafer, that does for power what a Pentium does for logic; the device is called "static," ironically, because it has no moving mechanical parts, and can thus open and close extremely fast. High-power silicon-based transfer switches were introduced in 1971, but the first such device with fully integrated microprocessor controls did not become available until 1994 (Figure 19).

An STS monitors the availability and quality of power from two (or more) independent sources of power, and instantly flips to a secondary feed upon detecting a sag or similar problem in the primary. The STS thus contains high-speed sensing and logic that allows switching behavior to be tailored, in real time, to meet the different optimum opening and closing of a circuit (a major factor in preventing upstream and downstream problems). Finally, an STS will typically incorporate redundant design features to ensure high, intrinsic reliability, and sensors to monitor conditions both inside the switch itself, and in the surrounding environment. An internal clock and on-board memory will log power events, and a communications channel (typically fiber-optic) provides links between the logic and the electronics that drive the switch, as well as with UPSs and off-site control systems.

Figure 19 Static transfer switch. *Source*: Danaher Power Solution.

Rack-mounted units can handle up to several kilowatts, while much larger car- and truck-sized devices route major power feeds running as high as 35 MW, and at 15, 25, and 35 kilovolts (kV). These devices can be configured for stand-alone operation. They may switch as needed between two or more primary "hot" power sources—two different grid feeds and a generator for example. Or they may stand alongside (or be integrated into) UPSs to coordinate power hand-offs among redundant UPS arrays when one unit fails, or to enable "hot swap" maintenance. Rack mounted solid-state switches can perform similar functions directly upstream of end-use devices—servers or telecommunications switches, for example—to select automatically among redundant power feeds so that the failure of any one feed is always invisible to the final load. And at ultra-high power levels—up to 100 MW—enormous arrays of solid-state switches are now being used to interconnect and isolate high-power transmission lines at about 50 grid-level interconnection points worldwide.

Coupled with large capacitors and power-conditioning high-power electronics, complex arrays of solid-state switches become a UPS. When grid power sags or fails altogether, they draw power from alternative sources; they likewise filter out surges and spikes in grid power that may be created by (to pick just one example) lightning strikes on transmission lines.

Generation and Transmission

Much of the critical-infrastructure literature refers to the grid as a single structure, and thus implicitly treats it as "critical" from end to end. But as discussed earlier, utilities themselves necessarily prioritize and rank the customers and loads they are expected to serve. Large power plants and high-voltage underground cables that serve densely populated urban areas obviously require more protection before they fail, and more urgent attention after, than small plants and rural distribution lines. In defining priorities and deploying new facilities, collaboration between utilities and critical-power customers is becoming increasingly important. Most notably, power is critical for the continued provision of other critical services—those provided by E911, air traffic control, wireline and wireless carriers, emergency response crews, and hospitals, among others.

The hardening of the grid begins at the top tier, in the generation and transmission facilities (Figure 20). Much of modern grid's resilience is attributable to the simple fact that interties knit local or regional grids into a highly interconnected whole, so that any individual end-user may receive power from many independent power plants, often located hundreds (or even thousands) of miles apart. Promoting development of this resilient architecture is the primary mission of NERC.

Thus, for example, New York's Marcy substation was upgraded in 2002 with high-power silicon switches that boosted the capacity of existing transmission wires by 200 MW.[90] California is now studying the feasibility of adding additional local interties to a major line that already runs from Oregon through Nevada. And after a 10-year wait for multi-state approvals, American Electric Power recently received permission to build a new 765-kV line linking a power-rich site in West Virginia to the power-hungry Virginia loop.

Initiatives like these were being pushed long before 9/11, because rising demand and increasingly strained supply were causing alarming increases in transmission congestion "events" (Figure 21). New interties provided an effective way to boost overall reliability and create capacity margins without building new plants; new interties also facilitated the wholesale trading that regulators had authorized in the 1990s. So long as there is sufficient margin in the generating capacity and redundancy in wires, and sufficiently fast and accurate control of key switches that

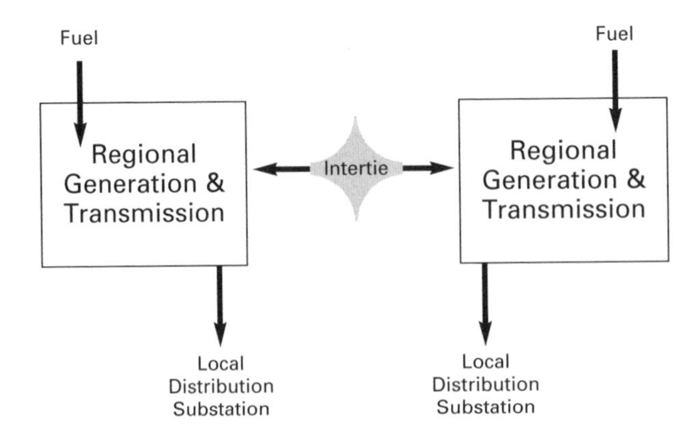

Figure 20

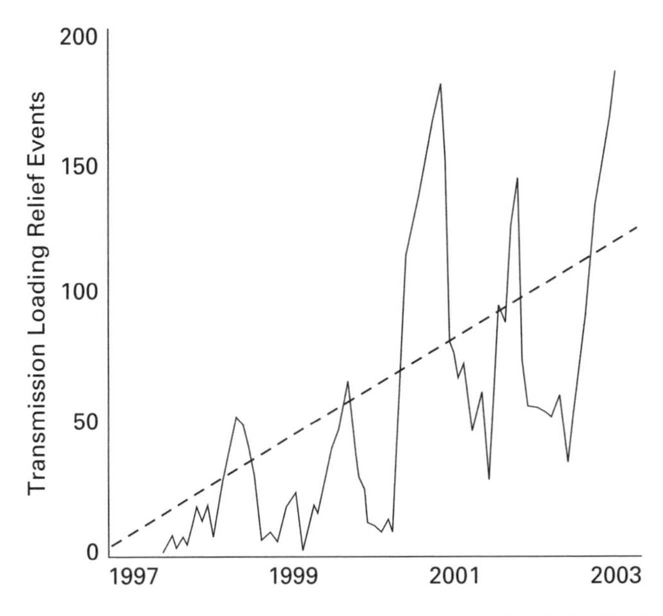

Figure 21 Electric grid congestion. *Source*: North American Reliability Council (NERC) Transmission Loading Relief Logs.

route power and isolate faults, the failure of any one power plant or line in the top tiers of the grid should never be discerned by end users at the bottom.

After 9/11, NERC expanded their recently created Critical Infrastructure Protection Advisory Group (CIPAG) to bring together the various public utilities responsible for securing the 680,000 miles of long-haul, high-voltage wires and the 7000 transmission-level substations.[91] Among other initiatives, CIPAG has formed a working group to inventory and develop a database of "critical spare equipment." The high-voltage transmission system uses high-power hardware—massive substation transformers, for example—that is often custom-built. Spares, if they exist at all, are rarely close at hand. Much can thus be done to assure overall continuity of operation through the intelligent sharing of standby assets.

Complementary discussions are addressing the possibility of creating a critical-equipment warehousing system, with spares warehoused at geographically dispersed locations, and the costs shared by the many potential beneficiaries of such planning. This solution has already been implemented for power line ("telephone") poles. As discussed further below, fleets of generators-on-wheels are now evolving as well.

Distribution and Distributed Generation

Very large end-users rely on similar intertie strategies one tier lower down in the grid to help secure their specific critical-power needs. At AOL's campus in Prince

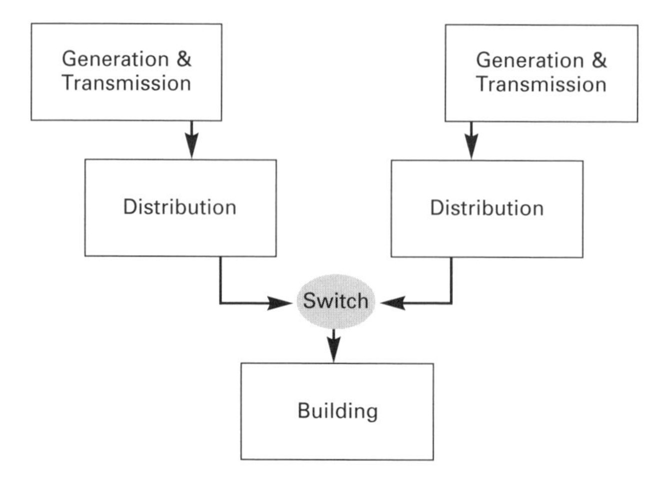

Figure 22

William County, Virginia, for example, a dedicated substation takes two separate 115 kV feeds from the grid. AOL also deploys backup generators of its own, but when critical-power loads are located in urban areas where on-site generation is infeasible, redundant connections to separate, electrically independent points on the grid are often the only practical approach for improving reliability (Figure 22)

In such configurations, the key "switch" controlling the dual feeds will often be a dedicated utility substation located on the doorstep of a factory, office park, or data center. The two separate high-voltage feeds to AOL's campus, for example, lead to a dedicated substation that steps the voltage down to 24 kV through redundant transformers, and feeds the power to two 25-kV sets of switchgear. By deploying additional substations in close collaboration with major critical-load customers, utilities shrink the footprint—that is, reduce the number of customers affected—by failures that occur elsewhere. And more substations create more points at which to interconnect independent parts of the grid, so that distant transmission lines and power plants effectively back each other up.

Substations can also serve as sites for utility deployment of distributed generation. With the addition of its own generating capacity, the substation is "sub" no longer–it becomes a full-fledged "mini-station" (Figure 23). Opportunities for deploying new generating capacity at this level of the grid—either permanently or when emergencies arise—are expanding as large electromechanical switches and related components are being replaced by new solid-state technologies (described in more detail below) that have much smaller footprints.

Utility-scale "generators on wheels"—either diesels or gas turbines—offer an important option for deployment in emergencies. Some substations already play host to small parking lots worth of tractor-trailers, each carrying 1–5 MW of generators powered by Cummins or Caterpillar diesel engines (Figure 24). Cummins' Kawasaki turbines, GE's Energy Rentals division, and Caterpillar's Solar Turbine

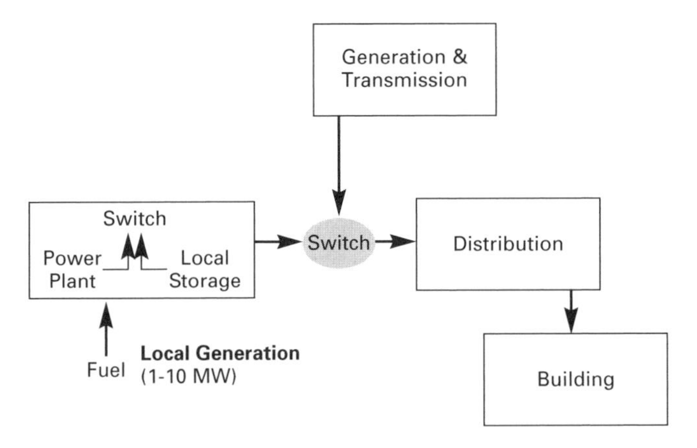

Figure 23

Figure 24 Substation diesel gensets. *Source*: Cummins Power Generation.

division offer mobile generating capacity in larger (10–25 MW) increments, powered by aeroderivative combined-cycle gas turbines, typically housed in a group of three or four trailers. Even larger turbines are being mounted on barges, for quick deployment at substations located near navigable waters and with suitable gas lines close at hand. Some 74 turbines, with 2000 MW of capacity, already float on power barges around Manhattan.

For the longer term, the DOE and many utilities are examining other possible sources of substation-level generation and storage. Large fuel cells present a potentially attractive alternative to gas turbines because they operate silently and with very low emissions.[92] Utilities have even tested megawatt-scale arrays of batteries for load-leveling and backup. In one trial a decade ago, Southern California Edison and other utilities assembled 8256 telecom-type lead–acid batteries in a massive 10 MW array, 50 miles outside of Los Angeles; the idea was to store grid power during off-peak hours of demand, and feed it back into the grid as needed. TVA is scheduled to bring on line this year a massive 12 MW flowing battery-type system called Regenesys (the electrochemistry is based on sodium bromide and sodium polysulfide); it can store 200 MW-hours of energy in its 170,000-ft^2 footprint.[93] As discussed further below, many of these studies (including two outstanding DOE reports published in 1999 and 2002)[94] have focused either on environmental objectives, or on smoothing out demand to lower prices by making more efficient use of the grid and other utility resources.

By such means, much can be (and is being) done to lower the likelihood of a loss of grid power needed by the most critical loads. As discussed further below, closer collaboration between utilities and their largest customers is now needed to advance such initiatives in the distribution tiers of the grid. Nevertheless, as discussed above, the grid is inherently frail, and there is only so much that feasibly can be done to make it less so. Mushrooming arrays of computer and telecom equipment in switching and data centers increasingly strain the utility's ability to provide sufficient quantities of power even when plants and the grid are all functioning normally. Guaranteeing supplies of critical power at locations like these ultimately means adding on-site generating capacity and storage to back up whatever is being done to improve reliability higher in the grid.

On-Site power

On-site power begins with on-site supplies of stored electrical, mechanical, or chemical energy. Engines, generators, and suitable arrays of power electronics are then required to generate power and condition it into suitable form (Figure 25).

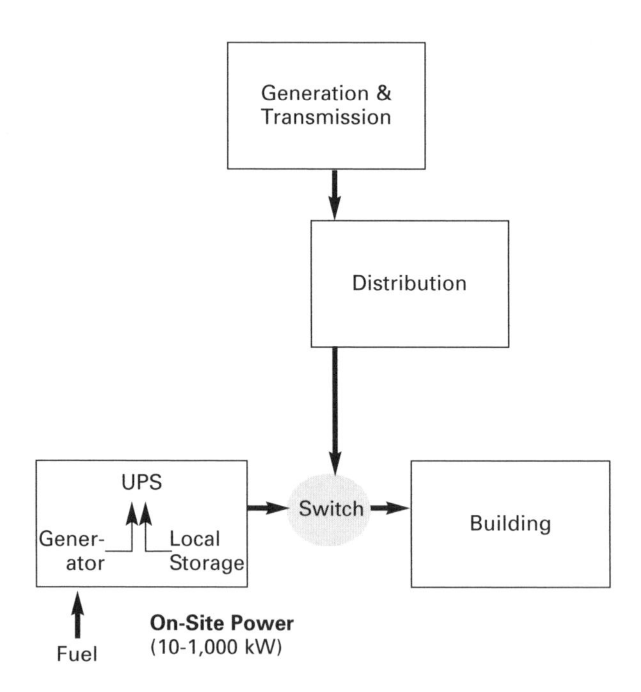

Figure 25

As noted, the boundaries between grid and on-site power are beginning to blur. Utilities build dedicated substations on the doorsteps of large customers, and may deploy distributed generation facilities at grid-level substations that serve the most critical loads. And while private contractors do most of the planning and installation of the facilities discussed below, utilities themselves are now actively involved in deploying on-site power to help guarantee what the grid alone cannot. According to recent surveys, almost 40% of utilities are now offering backup-power systems to commercial and industrial customers; most of the rest plan to begin offering such product/service contracts within the next few years, in collaboration with genset, microturbine, and fuel-cell manufacturers.[95]

A large technical literature already addresses the elements of on-site power systems. The IEEE's Recommended Practice for Emergency and Standby Power Systems for Industrial and Commercial Applications provides comprehensive technical and engineering evaluations of, and guidelines for, engineering, reliability, and operational aspects of on-site power systems. The National Fire Protection Association's (NFPA) Standard for Emergency and Standby Power Systems provides a comprehensive technical analysis of on-site power alternatives.[96] The first edition of the Standard emerged after the NFPA convened a Technical Committee on Emergency Power Supplies in 1976, to address "the demand for viable guidelines for the assembly, installation, and performance of electrical power systems to supply critical and essential needs during outages of the primary power source." The 2002 edition details hardware choices for batteries, generators, fuel tanks, cable connectors, switches, fuses, controls, mechanical layouts, maintenance, testing, and virtually all other aspects of hardware options, installation requirements, and operational capabilities.

Some part of the growth in demand for on-site power can be attributed to lack of adequate transmission-and-distribution capacity. With suitable engineering of the public-private interfaces, private generators not only reduce demand for power, they can also feed power back into the public grid. Some additional part of the rising demand for on-site power is attributable to growing interest in "alternative fuels" (such as fuel cells and photovoltaics) and co-generation. Certainly there has been a long-standing recognition of the role of on-site power for emergencies (caused, for example, by natural disasters) as is implicit in the NFPA document noted above. But in the aftermath of 9/11, the principal imperative for deploying on-site power is to assure continuity of critical operations and services.

Stored Energy

Energy can be stored electrically (in capacitors, for example), electro-chemically (batteries), mechanically (flywheels) and chemically (diesel fuel). By and large, however, these storage technologies fall into two groups that perform distinct functions (Figure 26). Batteries, flywheels, and ultra-capacitors are mainly "ride-through" technologies. They store quite limited amounts of energy, but can provide power quickly, to cover sags and outages that run from milliseconds to minutes or

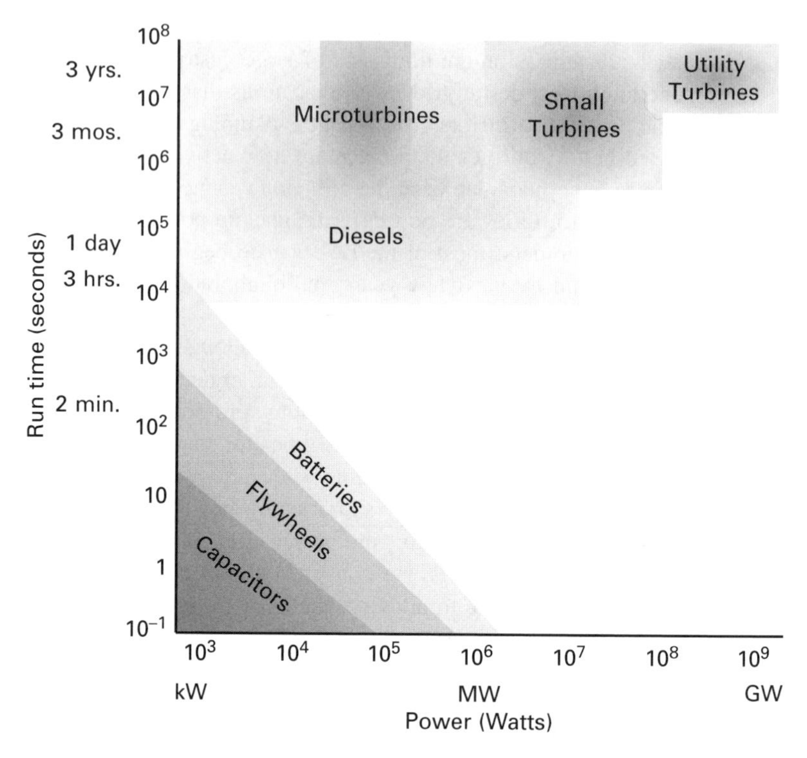

Figure 26 Generating and storing electricity. *Source*: Digital Power Group.

(at the outside) hours. Liquid fuels, by contrast, store large amounts of energy—enough to run backup generators for much longer periods. Diesel fuel and backup generators can thus provide what ride-through technologies cannot—critical-power "continuity" when grid power fails for many hours, days or longer.

Batteries

Rechargeable batteries remain the overwhelmingly dominant second-source of power. Batteries are widely used to provide ride-through power for the most common grid interruptions and outages, which is to say, the relatively short ones.

Portable devices rely on the exotic, expensive and often unstable battery chemistries of lithium, cadmium, nickel, silver, and zinc. These materials offer very high power densities—so the batteries are commensurately compact and light—but they are very expensive. They store far more energy per pound, but far less per dollar. Lead and acid are comparatively heavy and cumbersome, but they are also affordable, and no other battery chemistry has yet come close to beating them for all around utility (Figure 27).

"Flooded" lead–acid cells store about 20% more power than other lead-acid designs, but they vent hydrogen and oxygen (an explosive mixture), and have to be

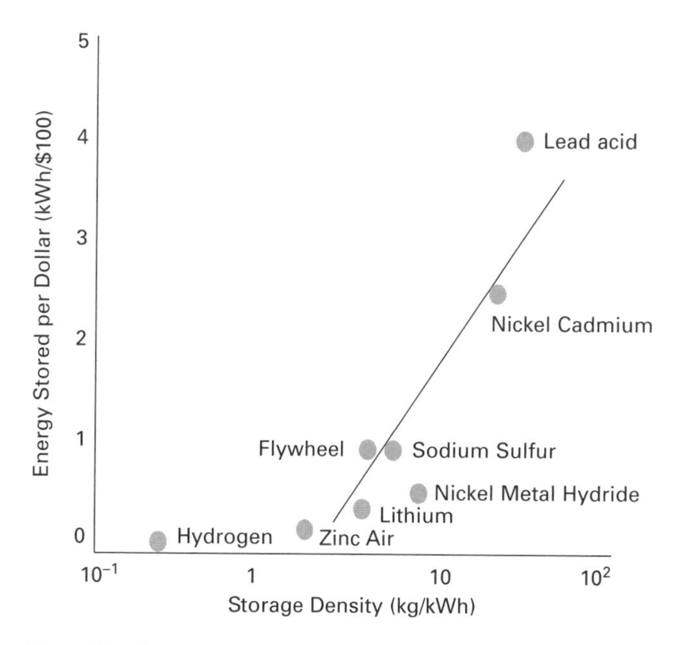

Figure 27 Cost to store grid power. *Source*: "Metal Fuel Cells," *IEEE Spectrum* (June 2001); "Portable Power," Buchmann, Cadex Electronics (2000); "Exoskeletons for Human Performance Augmentation," DARPA Workshop (1999), Defense Sciences Office; Electricity Storage Association.

watered periodically, either manually or (increasingly) by way of automated systems. Absorbed-glass-mat and gel-based batteries run sealed—they recombine the electrolytic gases within the battery, rather than vent them. These low-maintenance units can go pretty much anywhere safely—in enclosed cabinets, office environments, and basements (Figure 28).

Figure 28 Telecom backup batteries. *Source*: Enersys.

Battery manufacturers continue to pursue new chemistries. The high-temperature sodium–sulfur battery is a candidate to fill the niche between lead–acid and expensive lithium for high-energy storage. In late 2002, American Electric Power deployed the first U.S. commercial installation; a 100-kW system with two massive batteries able to provide seven hours of run time in about one-third the floor space of a lead–acid array.[97] But neither sodium–sulfur nor the sodium–bromide batteries noted earlier, are yet close to matching the venerable lead–acid technology's economical performance or the high reliability that comes with a huge market and a long history of operation.

Flywheels

For the most part, flywheels are even more limited than batteries in their practical ability to address the energy deficits created by any extended grid outage. They can and do substitute for batteries, however, to ride-through the short gap between the time when grid power fails and backup generators get up to speed. One of the commercial flywheel-based ride-through power units on the market today consists of two 600-pound steel flywheels, stacked vertically, and spinning silently in a vacuum at 7700 rpm; the wheels are connected to integrated motor/generators on a single shaft, with an array of electronics beyond the generator that converts the highly variable power generated by the spinning flywheel to the steady AC or DC outputs required by the loads.[98] Such designs can produce as much as 250 kW of power for 15 seconds, or 1 kWh of total energy.

Another emerging configuration uses a blend of static (battery) and active (flywheel) backup systems, with the flywheel used to handle the shortest and most frequent power dips (which range from milliseconds to fractions of a minute), with the batteries taking over for somewhat longer outages. In this configuration, the flywheel's main function is to extend the life-span of the batteries, which are substantially degraded when they are required to respond repeatedly to short-term dips in grid power.

Ultracapacitors

Ultracapacitors are increasingly being used to perform a similar function—to ride-through the gap between the failure of grid power and the start-up of a backup generator. The ultracapacitor's energy storage and performance characteristics are very similar to a flywheel's, but these devices contain no moving parts. Through advanced thin-film technologies, micro-material engineering, and automated production lines, a handful of manufacturers now make one-pound, soda-can-sized capacitors that deliver 2500 farads of capacitance (thousands of times more than conventional capacitor technology), and both price and size continue to fall steadily. Arrays of such capacitors are used, for example, to provide ride-through power in 10-kW computer servers. Once again, ultracapacitors don't eliminate batteries in most applications, but they can greatly improve battery functionality and expand battery markets overall.

Diesel Fuel

With or without the assistance of complementary flywheels and ultracapacitors, batteries store far less energy per unit of volume or weight than liquid hydrocarbon fuels. The battery banks in telecom central offices, which typically provide the office with a reserve time of four to eight hours, push the outer limits of battery backup. Cell tower base stations were originally designed around four-hour battery backup systems, but loads have risen to the point where few towers can run longer than an hour when grid power fails, and an hour of run time is achieved only if the batteries have been well-maintained. With rare exceptions, few facilities can economically rely on battery power for even that long; batteries are too bulky and too expensive to provide power for the even longer grid outages that must now be contemplated in critical-infrastructure planning (Figure 29).

Generators powered by liquid fuels will therefore play the key role in maintaining continuity through major grid outages. They can do so because liquid fuels store huge amounts of energy in very small volumes, because these fuels power the transportation sector for just that reason, and because the United States therefore has in place a huge, distributed infrastructure of trucks and tanks that transport and store primary fuel in quantities sufficient to keep key electrical loads lit for weeks or more. The far-flung, highly distributed infrastructure of diesel storage tanks is effectively invulnerable to the kinds of catastrophic failures that could incapacitate major power lines or gas pipelines.

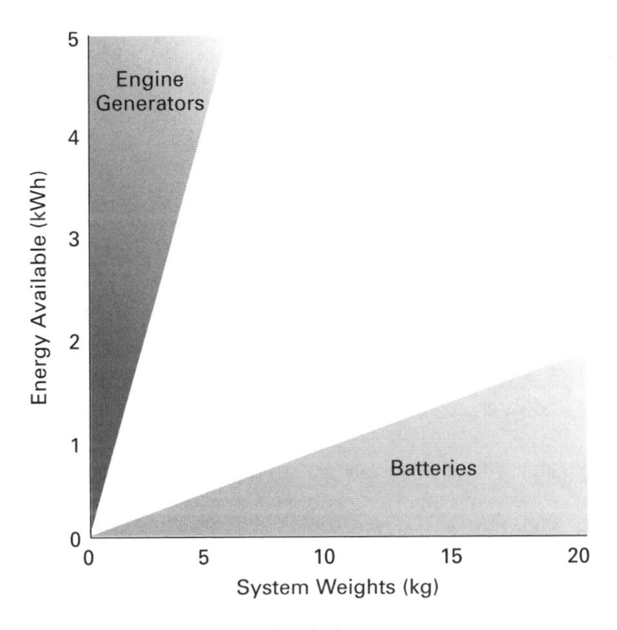

Figure 29 Storing on-site electrical energy.

Most backup diesel generators burn distillate fuel oil, the same fuel used for heating, and for aircraft. Trucks account for about half of U.S. distillate fuel consumption, and distillate fuel storage tanks are therefore already dispersed wherever trucks travel. Nearly 200,000 filling stations, for example, have underground storage tanks and many store diesel fuel specifically. Aviation accounts for another one-third of U.S. consumption; some 18,000 small (and large) airports thus provide a second major tier of dispersed storage. In addition, some 400,000 commercial buildings and eight million homes have storage tanks for their heating oil. And the United States has some 9000 regional oil distribution centers.[99] Many diesel generators are also configured for dual-fuel operation, and can thus also burn natural gas or propane, and stores of these fuels are extensive and dispersed, as well.

A standard 275-gallon residential heating oil tank, for example, contains enough energy to generate 4000 kWh, or some 40 times as much as a comparable volume of lead–acid batteries. And fuel tanks can be refilled by truck. If (say) 10% of the nation's electric loads are viewed as "critical," then one week's supply of our national fuel oil consumption would provide roughly one month of critical power.

Backup Generators

Backup diesel generators located in basements and parking lots, and on rooftops, now account for some 80,000 MW of backup generating capacity deployed near critical loads across the United States. There are, by contrast, only about 100 MW of natural gas-powered microturbines and 25 MW of fuel cells (100 units) installed. Diesel gensets are available in a wide range of sizes; on balance, they are cheap, readily available, and reliable, and modern designs run remarkably efficiently and cleanly *(Table 15)*. While all of these technologies are improving year by year, the marketplace has made clear that the diesel genset remains, by a wide margin, the most practical and affordable alternative for most backup applications.

Table 15 On-Site Power Sources

Type	Power (kW)	Cost ($/kW)	Efficiency (%)	Emissions (lb NO-/1000 kWh)	Fuel
Reciprocating engine	10–10,000	300–800	25–50	17/6	Oil/gas
Aeroderivative turbine	10,000–60,000	500–1700	30–60	1	Gas
Microturbine	20–100	900–1800	25–30	0.3–0.5	Gas
Fuel cell	10–250	5000–10,000	50	0.02	H (gas)

Source: "Using Distributed Energy Resources," www.eren.doe.gov/femp/

Diesel Gensets

To provide critical power for longer periods, the backup system of choice is the stand-by diesel generator. Sized from 10s to 1000s of kilowatts, diesel gensets can provide days (or more) of backup run time—the limits are determined by how much fuel is stored on-site, and whether or not supplies can be replenished during an extended outage (Figure 30).

Diesel generators are strongly favored over other options because they strike the most attractive balance between cost, size, safety, emissions, and overall reliability. Large power plants generate much cheaper power with coal and nuclear fuel, but only because they are very large, and are generally built where land is cheap, far from where most of the power they generate is consumed. They therefore depend on a far-flung grid to distribute their power—and the grid is exposed, and therefore vulnerable. Large gas turbines are attractive alternatives that are already widely used by utilities, but they come in 15- to 150-MW sizes—far too big for all but a tiny fraction of critical loads that require on-site power. And they are no more reliable than the gas pipelines that deliver their fuel.

For these reasons, diesel gensets under 3 MW have emerged to define a huge installed base of generating capacity (see Figure 6). Thus the FAA as earlier noted, for example, relies on nearly 3000 diesel generators to ensure back up power for its air traffic control centers, and tens of thousands of other diesel gensets are used to backup airport towers, hospitals, military bases, data centers, and other critical-power nodes. If both of the two primary grid feeds fail at AOL's campus in Prince William County, Virginia, the substation turns to power from a 13-unit string of 2 MW diesel generators, sitting on five days of fuel oil. Company-wide, AOL alone has 74 MW of backup generating capacity at its current facilities, with 26 MW destined for facilities under construction. In response to serious problems created by a power outage caused by Tropical Storm Isidore in 2002, Jefferson Parish, Louisiana (as just one example of such actions) approved installation of 17 diesel generators at its drainage pump stations.

Figure 30 Diesel genset. *Source*: Cummins Power Generation.

As noted above, power plants-on-wheels are of increasing interest to critical-power planners, because they offer the economies of sharing in much the same manner as fire engines, rescue vehicles, and the "critical spare equipment" program being developed by utilities. Companies like Cummins and Caterpillar have built (still relatively modest) fleets of 2-5 MW, trailer-mounted diesels, and these can be used to generate private, on-site power, just as they can be used by utilities for emergency generation at the substation level of the grid. In the aftermath of power outages caused by floods and hurricanes, the Army Corps' 249th Engineer Battalion (Prime Power) installs emergency trailer-mount generators at hospitals and other critical sites. Given the many obstacles—both cost-related and regulatory—that impede permanent deployment of backup generators, fleets of strategically positioned mobile power units offer the most practical assurance of power continuity for many critical applications.

Microturbines

Refrigerator-sized, gas-fired, air-cooled, very-low-emissions microturbines now come in sizes ranging from 30 to 100 kW. Capstone is the dominant player in this power niche. For tens of thousands of small electric loads in urban areas, these gas turbines represent an attractive alternative option for on-premises power. With very high power density they are two to five times as compact as the only commercial fuel cell being sold today. And when running on natural gas, just about as clean. They are lighter and quieter than diesel engines—that is why turbines are used to power aircraft—and they can be configured to run on either gas or diesel fuel. With some exceptions, however, they are not yet price-competitive with diesel gensets for most backup applications.

Fuel Cells

Fuel cells present another attractive, but even longer-term alternative. Their main virtue is that they can run very clean and quiet, and can thus be deployed directly on commercial premises. Two 200-kW ONSI units, for example, situated on the fourth floor of the Conde Nast Building, at 4 Times Square in the heart of Manhattan, power a huge sign on the building's façade. Other units under development by the likes of Ballard and Plug Power span the 5- to 250-kW range. Those under development by Siemens Westinghouse Power, and Atek range from 25 kW to 25 MW. FuelCell Energy and the United Technologies' ONSI build for the 200-kW to 2-MW space. But fuel cells remain, for the most part, a novel, relatively unproven, and comparatively expensive technology, and absent large on-premises gas storage tanks, they cannot offer any more continuity assurance than the gas lines that feed them. They are also quite a lot bulkier than diesels—a complete two-module, 2-MW fuel cell setup from FuelCell Inc. occupies about 4500 square feet; a 2-MW diesel setup, by comparison, requires only a $1200 - \text{ft}^2$ footprint.

Alternative Fuels and Technologies

Other generation technologies run on the much-promoted "alternative" fuels—sun and wind, most notably. But for now, at least, these alternatives are no more reliable than the fuels themselves, or the backup batteries deployed alongside; they are also uneconomical and—because they rely on such thin fuels—require a large amount of space to generate comparatively small amounts of power. With 2000 square feet set aside for on-site power, a diesel generator together with ancillary power conversion electronics and a buried fuel tank can provide a megawatt of power for a week (i.e., 100 MWh of total electrical energy). On the same footprint, a solar array with its essential backup batteries can provide only 1/100th as much power, and at roughly 100 times the capital cost.

Over the long term (and discussed briefly later), a more promising alternative to conventional gensets will emerge with the maturation of hybrid-electric trucks, buses, and cars, offering the possibility of linking the transportation sector's mobile, and highly distributed infrastructure of fuel tanks, engines, and generators directly to residences, small offices, and larger buildings.

"Uninterruptible Power"

Dual feeds to the grid, on-site batteries and generators, and the static transfer switches used to knit them together are all deployed to ensure the uninterruptible flow of power to critical loads. But the definition of the word itself—"uninterruptible"—depends a great deal on what kind of load is being protected. As discussed earlier, even a short loss of power can trigger a much longer interruption of business or manufacturing operations, because of the time it takes to reboot computers and restart machines. The purpose of a UPS is to isolate critical loads from even the momentary power interruptions that occur when grid power fails and batteries, generators, or other alternatives kick in, and to provide intelligent mediation between these alternative and redundant sources of power.

UPS typically refers to an AC device, used mainly to power computers; DC power plants perform a similar function in the telecom world. The power plants come in modular units, up to about 100 kW (DC) and 720 kW (AC). They are deployed in scalable architectures, with multiple units to accommodate loads running as high as megawatts. End-users have deployed some 20 GW of UPS capacity, which represents an aggregate capital investment of over $4 billion (see Figure 5).

A UPS performs two basic, complementary functions. It conditions power continuously, smoothing out the sags and spikes that are all too common on the grid and other primary sources of power. And by drawing on limited reserves of stored energy in large capacitors and on-board batteries, the UPS provides ride-through power, to cover for sags or complete power failures typically for up to (but rarely longer than) about 30 minutes. The power conditioning is performed by large arrays of digitally controlled power electronics; for ride through, the UPS dynamically selects and draws power from grid, batteries, backup generators, and other available sources. High-power applications may require multiple UPSs, and ensuring proper electrical synchronization is then a major technical challenge. Architectures must be designed to permit individual modules to be taken off-line for maintenance without removing the load from the conditioned power. The UPS will also monitor battery conditions and prevent destructive power flows (harmonics) among these various sources of power. Efficiency, economy of operation, and inherent reliability are essential. Sophisticated communications channels link the best UPSs to downstream loads, upstream power sources, and the static transfer switches that perform the higher-power switching functions.

At the heart of the UPS is its power conversion electronics—the silicon-based hardware that converts electricity from one form to another. Only in the past decade or so have the core high-power electronics matured to the point where they can provide cost-effective, efficient, reliable, digital-quality performance. The key enablers were the advent of high-performance high-power power chips, on the one hand, and, on the other, sophisticated, low-cost digital logic to provide intelligent control. UPS efficiencies have risen substantially—a significant factor in itself in terms of both heat management and cost of 24/7 operation. And superior power chips and designs have more than doubled the overall reliability of high-power power-conversion electronics.

In top-of-the-line devices, monitoring and software systems now play as crucial a role as the hardware. The software ensures smooth selection of power sources and load hand-offs. It continuously monitors and diagnoses the state of the grid, batteries, and sources of power, together with the condition of the UPS's own internal electronics. It provides predictive analysis of downstream problems—for example, current leaks that foreshadow the imminent failure of a capacitor or the insulation on a wire. And it provides automated notification and alarms, e-mails, paging, Web-based alerts, interfaces, and so forth (Figure 31).

Figure 31 Uninterruptible power supply. *Source*: Powerware.

Monitoring, Control, and Reliability-Centered Maintenance

Monitoring and maintenance already play a key role in maintaining power reliability, from the gigawatt-scale tiers at the very top of the grid, down to the UPS and individual loads at the very bottom. Such systems play even more essential roles in (a) the stabilization of still-functioning resources and (b) the rapid restoration of power to critical loads after a major failure of any significant part of the grid.

Grid-Level Monitoring and Control (SCADA)

At the grid level, SCADA systems are used by utilities and regional transmission authorities to monitor and manage power distribution grids and substations. A control center monitors a utility's generating plants, transmission and subtransmission systems, distribution systems, and customer loads. It oversees the automatic control of equipment in the field, and dispatches trucks as needed for manual intervention. Communication with the field equipment typically occurs over dedicated, utility-owned communications networks—analog and digital microwave and radio systems, and fiber-optic lines. Remote terminal units in the field collect data and communicate with control centers via these networks. New substations and equipment are beginning to incorporate "intelligent electronic devices" that push some of the intelligence and decision making into the field, closer to the action.

SCADA greatly improves reliability and provides the essential control infrastructure for the orderly restoration of power after a major outage in the higher tiers of the grid. But the existence of the control system itself creates new points of vulnerability. The SCADA sensors and control centers themselves have to remain powered, and thus define new critical-power nodes that need exceptionally robust and reliable backup power systems of their own. And the cyber-security of the SCADA computers and communications channels has become a major concern in its own right: The number of cyber attacks on the utility SCADA system has been rising rapidly. Sandia National Laboratories has been designated as the federal entity in charge of studying, providing solutions for and promoting SCADA security.[100]

Monitoring and Controlling On-site Power Networks

Grid feeds, static switches, batteries, backup generators, and UPSs likewise depend increasingly on embedded sensors and software to monitor their state and coordinate their operation with other components of a power network. Combined with GPS systems and extremely precise clocks, sophisticated analytical engines can determine the location, nature, and trajectory of failures at nearly every level, from specific pieces of equipment on up to the level of the building and beyond. In the past, many customers with critical power loads have been reluctant to let information of this kind leave their premises. But off-site monitoring services bring economies of scale and scope to these information-centered services, and with the advent

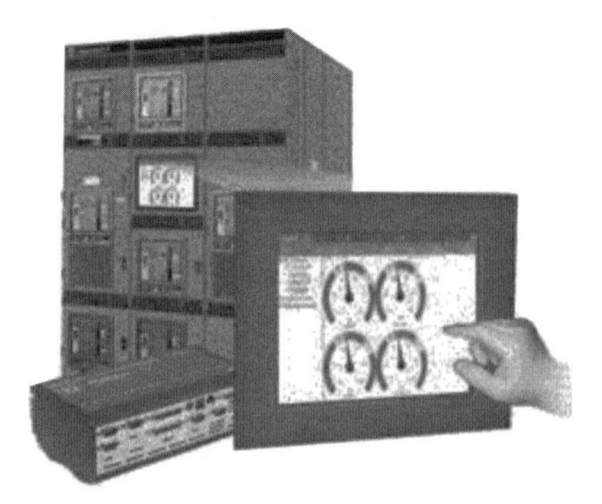

Figure 32 Power management software. *Source*: SquareD.

of highly secure communications networks, use of such services is now growing rapidly Figure 32.

Power-management software is increasingly being used as well, to provide overarching supervision and control. At the very least, such products can orchestrate the graceful shutdown of critical systems when power outages extend beyond the limits that on-site backup systems can handle. More sophisticated systems can direct the selective, sequential shedding of on-site loads, so that the most critical functions are the best protected. Recently commercialized solid-state circuit-breaker boxes, for example, permit a highly granular and dynamic triage of downstream loads, so that in the case of an extended outage limited battery power can be reserved for the most critical subsystems, with all others shut down—bringing aerospace levels of power control to building systems.

Extensive sensing and advance wired and wireless communications capabilities support the remote monitoring that is essential for all larger power networks. Such networks permit power-network managers to remotely monitor and control a UPS and all the equipment it protects. On-site power-control networks can link the UPS to power adapters deployed at critical loads, allowing the UPS to monitor key loads and optimize the distribution of power among them. These same power-control networks can communicate with the sensors and microprocessors embedded in static transfer switches, to monitor upstream sources of power, synchronize alternative power inputs, and react immediately to fluctuations or interruptions in supplies of power. The software can then control, stabilize, and isolate problems in real time.

For now, however, the networks that monitor and control on-site power operate entirely apart from those that monitor and control the grid. Given how heavily on-site power networks both depend on, and can interact with, the public grid, a key, long-term objective must be the integration of public and private power-control networks. We return to this issue later.

Reliability-Centered Maintenance

One of the most refractory problems in the assurance of critical power centers on whether or not backup equipment will actually work when a crisis hits. Batteries, for example, are notoriously unreliable unless meticulously maintained, and a key function for power management software is to provide both manual and user-defined battery tests for each UPS on the network. To address similar problems, the aviation industry relies heavily on "reliability centered maintenance" (RCM) to reduce the risks of catastrophic in-flight failures. RCM is, however, a relatively new concept in the critical-power industry, and it is very much more difficult to implement effectively than one might suppose.

Ironically, many reliability problems are created by maintenance itself, when, for example, technicians neglect to reconnect wires or flip switches after testing systems to establish that they are still working properly. A building's "power distribution unit" (PDU), for example, is an array of wires, mechanical clamps, switches, and circuit breakers that receive electric power from one source and distribute it to multiple loads. One common preventative maintenance policy centers on periodically measuring power flows through circuit breakers. To do so, a technician must open the PDU and clamp a meter to wires leading from the breakers. But that activity itself may unintentionally trip a breaker. Other maintenance-created problems are more pernicious, and will not be revealed until a major outage upstream exposes the concealed vulnerabilities.

Some of the most useful critical-power investments thus center on seemingly routine upgrades that swap older hardware with state-of-the-art replacements, which have built-in digital intelligence and monitoring capabilities. New PDU circuit breakers, for example, integrate sensors on to each wire, allowing the continuous, remote monitoring of current flows through each breaker. Reliability-centered maintenance can then be atleast partially automated, with the human element removed to a distance, where it is much more likely to add to reliability than to subtract from it.

Changes as seemingly simple as speeding up the performance of circuit breakers can greatly lower the likelihood of serious continuity interruptions precipitated by the power-protection hardware itself. Standard burn-out fuses—rather than circuit breakers—are the last line of defense in all circuits; when all else fails, power surges must be kept from destroying critical loads. Fuses are used, however, because ordinary circuit breakers flip open relatively slowly in many high-power emergency situations. And, the breaker in this case also creates an "arc flash" producing damaging electrical noise. A standard fuse creates no such noise because it burns out much faster—but the power then stays off until a technician manually replaces it. Recently developed fast-acting circuit breakers that eliminate the arc-flash can replace fuses in such applications, and can be remotely reset when sensors report the problem is clear (Figure 33).

Finally, sensor- and software-driven predictive failure analysis is now emerging, and will certainly become an essential component of next-generation RCM. By continuously monitoring the power waveforms, at every critical node in a system, unique signatures of many emerging problems can be recorded before failures occur.

Figure 33 Intelligent circuit breaker. *Source*: SquareD.

These algorithms cannot, of course, predict deliberate assaults on the network, but by making on-premises grids and backup systems much more robust, they can greatly increase the likelihood that such assaults will not in fact interrupt the delivery of power to critical loads, and they can greatly improve the mean-time-to-repair or recovery.

Resilient Design

There are definite limits to how much reliability can be added by hardware alone, and when systems are poorly designed, monitored, or maintained, more hardware can in fact reduce reliability rather than raise it. One of the most important—and least appreciated—challenges is to determine just how robust and resilient a design really is. It is far easier to declare a power network "reliable" or "robust" than to ascertain with confidence that it really is.

Standby diesel generators, for example, fail with some regularity. Some of the most pampered, carefully maintained backup diesel generators in the world reside at nuclear power plants. Yet about 1% of all nuclear-plant diesels fail to start when required, and fully 15% of the units will fail if run for 24 hours.[101] The operators and regulators of nuclear power plants are well aware of these limitations, and most nuclear plants have three separate, independent emergency power systems for just that reason. Because they are much less well maintained, diesel generators at hospitals and many other sites have failure rates 10 times higher. The May 2000 FAA report (noted earlier) identified failure rates in some of their diesel-generator-based systems at air traffic control centers that approached the grid's failure rates. More importantly, the same study showed a doubling in the past decade of the mean-time-to-repair for standby power systems.[102]

"Common mode" failures present a second refractory problem, particularly in the post-9/11 environment. For example, the high-voltage power lines and gas pipelines both present very inviting targets for terrorist attack; a simultaneous attack on both could cripple grid power and gas-fired backup generators. In the past, backup systems for commercial premises were often engineered to protect against common-mode failures caused by ordinary equipment failures, and by weather, but rarely engineered to protect against sabotage—and particularly not sabotage of the public infrastructure. A bank or data center might thus engineer a dual-feed to two independent high-voltage transmission lines—but would not plan for the further possibility that a deliberate attack might target both lines simultaneously. In the post-9/11 environment, the risk of more serious forms of common-mode failure must be taken more seriously.

Common-mode failures point to the more general challenge of analyzing risks of failures in complex, highly interdependent systems. The aviation and nuclear industries have spent many decades developing systematic, quantitative tools for analyzing the overall resilience of alternative architectures, and continuously improving the best ones. As those industries have learned, complex, probabilistic risk analysis is required for a rigorous assessment of reliability and availability. But these analytical tools are still relatively new and widely underused in the analysis of power supplies.

Used systematically, they require power engineers, statisticians, and auditors to physically inspect premises, analyze multiple failure scenarios, and draw on statistical databases to predict likely failure rates of the hardware, the human aspects of operation and maintenance, and external hazards. They must systematically take into account the key (though frequently overlooked) distinction between the reliability of the power itself, and the *availability* of the systems it powers. They must use as their basic inputs the mean-time-to-failure of individual subsystems and components, along with estimates of the risks of entirely human failures, which are often the most difficult to quantify. They must then build elaborate models for how components may interact to aggravate or abate problems.

They must then take fully into account not only the failure, but also the mean-time-to-repair. As discussed earlier, a 1-second power failure can entail a 10-minute reboot of a computer, or a week-long restart of a chip-fab; the availability of what really matters is then far lower than the nominal availability of the power behind it. The most common metric of *reliability* measures the probability of failure, ignoring the length of the ensuing down time entirely (Table 16). *Availability* metrics are far more difficult to ascertain, but they are much more useful, in that they attempt to include the time it takes to effect repairs and restart systems once a process is interrupted by a power failure.

Table 16 Typical Power Equipment Failure Rates*

Equipment	Failure Rate (per 1000 units)	Downtime per Failure (hours)
Generator	170	450
Small transformer	6	300
Large transformer	15	1000
Motor	7	70
Motor starter	14	60
Battery charger	30	40
Switchgear	1	260
Circuit breaker	4	5
Disconnect switch	6	2
Cable joint	0.8	30
Cable termination	4	10

* Representative examples.
Source: "IEEE Recommended Practice for Emergency and Standby Power Systems for Industrial and Commercial Applications" (1995).

With this said, both the analytical tools and the technologies required to engineer remarkably resilient, cost-effective power networks are now available. The challenge going forward is to promote their intelligent use where they are needed.

PRIVATE INVESTMENT AND THE PUBLIC INTEREST

Private-sector spending on homeland security is forecast in the range of $46–76 billion for fiscal year 2003.[103] But it is difficult to promote private investment in public security. Surveys of corporate security executives conducted since 9/11 have reported only modest increases in such spending in the first year after the attack.[104] Spending on security is widely viewed as pure expense.[105] New capital investment in electronic screening systems for detecting weapons or explosives in people or packages may be essential, but it neither generates revenue nor improves operational efficiency.

Power is different. The private sector was making huge investments in backup power long before 9/11, because electricity is essential for operating most everything in the digital age, and because the grid cannot provide power that is sufficiently reliable for many critical or merely important operations. Backing up a building's power supplies can be far more expensive than screening its entrances, but improving power improves the bottom line, by keeping computers lit and the assembly lines running.

Likewise, in the public sector, secure power means better service. At the very least, public safety and emergency response services require robust power supplies for their communications systems. Large city governments are run much like large businesses, but small communities are often even more vulnerable to outages, because their grids are more exposed, and because emergency service centers are more thinly dispersed. The city of Rushford, Minnesota, for example, recently approved the installation of four 3000-hp diesel backup generators to use during emergencies and when demand peaks strain available supplies.[106] The small town of McMinnville, Tennessee, installed 20 MW of diesel engines at a TVA substation for the same reasons.

The upshot, as discussed earlier, is that some 10% of the grid's capacity (80 GW) is now covered by backup generators. In recent years, roughly 1 MW of off-grid backup capacity has been added for every 6–10 MW of central-power-plant capacity brought on line. In addition, we estimate that approximately 3% of the grid's capacity (25 GW) is complemented by large UPS systems, with another 2% (10–15 GW) covered by smaller, desktop-sized units. Sales of long-life, high-performance backup lead–acid batteries have also risen sharply over the past decade; there is now an estimated cumulative installed base of some 30 million heavy-duty stationary backup lead–acid batteries. Most of this investment was made well before 9/11, and much of it falls well short of what is needed to provide adequate assurance of continuity of operations in the new environment. But these capital outlays do nevertheless confirm that the private sector has strong incentives to invest in critical-power infrastructure quite apart from any considerations related to homeland security.

Equally important is that such investments, though undertaken for private purposes, directly increase the reliability and resilience of the public grid as a whole. Larger users with their own on-site generating capacity can—and already do—sign "interruptible" power contracts with utilities; such contracts allow utilities to reduce peak demand by selectively shedding certain loads, rather than "browning out" (lowering the voltage to) large regions, or blacking out smaller ones entirely. In the event of a major assault on the grid, the process of restoring power to all will be speeded up and facilitated by the fact that some of the largest and most critical loads will be able to take care of themselves for hours, days, or even weeks.

Moreover, and even more important, the process of restoring power system-wide has to begin with secure supplies of power at the most critical nodes. Coordinating the response to a major power outage requires functioning telephone switches, E911 centers, and police communications, and the grid itself can't be re-lit unless its SCADA network remains powered. The most essential step in restoring power is not to lose it—or at worst, restore it very quickly—at key nodes and small, subsidiary grids from which the step-by-step restoration of the larger whole can proceed. Many of these nodes and grids are privately owned and operated, and securing their critical-power supplies thus depends, in the first instance, on private investment. Many of the rest are operated by local and state governments, and thus depend on investments made far down in the hierarchy of public sector spending.

Finally, in times of crisis, private generators can not only reduce demand for grid power, they can—with suitable engineering of the public-private interfaces—feed power back into limited segments of the public grid. Options for re-energizing

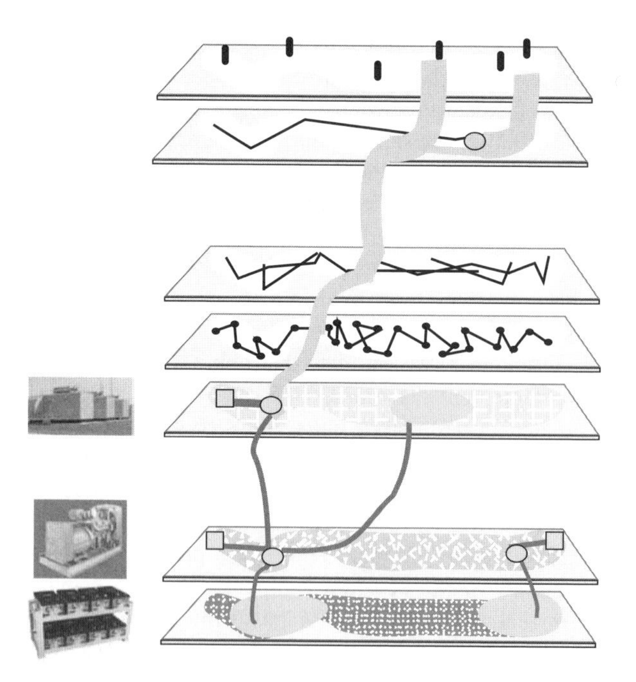

Figure 34 Adding resilience from the bottom.

the grid from the bottom-up are increasing as distributed generation expands, and as the grid's switches, substations, and control systems improve (Figure 34). As discussed further below, such options will multiply rapidly if hybrid-electric power plants come to be widely adopted in the transportation sector.

In sum, the single most effective way for government to secure the critical power infrastructure is to encourage private sector investment—not just by the relatively small numbers of quasi-public utilities and large federal agencies, but by private entities and state and local governments. Dispersed planning and investment is the key to building a highly resilient infrastructure of power.

Assess Vulnerabilities

Policy makers should be leading and coordinating the efforts of user groups, critical power providers, and utilities to conduct systematic assessments of critical-power vulnerabilities, for specific industries, utility grids, and configurations of backup systems.

As discussed above, planning for infrequent but grave contingencies is exceptionally difficult. Many critical power needs remain unaddressed simply because they have never been systematically examined. The most effective way to promote new private investment in critical power is for policy makers to help analyze and draw attention to limits and vulnerabilities of the grid, the types of loads that most require assured power continuity, and the types of on-site hardware that are capable of providing it.

Utility Protocols for "Electric Service Priority"

Major utilities already make a first—though often unsystematic—attempt to perform part of this assessment when they establish the ESP protocols noted earlier, to prioritize power restoration efforts after a major outage, typically targeting hospitals, emergency services and the like.[107] Such programs implicitly acknowledge that certain users and uses are atypically dependent on supplies of electric power, and suffer unusually serious consequences when their power fails. All such priorities are swept aside, however, when a high-level failure cuts off power to an entire region. Then, the focus is almost entirely on the systematic restoration of power from the top-down, beginning with the highest-power stations, trunks, and switching centers, in a process structured largely to minimize damage to the utility's own, most essential equipment.[108] In such circumstances utilities must—above all—maintain power supplies to their own control centers, communications and SCADA networks.

Thus, a utility's power-restoration priorities provide only limited information about critical-power requirements. Properly assessing underlying end-user vulnerabilities to power outages requires systematic assessment not just of "how important" the user is, but of the likelihood of (a) losing grid power at that user's specific location(s), (b) any backup actually starting/operating, and (c) losing or exhausting one or more of the on-premises components the user has in place to provide backup power.

The Chicago/DOE "Municipal Electric Power Vulnerability Assessment"

A joint study completed by the City of Chicago and the DOE in 2001 maps out the necessary elements of a comprehensive "municipal electric power vulnerability assessment" along just these lines. The study took as its starting point the cause and consequences of the April 1992 events that shut off utility power for weeks in the heart of Chicago. As that outage taught, and as the Chicago/DOE report points out, large cities have grown far more dependent on electric power than they used to be, in significant part because of urban society's ubiquitous dependence on digital and information hardware.

The Chicago/DOE report starts from the premise that a power vulnerability assessment "combines the information on the status of the electric power system. . . with information on the critical facilities and power-outage-sensitive individuals." The objective isn't merely to assess "the probability that the power will go out," it is to assess a community's "reliance on the electric power infrastructure and to project what the impacts might be if parts of it were disrupted."

This, the Chicago/DOE report emphasizes, requires a systematic, three-part analysis. The vulnerability assessment must begin with an analysis of the condition of the public grid, including the feeders and substations that serve the community, and an assessment of the extent to which "failures at one or a few substations could significantly affect" the community.

Equally important, is the identification of "critical facilities and power-outage-sensitive individuals," and the analysis of how much they depend on specific feeder connections or substations. In this regard, planners must determine, site by site, "whether several critical facilities and/or power-outage-sensitive individuals are connected to the same feeder," such that "the loss of one feeder might disrupt a number of facilities and sensitive individuals simultaneously."

Finally, the vulnerability assessment must analyze the adequacy of "backup measures in place for critical facilities and power-outage-sensitive individuals"—battery systems, on-site backup generators, portable generators, quick-connect circuit boxes, and so forth. Municipalities will generally perform this review for their own facilities; some municipalities "might find it useful to provide this review as a service to privately owned facilities as well." The Chicago/DOE report includes an appendix containing model forms for gathering basic information to systematize this type of review, but recognizes, as well, the importance of "on-site inspection . . . to clarify the actual status and condition of the backup measures."

The FAA's Power Systems Analysis

The comprehensive May 2000 FAA analysis, cited earlier, was conducted along similar lines, for the backup-power requirements at the Agency's roughly four dozen major control centers.[109] The report analyzed the historic record of grid power availability at those facilities, and data on the reliability of the backup systems currently in place, and thus the likelihood that the Agency's aging backup systems would perform as required when future outages occur. Subsequently, the FAA initiated a staged program to upgrade the most important and vulnerable systems. As part of that process, the Agency is deploying a network of extremely sensitive instruments for remote monitoring of key power equipment, together with software (originally developed for the U.S. Navy) for real-time predictive failure analysis.

Critical-Power for Telecommunications

Alongside the FAA, telecom regulators and carriers have progressed further than many other sectors in focusing on their critical-power needs and establishing service priorities in consultation with electric utilities.

In 1988, for example, the Federal Communications Commission (FCC) established its Telecommunications Service Priority program (TSP), to identify and prioritize telecommunication services that support national security and emergency preparedness missions. More than 50,000 phone lines have since been identified as critical, and the switching centers that service them have been placed on a master priority list for service restoration during an emergency.[110] Coverage falls into five categories. The top two are for national security; the third covers centers involved in public health, safety, and maintenance of law and order. The nearly 7000 local dispatch centers ("public-safety answering points") that handle E911 calls, for example, are eligible for listing in this category. The fourth category covers "public welfare and maintenance of national economic posture" and includes, among others, banks and other financial institutions under the sponsorship of the Federal Reserve Board. The fifth is a catchall category for the provision of new "emergency" services.

The DOE and the Office of National Communications Services subsequently established the Telecommunications *Electric* Service Priority (TESP) program, which prioritizes power restoration to critical telecommunications assets—which was subsequently reconstituted in 1994 as the National Electric Service Priority Program for telecommunications.[111] Some 230 telephone companies, more than 500 electric utilities, and regulatory authorities in all 50 states participate. While voluntary, the program maintains a database of the operational status of some 3500 "critical" assets nationwide, and electric utility emergency priority restoration systems have been revised accordingly.

Yet even in connection with critical telecom services, which depend entirely on their concomitantly "critical" supplies of power, there remain few standards specifying the minimum equipment required to maintain operational continuity in the event of serious grid outages. When backup power requirements are noted at all, they are often given short shrift. The National Emergency Number Association (NENA), which represents operators of E911 call centers, has promulgated backup-power standards that suggest "a minimum of 15 minutes of emergency power for full functionality" and "if budget permits, it is desirable to extend the 15 minutes to as much as 1 hour." Beyond that, NENA merely urges its members to plan for more "prolonged power outages," and recommends (without further specifics) that centers "be equipped with a source for long-term emergency power," which "may consist of a redundant utility power feed or a generator sized appropriately." Members are then advised to consult with "the local utility provider and a qualified power conditioning professional."[112]

According to the FCC's Director of Defense and Security (in the FCC's Office of Engineering and Technology) many E911-center administrators are not even aware that they can enroll in the National Communication System's (NCS) priority service list for access to communications lines and systems. The FCC draws a contrast with the financial industry, which has aggressively pursued priority restoration agreements.[113]

Vulnerability Assessments in Other Sectors

As summarized in *Table 17*, various entities in both the communications and financial industries have indeed completed a number of comprehensive threat assessments—among the most thorough we have seen. Alarmingly, however, most of those studies appear to either assume continuity of the supplies of power required to operate the digital equipment on which those sectors so completely depend, or simply take the position that the local utility is entirely responsible for securing the supply of power. Many of the entities responsible for ensuring critical-facility continuity are lagging well behind in assessing and addressing their underlying need for power, and also in plans to address the statistical certainty of local grid failure, extended outages, in particular.

Table 17 Electric Power Vulnerability Reports and Analyses with Vulnerability Assessments that Note or Focus on Electric Power. (First column indicates whether the indicated report includes or notes the role/importance of on-site critical power.)

☑	Title	Source/Author	Date	Selected Quote(s)
	Interagency Paper on Sound Practices to Strengthen the Resilience of the U.S. Financial System	Federal Reserve; Securities & Exchange Commission	April 7, 2003	"The agencies believe that it is important for financial firms to improve recovery capabilities to address the continuing, serious risks to the U.S. financial system posed by the post-September 11 environment." "Back-up sites should not rely on the same infrastructure components (e.g., transportation, telecommunications, water supply, and electric power) used by the primary site."
√	Potential Terrorist Attacks: Additional Actions Needed to Better Prepare Critical Financial Market Participants	General Accountig Office (GAO)	February 2003	"To begin work necessary to resume financial market operations, telecommunications carriers then had to obtain generators and use emergency power to support network operations and to coordinate with financial institutions to facilitate the resumption of stock exchange activities by September 17, 2001."
	The National Strategy For The Physical Protection of Critical Infrastructures and Key Assets	The White House	February 2003	"Almost every form of productive activity-whether in businesses, manufacturing plants, schools, hospitals, or homes-requires electricity." "Re-evaluate and adjust nationwide protection planning, system restoration, and recovery in response to attacks"
	Critical Infrastructure Protection: Efforts of the Financial Services Sector to Address Cyber Threats	General Accounting Office (GAO)	January 2003	". . . financial services sector is highly dependent on other critical infrastructures. For example, threats facing the telecommunications and power sectors could directly affect the financial services industry." ". . . the widespread and increasing use of supervisory control and data acquisition (SCADA) systems for controlling energy systems increases the capability of seriously damaging and disrupting them by cyber means."

√	Distributed Energy Resources Interconnection Systems: Technology Review and Research Needs	Department of Energy, National Renewable Energy Laboratory	September 2002	"[Distributed energy resources are] playing an increasing role in providing the electric power quality and reliability required by today's economy."
	Banking and Finance Sector National Strategy	National Strategy for Critical Infrastructure Assurance	May 13, 2002	"The banking and finance sector increasingly depends on third-party service providers for. . . telecommunications and electrical power."
	Information and Communications Sector: National Strategy for Critical Infrastructure and Cyberspace Security	Telecom consortium: CTIA, ITAA, TIA, USTA*	May 2002	". . . many customers in New York found that their communications problems stemmed not from destroyed telecommunications hardware but from power failures and stalled diesel generators." "The Telecommunications Electric Service Priority (TESP) initiative requests that electric utilities modify their existing ESP systems by adding a limited number of specific telecommunications critical facilities."
	The Electricity Sector Response to The Critical Infrastructure Protection Challenge	North American Electric Reliability Council (NERC)	May 2002	". . . work cooperatively with government, those within our industry, and other business sectors to identify and address roles, interdependencies, obstacles and barriers."

(*continued*)

Table 17 (continued)

☑	Title	Source/Author	Date	Selected Quote(s)
				"Interdependencies between other infrastructures and the electricity sector are complex and require continued review and assessment."
√	Analysis of Extremely Reliable Power Delivery Systems:	Electric Power Research Institute (EPRI)	April 2002	"develop a framework for understanding, assessing, and optimizing the reliability of powering new digital systems, processes, and enterprises"
√	Security Guidance for the Petroleum Industry	American Petroleum Institute	March 2002	"organizations increasingly rely on networked computer systems. . . . Computer systems have unique security issues that must be understood for effective implementation of security measures." ". . . refinery facilities or assets that may be subject to potential risk include:. . . Electrical power lines (including backup power systems)."
	Making the Nation Safer: The Role of Science & Technology in Countering Terrorism	National Academy of Sciences, National Research Council	2002	"The impact of a prolonged interruption in the electric power supply to any region of the country would be much larger than the economic loss to the energy sector alone." "Simultaneous attacks on a few critical components of the grid could result in a widespread and extended blackout." "While power might be restored in parts of the region within a matter of days or weeks, acute shortages could mandate rolling blackouts for as long as several years."

√	National Fire Protection Association (NFPA) Standard for Emergency and Standby Power Systems	National Fire Protection Association (NFPA)	2002	"Organized in 1976 by the NFPA in recognition of the demand for viable guidelines for the assembly, installation, and performance of electrical power systems to supply critical and essential needs during outages of the primary power source."
	Critical Infrastructure Interdependencies: Impact of the September 11 Terrorist Attacks on the World Trade Center	U.S. Department of Energy	November 8, 2001	
√	Chicago Metropolitan Area, Critical Infrastructure Protection Program: Critical Infrastructure Assurance Guidelines For Municipal Governments Planning For Electric Power Disruptions,	DOE, Metropolitan Mayors Caucus, City of Chicago	February 2001	
√	Power Systems Sustained Support, Investment Analysis Report	Federal Aviation Administration (FAA)	May 23, 2000	"The FAA cannot rely solely on commercial power sources to support NAS facilities. In recent years, the number and duration of commercial power outages have increased steadily, and the trend is expected to continue into the future."

(*continued*)

Table 17 (continued)

☑	Title	Source/Author	Date	Selected Quote(s)
√	National Information Assurance Certification and Accreditation Process (NIACAP)	National Security Telecommunications and Information Systems Security Committee	April 2000	"The contingency plan evaluation task analyzes the contingency, back-up, and continuity of service plans to ensure the plans are consistent with the requirements identified in the SSAA."
√	Interconnection and Controls for Reliable, Large Scale Integration of Distributed Energy Resources; Consortium for Electric Reliability Technology Solutions, Grid of the Future White Paper	U.S. Department of Energy	December 1999	"Customers will use distributed resources in the near term to improve the quality of power to sensitive equipment, to firm up poor reliability, to reduce their demand charges, and to take advantage of "waste" heat associated with on-site power generation, thus increasing cost effectiveness."
	Electric Power Risk Assessment	National Security Telecommunications Advisory Committee, Information Assurance Task Force	1998	". . . few utilities have an information security function for their operational systems." "A clear threat identification, combined with an infrastructure vulnerability assessment and guidelines for protection measures, is critical to stimulating effective response by individual utilities."
√	1998 New York Ice Storm: Mitigation Issues & Potential Solutions	Federal Emergency Management Agency (FEMA)	1998	There is no requirement to maintain a secondary redundant power supply.

				"Local news updates were not available because many stations did not have sufficient backup generator capacity." "Hospitals are not currently required to have on-site auxiliary power capacity sufficient for maintaining all their power systems" "Many pre-1978 wastewater treatment plants and pumping stations do not have alternative emergency power sources." "There is no inventory of existing generators at these facilities."
	President's Commission on Critical Infrastructure Protection, Critical Foundations	The White House	October 1997	"The significant physical vulnerabilities for electric power are related to substations, generation facilities, and transmission lines." "While the transportation system has long been dependent on petroleum fuels, its dependency on other infrastructures continues to increase, for example, on electricity for a variety of essential operations and on telecommunications to facilitate operations, controls, and business transactions." "As farsighted and laudable as these [earlier infrastructure vulnerability] efforts were, however, interdependencies within the energy infrastructure and with the other infrastructures were not studied, nor was the energy sector's growing dependence on information systems."
✓	Generic Standards for E9-1-1 PSAP Equipment	National Emergency Number Association (NENA)	June 20, 1996	"[plan for] prolonged power outages"

(*continued*)

Table 17 (continued)

☑	Title	Source/Author	Date	Selected Quote(s)
				"[centers to] be equipped with a source for long-term emergency power," which "may consist of a redundant utility power feed or a generator sized appropriately"
√	IEEE Recommended Practice for Emergency and Standby Power Systems for Industrial and Commercial Applications	Institute of Electrical and Electronics Engineers (IEEE)	December 12, 1995 (Revision of IEEE Std 446-1987)	"The nature of electric power failures, interruptions, and their duration covers a range in time from microseconds to days." "In the past the demand for reliable electric power was less critical." ". . . industrial and commercial needs contained more similarities than differences."

* Cellular Telecommunications and Internet Association (CTIA); Information Technology Association of America (ITAA); Telecommunications Industry Association (TIA); United States Telecom Association (USTA).

For the most part, other government, industrial, and commercial sectors have done very much less to assess their power vulnerabilities. Many older wastewater treatment facilities still lack the emergency power sources required to maintain pumping capability when grid power fails. In 1997, a number of public utility commissions called for backup power standards for newer water facilities, or for sufficient on-site fuel to cover extend outages.[114] The FEMA report cited earlier (which addresses the crippling impact of the 1998 Northeast ice storms),[115] pointed to across the board deficiencies in on-site auxiliary power capacity. Because of the outage duration, lack of on-site power even forced closures of many of the schools frequently designated as disaster shelters.

As the FEMA report emphasizes, the longer the outage, the more "critical" the lack of power tends to become. Many sites—including especially emergency response sites—that can ride through grid outages that last an hour or day become increasingly dysfunctional when power outages persist for longer periods. Again and again, the FEMA report emphasizes the need to assess power requirements and weigh the need for backup power where there was none at all, redundant backup systems where they found critical facilities depending on fundamentally unreliable backup generators (calling for the classic redundancy, defense-in-depth, common in military, aviation and nuclear industries), and most particularly for capability to operate for extended periods of time (beyond the minutes and hours typical where there was any backup at all).

Federal Initiatives and Oversight

Many of the reports just described provide useful models for what is required, but they also serve to highlight what is missing. In the current geopolitical environment, the analyses required to secure critical-power infrastructure can no longer wait until after the major outages materialize. In the aftermath of 9/11, critical-power infrastructure must be analyzed and secured in anticipation of significant threats, not just in reaction to their actual occurrence. The National Strategy for Homeland Security announced by the White House in July 2002, and substantially expanded in February 2003,[116] specifically directs federal agencies to "[f]acilitate the exchange of critical infrastructure and key asset protection best practices and vulnerability assessment methodologies." The National Strategy notes the "modeling, simulation, and analysis capabilities" on which "national defense and intelligence missions" rely, and declares that such tools must now be applied to "risk management, and resource investment activities to combat terrorism at home," and particularly with regard to "infrastructure interdependencies."

The responsibility to pursue vulnerability assessments would now logically lie with the Infrastructure Analysis, Information Assurance (IAIP) directorate of the DHS. The IAIP includes the CIAO (originally created in 1996), the former National Infrastructure Protection Center (NIPC), and representatives from four other federal government agencies. It coordinates the efforts of federal, state and local government officials and the private sector in protecting the nation's critical infrastructures. To that end, it brings together the capabilities needed "to identify and assess current and future threats, to map those threats against existing vulnerabilities, issue timely warnings and take preventive and protective action."[117]

TISP should be engaged in this effort as well.[118] Members include local, state, and federal agencies, the Army Corps of Engineers, professional associations, industry trade groups, code and standards organizations, professional engineering societies, and associations that represent the builders and operators of infrastructure facilities, among others. Among other goals, the organization seeks to "encourage and support the development of a methodology for assessing vulnerabilities," and to improve "protection methods and techniques" relevant to "domestic infrastructure security."

ESTABLISH CRITICAL-POWER STANDARDS FOR FACILITIES USED TO SUPPORT KEY GOVERNMENT FUNCTIONS

Federal and local organizations should work with the private sector to establish guidelines, procedures, and (in some cases) mandatory requirements for power continuity at private facilities critical to government functions.

Much of the private sector already has strong incentives to secure its critical-power requirements; for the most part, the government's role should be to inform and facilitate voluntary initiatives. At the same time, however, many federal, state, and local agencies have independent responsibility to ensure that private-sector facilities used to support key government functions can be relied on in times of crisis. This necessarily implies some form of governmental auditing or standard-setting for the backup-power systems that ensure continuity of operation of the chips, fiber-optics, radios and broadcast systems within privately owned telecom, data, computing, and financial networks on which key agencies at all levels of government rely.

Though they have yet to issue any specific, power-related standards, federal financial agencies have repeatedly noted the importance of power in maintaining the continuity of financial networks in times of crisis, and have affirmed their authority to audit the physical infrastructure on which key institutions rely. Federal regulators have general authority to examine all aspects of risk management at federally insured banks, including "use of information technology and third party service providers," and the evaluation can include systems in place to assure "business continuity." A January 2003 report by the GAO,[119] for example, notes that the "financial services sector is highly dependent on other critical infrastructures," particularly telecommunications and power. A follow-up GAO report concludes that the "business continuity plans" of many financial organizations, have "not been designed to address wide-scale events."[120] An April 2003 paper by the Federal Reserve stresses the importance of "improv[ing] recovery capabilities to address the continuing, serious risks to the U.S. financial system posed by the post-September 11 environment," with particular attention to the threat of "wide-scale disruption" of "transportation, telecommunications, power, or other critical infrastructure components across a metropolitan or other geographic area."[121]

Within the financial services community itself, however, securing power supplies is still—all too frequently—viewed as someone else's responsibility. The technology group of the Financial Services Roundtable and the Financial Services Information Sharing and Analysis Center (FS/ISAC) fully recognizes that the financial sector "increasingly depends on third-party service providers for. . . telecommunications and electrical power,"[122] and accepts that "the core infrastructure of the banking and finance sector must be examined to identify and assess areas and exposure points that pose systemic risk." Nevertheless, many of this sector's business-continuity efforts still, essentially, assume grid power and proceed from there.

If such attitudes are still encountered even in sophisticated, well-funded financial circles, they dominate elsewhere, in planning for service continuity in local government agencies, wastewater treatment plants, schools that double as disaster recovery centers, broadcast facilities, hazardous chemical storage, and even many medical facilities. Government agencies at all levels have significant financial stakes in these services, and in the underlying facilities used to provide them, and thus share responsibility for ensuring that adequate preparation is made for emergencies.

In this regard, the Chicago/DOE study, discussed above, provides a model for the analysis, guidelines, and standards that municipal governments can play a key role in developing. It makes no attempt to tell utilities how to make the public grid more reliable; rather, it focuses on identifying critical-power users, conducting site-specific vulnerability assessments, and assessing technology alternatives—batteries, generators, dual feeds, and so forth—that can keep key loads lit when the grid fails. The Chicago/DOE report does not go on to establish mandatory standards for the private facilities on which the City of Chicago itself relies. But power-related mandates of that character must emerge in due course to define the private facilities required to ensure operational continuity of public services in times of widespread disruption to the public infrastructure.

Share Safety- and Performance-Related Information, Best Practices, and Standards

Utilities, private suppliers, and operators of backup power systems should develop procedures for the systematic sharing of safety- and performance-related information, best practices, and standards. Policy makers should take steps to facilitate and accelerate such initiatives.

The federally run Centers for Disease Control (CDC) complies epidemiological databases and performs the analyses that public health officials can rely on in formulating responses. To that end, the CDC obtains information from state agencies, private entities, and individuals, with suitable protections in place to protect both privacy and proprietary information. In similar fashion, aviation safety has been greatly enhanced by a systematic government process for investigating all significant equipment failures and major accidents, and sharing the information industry wide. The National Transportation Safety Board investigates and maintains a comprehensive database of accidents and incidents that extends back to 1962; industry participation is mandatory, but the information generated in the process is used only for improving safety system wide; it may not, for example, be used in civil litigation.[123] The Aviation Safety Reporting System (ASRS) is a complementary, voluntary program that collects, analyzes, and responds to less serious incident reports. The program focuses mainly on human factors and guarantees confidentiality; more than 300,000 reports have been submitted to date.[124]

The government's National Communications System works closely, and in similar fashion, with the private National Security Telecommunications Advisory Committee to ensure the continuity of the telecommunications services on which the most important government users depend.[125] The two groups are jointly charged with ensuring the robustness of the national telecommunications grid. They have been working together since 1984 and have developed effective means to share information about threats, vulnerabilities, operations, and incidents, which improves the overall surety of the telecommunications network. Comparable programs should be developed to analyze—and learn from—failure at every tier of the electric infrastructure, including distributed generation and on-site power.

In the electric power sector, the nuclear industry developed a similar, comprehensive, information-sharing program in the aftermath of the 1979 events at Three Mile Island. The Institute for Nuclear Power Operations established a comprehensive system for the sharing of safety-related information, best practices, and standards among utilities, equipment vendors, architect/engineers, and construction firms.[126] The Institute's information sharing programs include an equipment failure database and a "Significant Event Evaluation and Information Network." These and other assets have sharply reduced duplicative evaluation of safety- and performance-related events.[127] Since the establishment of these programs, the nuclear industry has maintained a remarkable, two-decade record of safe operations, while also dramatically improving the availability—and thus the economic value—of its facilities.

In light of the growing importance of on-site power, the providers and users of this equipment should now seriously consider establishing a national advisory group comparable to the utility sector's NERC/CIPAG (discussed above) and empowered to work with CIPAG to coordinate the complementary development of on-site power-protection infrastructure. Independently, the NFPA Standard for Emergency and Standby Power Systems, noted earlier, sets out what amounts to a best-practices guide to on-site power; it also classifies different backup systems and configurations by run time. It is difficult to see how CIPAG can properly fulfill its stated mission—"to advance the physical and cyber security of the critical electricity infrastructure of North America"—without a systematic process for coordinating grid—level and on-site power-protection initiatives. The same holds for other NERC activities, including those of the Electricity Sector Information Sharing and Analysis Center (which gathers and interprets security-related information and disseminates it within the industry and the government) and NERC's best-practices advisories for the business community.[128] On-site power has emerged as a new, essential tier of the power infrastructure, and the security of the system as a whole can no longer be analyzed without due consideration of on-site power facilities.

More generally, as discussed in the 1997 Presidential Commission report, Critical Foundations,[129] the hardening of critical infrastructure will depend on "creation of a trusted environment that . . . allow[s] the government and private sector to share sensitive information openly and voluntarily." To that end, the 1997 report proposed changes in various laws that currently inhibit the protection of confidential information, and thus discourage participation in information sharing. The report specifically flagged, as areas of potential concern, the Freedom of Information Act, insufficient protection of trade secrets and proprietary information, classified information, antitrust laws, civil liability, and potential national security issues arising from participation by foreign corporations in information sharing arrangements. Little was done to follow up on these proposals in 1997; they should be acted upon now.

Interconnect Public and Private Supervisory Control and Data Acquisition Networks

The supervisory control and data acquisition networks operated by utilities and the operators of backup power systems should be engineered for the secure exchange of information, to facilitate coordinated operation of public and private generators and grids. Policy makers should take steps to facilitate and accelerate that development.

As described earlier, SCADA systems are used by utilities and regional transmission authorities to monitor and manage power distribution grids and substations. Extensive arrays of sensors and dedicated communications links feed information to the major control centers that monitor the overall state of the grid, and control its constituent parts. Very similar data networks and supervisory systems perform the same functions on private premises, supervising and controlling grid feeds, static switches, batteries, backup generators, and UPSs.

At present, there is very little direct, electronic linkage between the public and private networks that supervise and control the flow of power. This is a serious deficiency, and one that should be comparatively easy to rectify. Better communication is essential if utilities are to collaborate more closely with the owners of private generators, to shed loads (or even to draw privately generated power back into the grid) when large power plants or major transmission lines fail. And the owner–operators of private generators and grids would be much better positioned to protect their facilities from major problems propagating through the grid (Figure 35).

Among other advantages, advanced networking of this kind would make possible dynamic updating of service-restoration priorities. At present, for example, the telephone-service priority program rules require service users to revalidate their priority status every two years. Most utility ESP protocols are equally static, and therefore much less useful than they might be. Critical-power needs can obviously change month to month, or even hour to hour. Responses to widespread outages

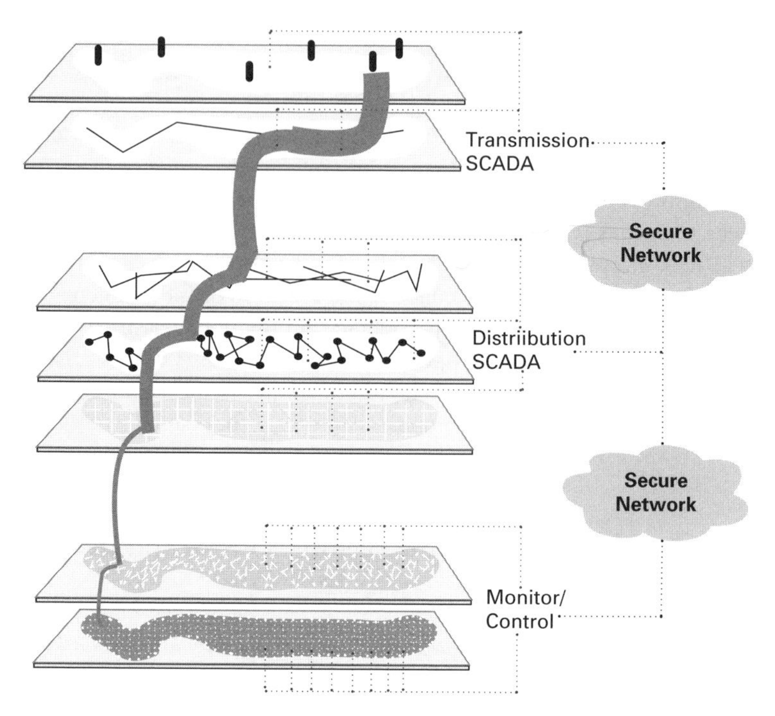

Figure 35 Networking the tiers.

cannot be optimized on the basis of two-year-old information. By way of analogy, some private companies rely on private weather reporting services to provide site-specific, real-time weather alerts to assist in power management and control. Thor Guard's service, for example, can provide localized forecasts of lightning hazards,[130] allowing enterprises with on-site power to disconnect from the public grid when the risk of dangerous transients is high. More efficient and dynamic exchanges of electronic information between the various public and private tiers of the grid could greatly improve the resilience of the whole.

Secure Automated Control Systems

The necessary integration of SCADA networks operated by utilities and the operators of backup power systems requires high assurance of cybersecurity of the networks in both tiers. Policy makers should take steps to advance and coordinate the development of complementary security protocols in the public and private tiers of the electric grid.

More efficient exchange of information between public and private power-control networks is clearly desirable, but at the same time, the very existence of highly interconnected control networks can create new vulnerabilities that must be assessed and addressed.

As part of its more general charge to improve the surety of the nation's energy infrastructure, Sandia National Laboratories has focused specifically on the cyber vulnerabilities of the electric power industry's SCADA networks.[131] In addition to identifying significant security issues specific to particular systems, Sandia has explored the more general problems inherent in the trend toward fully automated, networked SCADA systems, increasing reliance on fully automated control, loss of human expertise, reliance on equipment of foreign manufacture, and the difficulty of performing meaningful security inspection and validation.

SCADA systems, Sandia notes, "have generally been designed and installed with little attention to security. They are highly vulnerable to cyber attack... [S]ecurity implementations are, in many cases, non-existent or based on false premises." Sandia also cites public reports that these systems have been the specific targets of probing by A1 Qaeda terrorists. Over 1200 cyber attacks were detected *per energy company* over a six-month period in 2002—20% more than against an average financial company, and twice as many as an average e-commerce company; moreover, a disproportionately high fraction (70%) of the attacks on energy company SCADA networks were classified as "severe."[132] (Figure 36).

To address some of these problems, Sandia is working with international bodies to develop open but secure standards for these networks. This initiative is, clearly, an essential complement to the further development of utility SCADA networks,

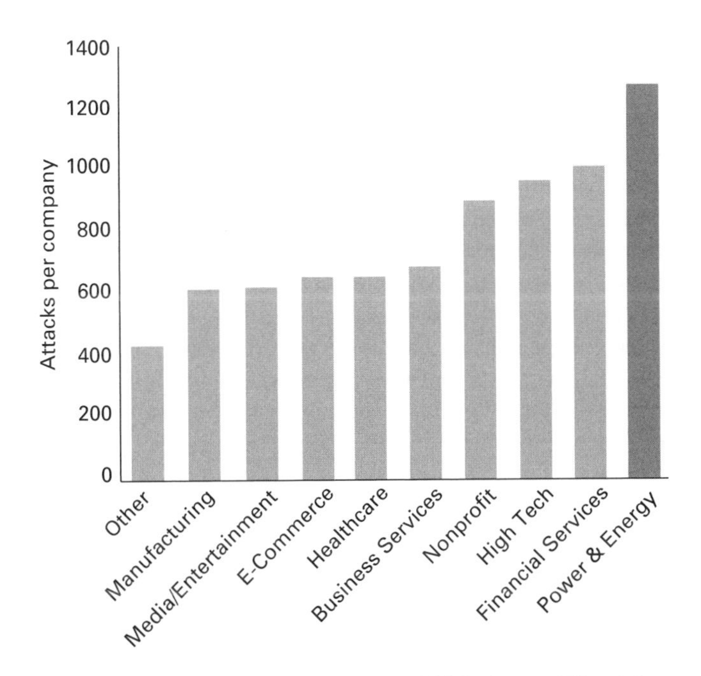

Figure 36 Cyber attacks (from 1/1/02 to 6/30/02). *Source*: "Riptech Internet Security Threat Report: Attack Trends for Q1 and Q2 2002," Riptech, Inc. (July 2002).

and an equally essential predicate to any interconnection of public and private control networks.

Share Assets

Policy makers and the private sector should take steps to promote sharing of "critical spares" for on-site generation and power-conditioning equipment, and to advance and coordinate the establishment of distributed reserves and priority distribution systems for fuel required to operate backup generators.

As noted earlier, the utility industry's CIPAG has formed a working group to inventory and develop a database of high-power "critical spare equipment" used in the high-voltage transmission system. Such programs address the straightforward, serious challenge of locating spares in times of emergency, and getting them quickly to where they are needed. The intelligent sharing of standby assets shortens response times and can greatly reduce the collective cost of preparing for rare but serious interruptions.

The CIPAG program was initially developed from a 1989 FBI request to NERC, to identify and locate equipment (specifically, the large and hard-to-replace extremely-high-voltage transformers) that might be available for loan in the event of a terrorist attack. NERC's spare equipment database, which is now accessible on a secure Web site, has since grown to include over 900 transformers.[133] The CIPAG initiative has recently included "emphasis on deterring, preventing, limiting, and recovering from terrorist attacks"[134] and the class of critical equipment is to be expanded to incorporate all other grid-critical equipment.

Comparable programs could, in similar fashion, significantly improve operational resilience in the lower tiers of the grid. Power plants on wheels (or on barges) offer the economies of sharing in much the same manner as fire engines and rescue vehicles, and bypass the many cost-related and regulatory obstacles that impede permanent deployment of backup generators.

Fuel supplies for backup generators present similar challenges, and similar opportunities. As recently noted in The National Strategy for The Physical Protection of Critical Infrastructures and Key Assets,[135] "[a]ssuring electric service requires operational transportation and distribution systems to guarantee the delivery of fuel necessary to generate power." The DOE addressed an analogous problem in a program initiated in 1998. Concerned about potential interruptions (or price run-ups) in supplies of home heating oil, the Department established four new regional heating oil storage terminals—in effect a mini Strategic Petroleum Reserve—containing a total of 2 million barrels of reserve fuel.[136]

Enhance Interfaces Between On-Site Generating Capacity and The Public Grid

Improved technical and economic integration of on-site generating capacity and the public grid can back up critical loads, lower costs, and improve the overall resilience of the grid as a whole, and should therefore rank as a top priority for policy makers and the private sector.

Many utilities are already offering backup-power services to commercial and industrial customers, or plan to begin doing so within the next few years. Additional impetus for utility and private development of "distributed" or on-site power has come from growing interest in fuel cells, photovoltaics, and co-generation. With suitable engineering of the public-private interfaces, and appropriate tariffs in place, on-site generators can back up critical loads, lower costs, and improve the overall resilience of the grid as a whole.

Wisconsin Public Power, Inc. (WPPI) for example, now offers large (>250 kW) customers long-term contracts for the installation and maintenance of on-site standby generators, together with remote monitoring and control. WPPI also fires up the units during periods of peak demand or high price.[137] In Orlando, Cirent Semiconductor backs up the most critical loads in its much larger (35 MW) facility with an array of 2 MW diesel gensets. In cooperation with the Florida Power Corporation (FPC), these generators are now being used for peak shaving as well: In return for lower rates, Cirent fires up the units during times of peak usage to displace up to 6 MW of demand from FPC's public grid.

Such opportunities have been systematically explored in two excellent papers written by the DOE: A 1999 white paper prepared in collaboration with a consortium of the national laboratories, Integration of Distributed Energy Resources,[138] and an outstanding September 2002 report, Distributed Energy Resources Interconnection Systems: Technology Review and Research Needs.[139] While surprisingly silent on the subjects of critical power and homeland security, the latter report includes a detailed survey of distributed-power technologies, a complete analysis of the technologies required to interconnect distributed capacity to the grid, and a thorough set of recommendations for both the technical and the policy-making communities with regard to interconnection standards (Figure 37).

Federal and state regulators, together with private standards-setting bodies, are now actively developing programs to facilitate such arrangements, and to make them profitable to all participants. As noted earlier, an IEEE publication sets out comprehensive definitions of technical standards for reliable on-site industrial and commercial power systems.[140] More recently, the IEEE approved a new Standard for Interconnecting Distributed Resources With Electric Power Systems.[141] California, New York, and a number of other states have interconnection tariffs and procedures in place; most of them defer to bilateral agreements between the utility and the owner of the distributed generating capacity, with disputes handled case-by-case by the public utility commission. At the federal level, FERC likewise relies heavily on voluntary arrangements, but the Commission is also developing a model interconnection agreement that would create a streamlined process for interconnection of under-20-MW facilities, and an even simpler process for facilities under 2 MW.[142]

Over the longer term, the emergence of hybrid-electric trucks, buses, and cars will offer the possibility of linking the transportation sector's mobile, and highly distributed infrastructure of fuel tanks, engines, and generators directly to

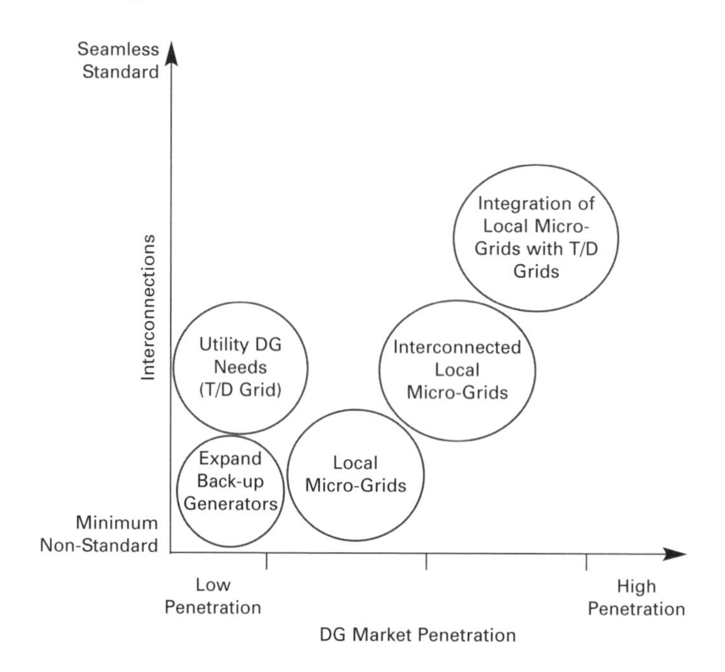

Figure 37 Distributed generation and the public grid. *Source*: "Grid of the Future White Paper on Interconnection and Controls for Reliable, Large Scale Integration of Distributed Energy Resources," Transmission Reliability Program Office of Power Technologies (December 1999).

electrical breaker boxes and UPSs in residences, small offices, and larger buildings. As discussed earlier, one of the diesel genset's main attractions is that it runs on the transportation sector's fuel, and can thus draw on that sector's huge, distributed, and therefore resilient infrastructure of fuel storage and delivery. The hybrid car designs already on the road—the Toyota Prius, Honda Insight and Civic, and shortly available Ford Escape—have 10- to 33-kW power plants, with primary fuel on board in the gas tank. Emerging larger vehicles will have 100 kW. If and when they eventually arrive, simple, safe bridging from the transportation sector's power plants to building-level electrical grids will offer potentially enormous improvements in the resilience and overall reliability of electrical power supplies.

One cannot underestimate either the technical or the economic challenges involved in connecting distributed generating capacity to the grid. At the technical level, generators must be very precisely synchronized with the grid; otherwise they can create destructive reverse power flows damaging grid equipment for all (or cause catastrophe on the small generator side). Isolation switches are essential to protect utility workers from the hazards of privately generated power moving up the lines during a grid outage. Economic integration is equally challenging. Large, centralized power plants and the grid itself are shared capital-intensive resources, and their costs must be allocated rationally. But with these important qualifications noted, the orderly physical and economic integration of utility and on-site power systems offers an enormous opportunity to secure critical-power at no net cost to either utilities or end-users.

Remove Obstacles

Private investment in critical-power facilities creates public benefits, and policy makers should explore alternative means to remove obstacles that impede private investment in these facilities.

Zoning and environmental regulations, and related issues of insurance, often present serious obstacles to deployment of on-site generating facilities. State and local fire codes may prevent on-site fuel storage. Companies storing fuel oil for backup generators or vehicle fleets may have to prepare written Spill Prevention Control and Countermeasure (SPCC) plans to comply with the Clean Water Act, as well as a maze of local, state and federal environmental regulations.[143] Municipal zoning and safety regulations affect where storage tanks can be sited, and how large such tanks may be.

As standby or emergency facilities, backup diesel generators are exempt from some emissions related regulations, but the exemptions typically limit annual operations to 80 or 120 hours per year, depending on the air district,[144] and thus eliminate the possibility of reducing costs by using the same facilities to generate power during high-price periods of peak demand. Noise abatement is becoming an issue as well.[145]

Some of these obstacles can readily be addressed: Insurance problems, for example, can be addressed by liability limits. Others will require a systematic reassessment of priorities—how to strike appropriate balances, for example, between air quality and homeland security in the new, post-9/11 environment. The various obstacles noted above often fall under the jurisdiction of separate federal, state, and local regulatory authorities. Regulators at all levels of government should work more closely to simplify and accelerate the process of providing regulatory clearance for private investment in critical-power facilities. Federal regulators should, as necessary, be prepared to assert preemptive authority here, to ensure that parochial regulatory concerns do not unduly impede investment in facilities that contribute materially to the overall resilience of the interstate power grid (Figure 38).

In striking these balances, it bears emphasizing, once again, that private investment in on-site generating facilities creates public benefits too: In the event of a major assault on the public grid, these facilities will relieve demand, and maintain the resilient islands of power required to restart the rest of the system. To the extent that competing public policies—environmental regulations, for example—reduce incentives to private investment here, policy makers should search for other mechanisms to increase them. Because power generation facilities are so capital intensive, allowing accelerated depreciation (or immediate expensing, or similar favorable treatment) for tax purposes would have a substantial impact. A proposal to that effect was included in both the White House's National Strategy and GAO's Critical Infrastructure Report.[146]

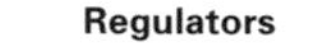

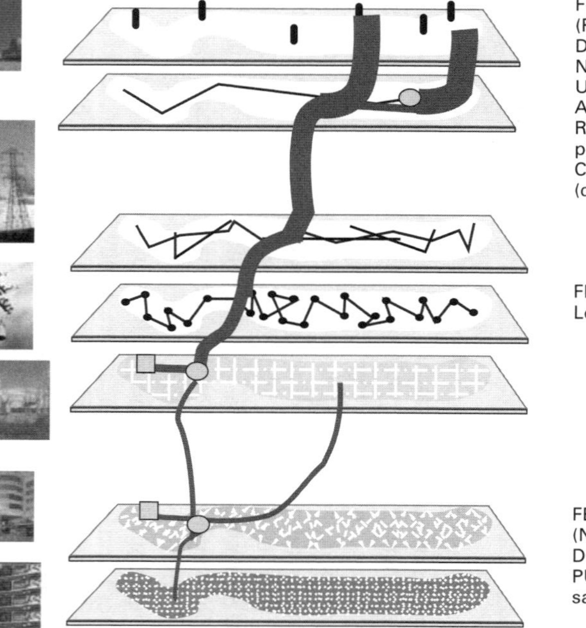

Regulators	Players & Standards
Federal Energy Regulatory Commission (FERC), Department of Energy (DOE), Department of Transportation (DOT), Nuclear Regulatory Commission (NRC), U.S. and State Environmental Protection Agencies (EPAs), National Association of Regulatory Commissions (NARUC) (e.g., pipeline safety), State Public Utility Commissions (PUCs), Local Governments (county and city) (e.g., zoning, siting)	Critical Infrastructure Assurance Office (CIAO- DHS), Infrastructure Analysis, Information Assurance (IAIO - DHS), Critical Infrastructure Protection Advisory Group (CIPAG - NERC)
FERC, DOE, DOT, EPAs, NARUC, PUCs, Local Governments (e.g., zoning, siting)	Electric Power Research Insitute (EPRI), Telecommunications Electric Service Priority (TESP), Insitute of Electrical and Electronics Engineers (IEEE), Electrical Generating Systems Association (EGSA), National Electrical Manufacturers Association (NEMA)
FERC, National Fire Protection Agency (NFPA), Underwriter's Laboratory (UL), DOE, Department of Labor (OSHA), EPAs, PUCs, Local Governments (e.g., health, safety)	Information Sharing and Analysis Center (ISAC), National Fire Protection Agency (NFPA), International Electrotechnical Commission (IEC)

Figure 38 Power regulators and players.

NOTES

1 Testimony of Kenneth C. Watson, President, Partnership for Critical Infrastructure Security, House Energy and Commerce Committee, Subcommittee on Oversight and Investigation, Hearing on Creating the Department of Homeland Security (DHS) (July 9, 2002).
2 *Critical Infrastructure Protection in the Information Age*. Executive Order 13231 (October 2001).
3 *Critical Infrastructure Interdependencies: Impact of the September 11 Terrorist Attacks on the World Trade Center*, Department of Energy (DOE) (November 2001).
4 *Information & Communications Sector: National Strategy for Critical Infrastructure and Cyberspace Security*, Cellular Telecommunications & Internet Association (CTIA), Information Technology Association of America (ITAA), Telecommunications Industry Association (TIA), United States Telecom Association (USTA) (May 2002).
5 Includes nearly 1000 stations affiliated with the five major networks – NBC, ABC, CBS, FOX, and PBS. In addition, there are about 9,000 cable TV systems. CIA World Databook (1997), http://www.cia.gov/cia/ publications/factbook/geos/us.html.
6 CIA World Databook (1998), http://www.cia.gov/cia/publications/factbook/ geos/us.html.
7 *Trends in Telephone Service*, Federal Communications Commission (FCC) (May 2002), http://www.fcc.gov/Bureaus/Common_Carrier/Reports /FCC-State_Link/IAD/trend502.pdf.
8 *Industry Statistics*, NCTA (November 2002), http://www.ncta.com/industry_overview/indStat.cfm?indOverviewID=2.
9 *Background on CTIA's Semi-Annual Wireless Industry Survey*, CTIA (December 2002), http://www.wow-com.com/industry/stats/surveys.
10 May not include pico cells. *CTIA Semiannual Wireless Industry Survey* (December 2002), http://www.wow-com.com/industry/stats/surveys.
11 Assumes 1 BSC per every 75 base stations. "... BSCs ... are responsible for connectivity and routing of calls for 50 to 100 wireless base stations." *VII GSM Call Processing*, http://www.privateline.com/PCS/GSMNetworkstructure.html#anchor 3381743.
12 There are approximately 4,500 land-based transponder sites worldwide, evenly divided between North America, Western Europe, and the Asia Pacific region, with television programming back-haul accounting for half of total use. *Teleports and Carriers Market Facts*, World Teleport Association.
13 *Internet Data Centers*. Salomon Smith Barney, (August 3, 2000). Eagle, Liam, *Tier 1 Releases Hosting Directory. Data Center Report*, (April 4, 2002), http://thewhir.com/ marketwatch/tie040402.cfm.
14 *The Internet – What Is It?* Boardwatch Magazine (2000).
15 *CIA World Databook*, http://www.cia.gov/cia/publications/factbook/geos/ us.html.
16 *Directory of Internet Service Providers*, Boardwatch Magazine (2000).
17 *Directory of Internet Service Providers*, Boardwatch Magazine (2000).
18 Estimate based on Energy Information Administration (EIA) commercial building data. Figure is 10 percent of $>100,000$ sq ft buildings.
19 *Critical Infrastructure Protection: Efforts of the Financial Services Sector to Address Cyber Threats*. General Accounting Office (GAO) report (GAO-03-173) (January 2003).
20 *Critical Infrastructure Protection: Efforts of the Financial Services Sector to Address Cyber Threats*, GAO (GAO-03-173) (January 2003).
21 *Potential Terrorist Attacks: Additional Actions Needed to Better Prepare Critical Financial Market Participants*, GAO (GAO-03-414) (February 2003).
22 *Critical Infrastructure Protection: Efforts of the Financial Services Sector to Address Cyber Threats*, GAO (GAO-03-173) (January 2003).
23 *Interagency Paper on Sound Practices to Strengthen the Resilience of the U.S. Financial System*. Federal Reserve (April 2003).
24 Registered hospitals as of December 2002, American Hospital Association, http://www.hospitalconnect.com/aha/resource_center/fastfacts/fast_facts_US_hospitals.html.
25 Estimate derived by multiplying nationwide hospital total by 4, a ratio we found in select U.S. cities.
26 Not a comprehensive count of all elderly care facilities, but nursing homes provide 24/7 care. *JAMA Patient Page: Nursing Homes*, The Journal of the American Medical Association, http://www.medem.com/MedLB/article_detaillb.cfm?article_ID= ZZZV2MOSIUC&sub_cat=392.

See also, American Association for Homes and Services for the Aging, http://www.aahsa.org/index.shtml.

27 *Facts and Figures*, National Emergency Numbers Association (NENA), http://www.nena.org/PR_Pubs/911fastfacts.htm

28 *A Needs Assessment of the U.S. Fire Service: A Cooperative Study*, Federal Emergency Management Agency (FEMA) U.S. Fire Administration (USFA). National Fire Protection Association (NFPA), FA-240 (December 2002).

29 Assumes an average of 5 important government buildings per 100,000 people.

30 *1998 New York Ice Storm: Mitigation Issues & Potential Solutions*, FEMA Region II (1998), http://www.appl.fema.gov/reg-ii/1998/ nyice4.htm#ELECTRIC.

31 *A Needs Assessment of the U.S. Fire Service: A Cooperative Study*, FEMA, USFA, NFPA, (FA-240) (December 2002).

32 Of the 12 automated systems that are considered mission-critical, one has already been made compliant, one will be upgraded, three have been retired, four more will be retired, and three will be replaced. Testimony of Kathleen Himing, Chief Information Officer Federal Energy Regulatory Commission, before the Subcommittee on Technology Committee on Science, U.S. House of Representatives (May 14, 1998), http://www.house.gov/science/himing_05-14.htm.

33 North American Electric Reliability Council (NERC), http://www. nerc.com.

34 NERC, http://www.nerc.com.

35 NERC, http://www.nerc.com.

36 EIA (2000), http://www.eia.doe.gov/cneaf/electricity/ipp/htmll/t17p01.html.

37 "Between 20 and 100 miles separate pumping stations, depending on the pressure at which the pipeline is operated and upon the terrain over or through which it runs." Association of Oil Pipe Lines, http://www.aopl.org/about/questions.html. We used a 60-mile separation to derive estimate.

38 Compressor station every 60 miles for high-pressure natural gas transmission pipelines. *PennWell*, http://www.pennwell.com.

39 *National Transportation Statistics 2002*. Bureau of Transportation Statistics (BTS), http://www.bts.gov/publications/national_ transportation_statistics/2002.

40 *National Transportation Statistics 2002*, BTS, http://www.bts.gov/publications/national_transportation_statistics/2002.

41 *Economic Reports: Operators & Producing Wells*. Independent Petroleum Association of America (IPAA) (2002), http://www.ipaa.org/info/ econreports/usps.asp?Table=Chart03.

42 *Economic Reports: Operators & Producing Wells*. IPAA (2002), http://www.ipaa.org/info/econreports/usps.asp?Table=Chart03.

43 *Security Guidance for the Petroleum Industry*, American Petroleum Institute (March 2002).

44 *Security Guidance for the Petroleum Industry*, American Petroleum Institute (March 2002).

45 *Security Guidance for the Petroleum Industry*, American Petroleum Institute (March 2002).

46 *Security Guidance for the Petroleum Industry*, American Petroleum Institute (March 2002).

47 *Security Guidance for the Petroleum Industry*, American Petroleum Institute (March 2002).

48 *Fueling the Future*, American Gas Association, http://www.fuelingthefuture.org/contents/ExpandingNaturalGasDelivery.asp

49 U.S. Geological Survey, http://ga.water.usgs.gov/edu/tables/mapwwfac.html.

50 American Water Works Association, http://www.awwa.org/Advocacy/ pressroom/waterfax.cfm, http://www.awwa.org/Advocacy/pressroom/ TrendsIssues/.

51 The Federal Aviation Administration (FAA) has 4 Air Route Traffic Control Centers (ARTCC) and 19 Air Traffic Control Towers (ATCT), http://www2.faa.gov/index.cfm/1042/. See http://www2.faa.gov/index.cfm/1043/ for ATCT list. FAA also has Automated Flight Service Stations, Automated International Service Stations, and Terminal Radar Approach Control Facilities, http://www2.faa.gov/index.cfm/1042; 19 Airport Traffic Control Centers (ATCC) http://www2.faa.gov/index.cfm/1043/; 24 Automated Flight Service Stations (AFSS); http://www2.faa.gov/index.cfm/1044; 3 Automated International Flight Service Stations (AIFSS) http://www2.faa.gov/index.cfm/1045; 6 Terminal Radar Approach Control (TRACON) Facilities http://www2.faa.gov/index.cfm/1046

52 There are nearly 20,000 airports nationwide, but only 5300 are designated for public use. We estimate that roughly 75% of these, or 4000, have control towers.

53 Class I rail. Number of rail control centers assumes one per 1,000 miles of track. *National Transportation Statistics 2002*, BTS, http://www.bts.gov/publications/national_transportation_ statistics/2002.

54 Municipal DOTs as well as private companies such as Metro Networks, which runs 68 centers nationwide, operate traffic control centers. Traffic data is gathered by electronic and human means, crunched, and then transmitted to radio and TV stations, Web sites, and wireless service providers. Dan Baum and Sarah Schmidt, *Free-Market Gridlock*, Wired (November 2001), http://www.wired.com/wired/archive/9.11/usa-traffic.html. There are, for example, 3 VDOT operated traffic control centers in VA http://www.virginiadot.org/infoservice/news/newsrelease-stcalerts.asp and 8 GDOT traffic centers in GA.

55 *Security Guidance for the Petroleum Industry*, American Petroleum Institute (March 2002).

56 See in particular the work at the Energy and Critical Resources program at the Sandia National Laboratories, http://www.sandia.gov/ programs/energy-infra.

57 Marsh, R. T., *Critical Foundations: Protecting America's Infrastructure*, President's Commission on Critical Infrastructure Protection (October 1997), http://cyber.law.harvard.edu/is03/Readings/critical_infrastructures.pdf.

58 *Information & Communications Sector: National Strategy for Critical Infrastructure and Cyberspace Security* (May 2002).

59 *Chicago Metropolitan Area, Critical Infrastructure Protection Program: Critical Infrastructure Assurance Guidelines for Municipal Governments Planning for Electric Power Disruptions*, DOE, Metropolitan Mayors Caucus, City of Chicago (February 2001).

60 *Energy: The First Domino in Critical Infrastructure*, Computer World (September 2002).

61 790 GW is net summer capacity at electricity-only plants. Total capacity for all sectors is 855 GW, which includes independent power producers, commercial plants, and industrial plants. *Annual Energy Review 2001*, EIA, http://www.eia.doe.gov/emeu/aer/contents.html.

62 Total additions to grid capacity were 12 GW in 1999, 31 GW in 2000, and 48 GW in 2001. *Annual Electric Generator Report - Utility* and *Annual Electric Generator Report—Nonutility*, EIA (1999, 2000, 2001).

63 *U.S. Industrial Battery Forecast*, Battery Council International (April 2002).

64 Hurricane Fran left 792,000 Progress Energy customers without power and was designated by FEMA as the "largest concentrated power outage caused by a hurricane in U.S. history." Progress Energy, http://www.progress-energy.com/aboutenergy/learningctr/stormtips/hurricanespast.asp. See also National Oceanic and Atmospheric Administration, http://www.noaa.gov/hurricanean-drew.html. Ice storms in the Northeast in 1998 and in the Carolinas in 2002 left hundreds of thousands without power.

65 *IEEE Recommended Practices for the Design of Reliable Industrial and Commercial Power Systems*, Institute of Electrical and Electronics Engineers (IEEE), (IEEE Std 493–1997).

66 *IEEE Recommended Practices for the Design of Reliable Industrial and Commercial Power Systems*, IEEE, (IEEE Std 493–1997).

67 *Making the Nation Safer: The Role of Science & Technology in Countering Terrorism*, National Academy of Sciences, National Research Council (2002).

68 *1999 Commercial Buildings Energy Survey: Consumption and Expenditures*, EIA.

69 *Manufacturing Consumption of Energy 1998*, EIA.

70 *IEEE Recommended Practices for Emergency and Standby Power Systems for Industrial and Commercial Applications*, IEEE (Revision of IEEE Std 446–1987) (December 1995).

71 *IEEE Recommended Practices for Protection and Coordination of Industrial and Commercial Power Systems*, IEEE (IEEE Std 242–2001).

72 *1999 Commercial Buildings Energy Survey: Consumption and Expenditures*, EIA.

73 Statistical average calculated from total sub-sector annual electric use based on assumption of 100 hr/wk operation of all buildings yielding an approximation not accounting for peak building demand or high (or low) duty cycle buildings.

74 *The Cost of Power Disturbances to Industrial & Digital Economy Companies*. Electric Power Research Institute (EPRI), prepared by Primen (June 2001).

75 *Manufacturing Consumption of Energy 1998*, EIA.

76 The Manufacturing Institute, *The Facts About Manufacturing*. http://www.nam.org/secondary.asp?TrackID=&CategoryID=679.

77 University of Delaware's Disaster Research Center quoted in *Disaster Recovery for Business, Operation Fresh Start*. DOE, http://www.sustainable.doe.gov/freshstart/business.htm.

78 See *Dig More Coal. The PCs are Coming, Forbes* (May 31, 1999), and more recently *Silicon and Electronics* (February 2003), http://www.digitalpowergroup.com/Downloads/Silicon%20and%20Electronics.html.

79 *Analysis of Extremely Reliable Power Delivery Systems: A Proposal for Development and Application of Security, Quality, Reliability, and Availability (SQRA) Modeling for Optimizing Power System Configurations for the Digital Economy*, EPRI, Consortium for Electric Infrastructure to Support a Digital Society (April 2002).

80 Huber, Peter and Mark Mills, *Silicon and Electrons* (February 2003), http://www.digitalpowergroup.com/ Downloads/Silicon%20and%20Electronics.html.

81 *Office Equipment in the United Kingdom, A Sector Review Paper on Projected Energy Consumption For the Department of the Environment, Transportation, and the Regions*," (January 2000).

82 *Electricity Use Soars*, Crain's New York Business (May 19–25, 2003). "New York's electric demand continues to rise and shows little sign of abating. Unless significant generating capacity is added to the system—and soon—demand is going to overwhelm supply and reliability will be at risk," said William J. Museler, NYISO President and CEO. "Because of the two to three-year lead time to build large baseload plants, if New York is to remedy this situation it needs to get a new siting law in place, plants approved and construction commenced immediately." *New York Independent Operator Announces Summer Electricity Forecast*, NYISO Press Release (February 25, 2003).

83 *Digital Economy 2002*. Economics and Statistics Administration, U.S. Department of Commerce (February 2002), http://www.esa.doc.gov/508/esa/DIGITALECONOMY2002.htm.

84 Oliner, Stephen D. and Daniel E. Sichel, *Information Technology and Productivity: Where Are We Now and Where Are We Going*? Federal Reserve Board (May 10, 2002), http://www.federalreserve.gov/pubs/feds/2002/200229/200229pap.pdf.

85 *Annual Energy Outlook 2003*, EIA, DOE (January 2003), http://www.eia.doe.gov/oiaf/aeo/pdf/0383(2003).pdf.

86 *Top Security Threats and Management Issues Facing Corporate America*, Pinkerton (2002), http://www.pinkertons.com/threatsurvey/default.asp.

87 *Revalidation of CIP F-11, Power Systems Sustained Support (PS 3) Program*, FAA (November 12, 1998).

88 *Power Systems Sustained Support, Investment Analysis Report*, FAA (May 23, 2000).

89 *Power Systems Sustained Support, Investment Analysis Report*, FAA (May 23, 2000).

90 For the "bible" on high-power silicon systems for grid-level upgrades, see: Hingorani and Gyugyi, *Understanding FACTS: Concepts and Technology of Flexible AC Transmission Systems*, IEEE Press (2000).

91 NERC was designated as the federal government's "Electricity Sector Information Sharing and Analysis Center," to gather and interpret security-related information and disseminate it within the industry and the government. NERC has developed a best-practices document: *Security Guidelines for the Electricity Sector* (June 2002), http://www.nerc.com/~filez/cipfiles.html.

92 Currently, United Technologies' 200-kW ONSI is the only commercially available fuel cell; roughly 100 of these molten carbonate systems are installed

93 *Environmental Assessment: The Regenesys Energy Storage System*, Tennessee Valley Authority (August 2001), http://www.tva.com/environment/reports/regenesys/chapter_2.pdf.

94 See for example: *Interconnection and Controls for Reliable, Large Scale Integration of Distributed Energy Resources; Consortium for Electric Reliability Technology Solutions, Grid of the Future White Paper*, DOE (December 1999); *Distributed Energy Resources Interconnection Systems: Technology Review and Research Needs*, DOE National Renewable Energy Laboratory, (September 2002).

95 *Onsite Power for C&I Customers*, Chartwell Inc. (2002)

96 *Standard for Emergency and Standby Power Systems*, NFPA (2002).

97 *AEP Dedicates First U.S. Use of Stationary Sodium Sulfur Battery*, PRNewswire (September 23, 2002).
98 See Powerware UPS systems incorporating the Active Power flywheel.
99 Data from DOE/EIA *Commercial Building data*, Petroleum Distributors Association and the FAA.
100 *Sandia SCADA Program: High-Security SCADA LDRD Final Report*, Sandia National Laboratories (April 2002), http://infoserve.sandia.gov/cgi-bin/techlib/access-control.pl/2002/020729.pdf.
101 Fairfax, Steven. *Credit Cards Lessons for Life-Support Systems*, Mtechnology (July 30, 2001).
102 *Power Systems Sustained Support: Investment Analysis Report*, FAA (May 23, 2000).
103 *Private Sector Spending*, Homeland Security & Defense (July 16, 2003) (From a June 2002 analysis by Deloitte Consulting).
104 *Skeptical of Attacks*. . . Homeland Security & Defense (October 23, 2002) (Less than half of 230 companies surveyed by the Council on Competitiveness are spending any more on security than they were a year ago).
105 Survey sponsored by ASIS International, an Alexandria, Va., organization of security professionals.
106 Winona Daily News (June 15, 2003), http://www.winonadailynews.com/articles/2003/06/15/news/03lead.txt.
107 Pacific Gas and Electric, for example, first targets restoration of generation and distribution facilities serving public service and emergency service agencies like hospitals, police, fire, water pumping stations, communication facilities, and critical service to small groups or individuals. (http://www.pge. com/004_safety/004c9_restoration_po.shtml) The Rockland Electric Company of Rockland New Jersey offers a 24-hour hotline and first priority status for restoration of service to residential customers who rely on life support equipment such as kidney dialysis machines, apnea monitors, oxygen concentrators, respirators, ventilators, and infusion feeding pumps. (http://www.oru.com/publications/RECO-RR2001.pdf) Tampa Electric restores power first to hospitals, disaster centers, and police and fire stations. It then concentrates on water and sewer installations, followed by telephone service and residential customers who depend on power for life-support systems. (http://www.tampaelectric.com/TENWRelease091499a.html)
108 See for example: *Blackstart Regional Restoration Plan*. Southeastern Electric Reliability Council (March 14, 2003).
109 *Critical Infrastructure Assurance Guidelines For Municipal Governments: Planning For Electric Power Disruptions*. Washington Military Department, Emergency Management Division (February 2001), http://emd.wa.gov/3-map/a-p/pwr-disrupt-plng/14-app-b-franchise.htm.
110 *Power Systems Sustained Support. Investment Analysis Report*, FAA (May 23, 2000).
111 See www.tsp.ncs.gov
112 See http://www.ncs.gov/Nstac/IssueReview98/PreviouslyIssues.html
113 *Generic Standards for E9-1-1 PSAP Equipment*, National Emergency Number Association (NENA) (June 20, 1996).
114 See www.fcc.gov/hspc/emergencytelecom.html. More information appears at tsp.ncs.gov.
115 The Wastewater Committee Of The Great Lakes–Upper Mississippi River Board of State And Provincial Public Health And Environmental Managers, http://www.dec.state.ny.us/website/dow/10states.pdf
116 *1998 New York Ice Storm: Mitigation Issues & Potential Solutions*, FEMA (1998), http://www.appl.fema.gov/reg-ii/1998/nyice4.htm#ELECTRIC.
117 *The National Strategy For The Physical Protection of Critical Infrastructures and Key Assets*, The White House (February 2003).
118 See http://www.nipc.gov/sites/newrelatedsites.htm.
119 See http://www.tisp.org.
120 *Critical Infrastructure Protection: Efforts of the Financial Services Sector to Address Cyber Threats*, GAO (January 2003).
121 *Potential Terrorist Attacks: Additional Actions Needed to Better Prepare Critical Financial Market Participants*, GAO (February 2003).
122 *Interagency Paper on Sound Practices to Strengthen the Resilience of the U.S. Financial System*, Board of Governors of the Federal Reserve System; Office of the Comptroller of the Currency; and Securities and Exchange Commission (April 7, 2003).

123 *Banking and Finance Sector National Strategy*. The National Strategy for Critical Infrastructure Assurance (May 13, 2002).
124 See http://www.ntsb.gov/aviation/report.htm.
125 See http://asrs.arc.nasa.gov/overview_nf.htm.
126 See generally: *National Strategy for Critical Infrastructure and Cyberspace Security*. CTIA; ITAA; TIA; USTA (May 2002).
127 See http://www.nei.org/index.asp?catnum=2&catid=57.
128 See http://www.nrc.gov/reading-rm/doc-collections/gen-comm/gen-letters/1982/gl82004.html.
129 See, for example, *Security Guidelines for the Electricity Sector*, NERC (June 2002), http://oea.dis.anl.gov/documents/Security_Guidelines_for_the_Electricity_Sector_June_2002.pdf.
130 *Critical Foundations*, President's Commission on Critical Infrastructure Protection (October 1997).
131 See http://www.weatherdata.com/products/index.php; http://www.thorguard.com/about.asp; http://www.meteorlogix.com/products/ mxinsight.cfm.
132 Statement of Dr. Samuel G. Varnado, Sandia National Laboratories, United States House of Representatives, Committee on Energy and Commerce, Subcommittee on Oversight and Investigations (July 9, 2002).
133 *Internet Security Threat Report: Attack Trends for Q1 and Q2 2002*. Riptech (July 2002).
134 See http://www.sercl.org/minutes/mic-0303/mic0303a.pdf.
135 *Critical Infrastructure Protection Advisory Group Scope*, NERC (January 17, 2003).
136 *The National Strategy For The Physical Protection of Critical Infrastructures and Key Assets*, The White House (February 2003).
137 *Report To Congress On The Feasibility Of Establishing A Heating Oil Component To The Strategic Petroleum Reserve*, DOE (June 1998).
138 http://www.newrichmondutilities.com/business_customers/default.asp?CategoryNumber=3&SubcategoryNumber=1.
139 *Interconnection and Controls for Reliable, Large Scale Integration of Distributed Energy Resources*, Consortium for Electric Reliability Technology Solutions, Grid of the Future White Paper, DOE (December 1999).
140 *Distributed Energy Resources Interconnection Systems: Technology Review and Research Needs*, National Renewable Energy Laboratory (September 2002).
141 *IEEE Recommended Practices for the Design of Reliable Industrial and Commercial Power Systems*, IEEE (December 1997).
142 See http://www.energy.ca.gov/distgen/interconnection/ieee.html.
143 See http://ferc.gov/Electric/gen_inter/small_gen/RM02-12-000.pdf
144 In 1995, the EPA conducted a survey of oil storage facilities potentially subject to the Agency's SPCC regulation. The survey found approximately 438,000 facilities, and estimated that there were well over 1 million underground storage tanks in the country subject to SPCC oversight.
145 See http://www.distributed-generation.com/regulatory_issues.htm.
146 Broadcast Engineering, http://broadcastengineering.com/ar/broadcasting_ ups_backup_power.
147 The Security Industry Association and Real Estate Roundtable are lobbying passage of a bill introduced this past March to amend federal tax laws to allow full deduction for homeland security expenses (Public Safety and Protection Act, HR 1259). The bill would provide for 100 % expensing of a wide variety of security-related costs ranging from hardening physical premises, to software, biometrics, and "computer infrastructure."

Appendix B

BITS Guide to Business-Critical Power

1 EXECUTIVE SUMMARY

The *BITS Guide to Business-Critical Power* (the *Guide*) provides financial institutions with industry business practices for understanding, evaluating, and managing risks associated when the predicted reliability and availability of the electrical system is disrupted. Further, it outlines ways financial institutions can enhance reliability and ensure uninterrupted back-up power. The *Guide* is written for interested parties—from CEOs to business managers, risk managers to business continuity professionals, procurement experts to facilities managers—as they analyze risks, conduct due diligence for critical power, and integrate evolving regulatory and building code requirements into business continuity plans.[1]

Business practices for the financial services industry mandate continuous uptime for computer and network equipment to facilitate around-the-clock trading and banking activities anywhere and everywhere in the world. Financial services institutions are appropriately and understandably intolerant of unscheduled downtime. Business-critical power is the power that an organization absolutely requires to achieve its business objectives. Today more than ever, financial institutions are demanding continuous 24-hour system availability.

The risks to the financial services industry associated with cascading power supply interruptions from the public electrical grid in the United States have increased

[1] This document is intended only to provide suggestions on business objectives, not to provide legal advice. An appropriate legal professional should be engaged to provide such advice on a case-by-case basis.
Appendix B is reprinted with permission from © BITS.

due to the industry's ever-increasing reliance on computer and related technologies. As the number of computers and related technologies continues to multiply in this increasingly digital world, demand for reliable quality power increases as well. Without reliable power, there are no goods and services for sale, no revenues, and no profits.

The financial services industry has been innovative in the design and use of the latest technologies, driving its businesses to increased digitization in this highly competitive business environment. Achieving optimum reliability is very challenging since the supply and availability of uninterrupted, conditioned power is becoming more and more critical to the industry. For example, data centers of the past usually required the installation of stand-alone protective electrical and mechanical equipment only for computer rooms. Data centers today operate on a much larger scale 24/7. The proliferation of distributed systems using hundreds of desktop PCs and workstations connected through LANs and WANs that simultaneously use dozens of software business applications and reporting tools makes each building a "computer room." The uninterrupted power needs for any single institution are formidable, but the power requirements are truly staggering when the entire interconnected financial services industry is considered.

To provide continuous operation under all foreseeable risks of failure is a non-trivial matter and requires a holistic, enterprise approach. Communication between managers of business lines, business continuity and facilities is vital. Only when all parties fully understand the three pillars of power reliability—design, maintenance and operation—can an effective plan be funded and implemented. The costs associated with reliability enhancements are significant and sound decisions can only be made by quantifying performance benefits against downtime cost estimates.

Financial institutions cannot develop a plan to protect against threats they do not envision. This *Guide* assumes the following:

- There will be power failures that affect your financial institution.
- Financial institutions may be exposed to regulatory or fiscal penalties (monetary or customer loss) as a result of these outages.
- The only way to ensure that your financial institution will be protected is to buy and install standby power generation and/or power protection systems (hereinafter referred to as critical power) so as to make the facility independent of the public power "grid" when needed.[2]
- Reliability and facility infrastructure health are not guaranteed simply by investing in and installing new equipment. Unexpected failures can compromise even the most robust facility infrastructure if appropriate testing, maintenance and due diligence techniques are not employed.
- Financial institution personnel need to be trained and records kept.

[2] For a discussion of the range of options for standby power generation and/or power protection systems, please see Section III (Needs Analysis/Risk Assessment) and Section IV (Installation) of this Guide.

The sections that follow address each of these assumptions in more detail and endeavor to address applicability from a variety of perspectives. Each financial institution will have its own organizational structure, so it will be up to each financial institution to interpret the terms in this document according to its respective structure.

In general, the provision of critical power will often fall under the purview of the business continuity professional working in concert with business managers, risk managers and practitioners. The following definitions can be used as guidelines:

- Business Continuity Professional refers to the individual responsible for preparing and coordinating the business continuity process. Potential titles: business continuity manager; disaster recovery coordinator; business recovery coordinator.
- Business Manager refers to the individual responsible for lines of business and ensuring that the financial institution is profitable. This individual is the ultimate decision maker and the level of executive can run the gamut from the most senior executive to a line manager. Potential titles: chief executive officer; executive vice president; senior vice president; vice president; chief technology officer; chief information officer.
- Risk Manager refers to the individual responsible for evaluating exposures, and controlling exposures through such means as avoidance or transference. There are various types of risk, including operational, credit and market risk. This *Guide* deals with operational risk. Potential titles: corporate risk officer; risk management officer; chief risk officer.
- Practitioner refers to the individual who will be responsible for the implementation and maintenance of critical power. Potential titles: facilities manager; chief engineer; event manager.

A list of questions is included at the end of each section. A consolidation of all the questions is included in Appendix B.5 and items are cross-referenced to their respective section. The questions are presented in a "worksheet" format providing space so that financial institutions can indicate whether the question is applicable and record comments germane to the question.

These questions are a starting point for a rigorous examination of a financial institution's business continuity strategy for critical power needs. They may also serve as considerations in procuring adequate levels of critical power.

II THE GRID

Electricity occupies a uniquely important role in all operations of financial institutions. The loss of power takes out data and communications capabilities and virtually all of the new systems and technologies being deployed for physical and operational security.

The public electric grid is inherently vulnerable. Relatively small numbers of huge power plants are linked to millions of locations by hundreds of thousands of miles of exposed wires. Nearly all high-voltage lines run above ground and traverse open country. A handful of high-voltage lines serve entire metropolitan regions. Serious problems may propagate rapidly through the grid itself.

Most accidental grid interruptions last barely a second or two, and many "power quality" issues involve problems that persist for only tens of milliseconds (one or two cycles). In most areas of the country, grid outages of an hour or two occur, on average, no more than once or twice a year, and longer outages are even rarer. Accidental outages tend to be geographically confined as well; the most common involve blown circuits in a single building (typically caused by human error, much of it maintenance related), or interruptions confined to the area served by a single utility substation.

There is normally very little risk that several high-voltage lines feeding a metropolitan area from several different points on the compass will fail simultaneously, and when just one such line fails, all the resources at hand can be mobilized to repair it. Deliberate assaults, by contrast, are much more likely to disable multiple points on the network simultaneously. A 2002 National Academy of Sciences report drove this reality home, observing starkly: "[A] coordinated attack on a selected set of key points in the [electrical] system could result in a long-term, multistate blackout. While power might be restored in parts of the region within a matter of days or weeks, acute shortages could mandate rolling blackouts for as long as several years."[3] Operations that can afford simply to shut down and wait out short blackouts may not be able to take that approach in response to the mounting threats of longer outages.

Understanding the Grid

The national electric grid is a vast, sprawling, multi-tiered structure that reaches everywhere and is used by everyone. Measured by route miles and physical footprint, the North American grid is by far the largest network on the planet.

Architecturally similar arrays of generators, wires, switches, and transformers appear within each of the grid's principal tiers. The generation and transmission tiers at the top have stadium-sized, gigawatt-scale power plants and commensurately high-voltage wires, building-sized transformers, truck-sized capacitors, and arrays of mechanical, electromechanical, and electronic relays and switches. The distribution tiers in the middle have tennis-court-sized, megawatt-scale substations, van-sized transformers, and barrel-sized transformers mounted ubiquitously on poles and in underground vaults. The bottom tiers transform, condition, and distribute power within factories, commercial buildings, and homes via power-distribution units, lower-voltage on-premise grids, and dispersed switches, batteries, and backup systems further downstream.

[3] *Making the Nation Safer: The Role of Science & Technology in Countering Terrorism*, National Academy of Sciences, National Research Council (2002).

The top tier of the grid is typically fueled by coal, uranium, water, or gas; each lower tier is typically "fueled" initially by the electric power delivered from the tier above. Generating stations in the top tier dispatch electrical power through some 680,000 miles of high-voltage, long-haul transmission lines, which feed power into 100,000 substations. The substations dispatch power, in turn, through 2.5 million miles of local distribution wires. Nearly all high-voltage lines run aboveground and traverse open country, and a handful of high-voltage lines serve entire metropolitan regions. At the same time, a couple of large power plants can provide all the power required by a city of a half-million. Many communities are served by just a handful of smaller power plants, or fractional shares of a few bigger power plants.

In the most primitive architecture, the grid includes only a power plant and wires. Power is generated at the top tier, consumed at the bottom, and transported from end to end by a passive, unswitched, trunk-and-branch network. This was the structure of the very first grid, from Edison's Pearl Street station in New York, in 1882. The higher up things fail, the more widely the failure is felt.

The modern grid is, of course, much more robust. Many different power plants operate in tandem to maintain power flows over regions spanning thousands of miles. In principle, segments of the grid can be cut off when transformers fail or lines go down, so that failures can be isolated before they propagate to disrupt power supplies over much larger regions. The effectiveness of such failure isolation depends on the level of spending on the public grid, which has been in decline for years. Identical strategies of isolation and redundancy are used on private premises to make the supplies of power to critical loads absolutely assured, insulating those loads from problems that may affect the grid.

Switches control the flow of power throughout the grid, from power plant down to the final load. "Interties" between high-voltage transmission lines in the top tiers allow even the very largest plants to supplement and back each other up. Distributed generation facilities in the middle tiers can power smaller segments of the grid and keep them lit even when power is interrupted in the highest tiers. When power stops flowing through the bottom tiers of the public grid, critical-power circuits on private premises are isolated and private, on-premises generators kick in.

Much of the critical-infrastructure literature refers to the grid as a single structure and thus implicitly treats it as "critical" from end to end. But utilities themselves necessarily prioritize and rank the customers and loads they are expected to serve. Large power plants and high-voltage underground cables that serve densely populated urban areas obviously require more protection before they fail, and more urgent attention after, than small plants and rural distribution lines. In defining priorities and deploying new facilities, collaboration between utilities and critical power customers is becoming increasingly more important. Most notably, power is critical for the continued provision of other critical services—those provided by E911, air traffic control, wireline and wireless carriers, emergency response crews, and hospitals, among others.

Hardening of the grid begins at the top tier, in generation and transmission facilities. Much of modern grid's resilience is attributable to the simple fact that "interties" knit local or regional grids into a highly interconnected whole, so that

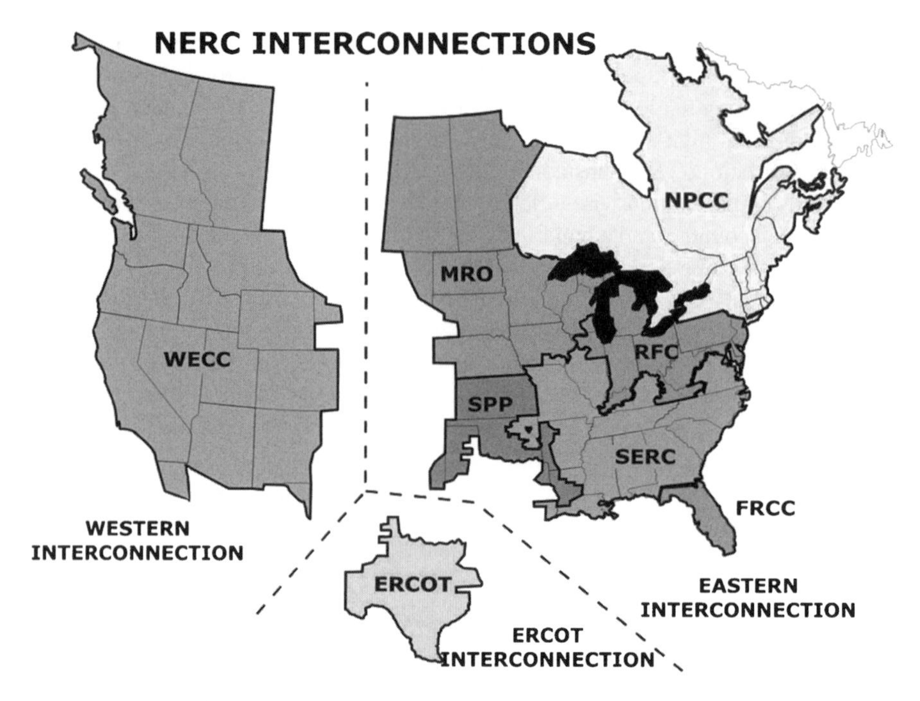

The main interconnections of the U.S. electric power grid and the 10 North American Electric Reliability Council (NERC) regions. (*Source*: North American Electric Reliability Council.)

any individual end-user may receive power from many independent power plants, often located hundreds (or even thousands) of miles apart. Promoting development of this resilient architecture is the primary mission of the North American Electric Reliability Council (NERC).

It is important to note that there is no "national power grid" in the United States. In fact, the continental United States is divided into three main power grids[4]:

- The Eastern Interconnected System, or the Eastern Interconnect
- The Western Interconnected System, or the Western Interconnect
- The Texas Interconnected System, or the Texas Interconnect.
- ERCOT—Electric Reliability Council of Texas
- FRCC—Florida Reliability Coordinating Council
- MRO—Midwest Reliability Organization
- NPCC—Northeast Power Coordinating Council
- RFC—Reliability *First* Corporation

[4] Reprinted from the Department of Energy's website.
http://www.eere.energy.gov/de/us_power_grids.html

- SERC—Southeastern Electric Reliability Council
- SPP—Southwest Power Pool
- WECC—Western Electricity Coordinating Council

The Eastern and Western Interconnects have limited direct current interconnections with each other; the Texas Interconnect is also linked with the Eastern Interconnect via direct current lines. Both the Western and Texas Interconnects are linked with Mexico, and the Eastern and Western Interconnects are strongly interconnected with Canada. All electric utilities in the mainland United States are connected with at least one other utility via these power grids.

The grid systems in Hawaii and Alaska are much different than those on the U.S. mainland. Alaska has an interconnected grid system, but it connects only Anchorage, Fairbanks, and the Kenai Peninsula. Much of the rest of the state depends on small diesel generators, although there are a few minigrids in the state as well. Hawaii also depends on minigrids to serve each island's inhabitants.

New interties provide an effective way to boost overall reliability and create capacity margins without building new plants; new interties also facilitated the wholesale trading that regulators authorized in the 1990s. So long as there is sufficient margin in the generating capacity and redundancy in wires, and sufficiently fast and accurate control of the key switches that isolate faults, the failure of any one power plant or line in the top tiers of the grid should not be discernible by end users at the bottom.

After September 11, 2001, NERC expanded their recently created Critical Infrastructure Protection Advisory Group (currently the Critical Infrastructure Protection Committee [CIPC]) to bring together the public utilities responsible for securing the thousands of miles of long-haul, high-voltage wires and the 7000 transmission-level substations. Among other initiatives, CIPC has formed a working group to inventory and develop a database of "critical spare equipment." Given the customization required in the high-voltage transmission system (which often uses custom-built, high-power hardware such as massive substation transformers), maintenance of spare equipment can be challenging. CIPC's database helps to assure overall continuity of operation through the intelligent sharing of standby assets.

Complementary discussions are addressing the possibility of creating a critical-equipment warehousing system, with geographically dispersed warehousing of spares, and cost-sharing by the potential beneficiaries of such planning. This solution has already been implemented for power line ("telephone") poles and is now being applied to fleets of substation-scale generators-on-wheels.

Very large end-users rely on similar intertie strategies one tier lower down in the grid to help secure their specific critical power needs. In such configurations, the key "switch" controlling the dual feeds will often be a dedicated utility substation located on the doorstep of a factory, office park, or data center. By deploying additional substations in close collaboration with major critical-load customers, utilities shrink the footprint (i.e., reduce the number of customers affected) by failures that occur elsewhere. More substations create more points at which to interconnect independent parts of the grid so that distant transmission lines and power plants effectively back each other up.

Substations can also serve as sites for utility deployment of distributed generation. With the addition of its own generating capacity, the substation is "sub" no longer—it becomes a full-fledged "mini-station." Opportunities for deploying new generating capacity at this level of the grid, either permanently or when emergencies arise, are expanding as large electromechanical switches and related components are being replaced by new solid-state technologies that have much smaller footprints.

By such means, much can be, and is being, done to lower the likelihood of a loss of grid power needed by the most critical loads. Closer collaboration between utilities and their largest customers is now needed to advance such initiatives in the distribution tiers of the grid. Nevertheless, the grid is inherently subject to external factors that can cause outages, and there is only so much that can be feasibly done to secure it against these possible problems. Guaranteeing supplies of critical power at high-demand, high-reliability locations, such as data centers and communications hubs, ultimately means adding increasingly sophisticated on-site generating capacity and storage to back up whatever is being done to improve reliability higher in the grid.

Power Restoration

The first essential step in restoring power after a major outage is to isolate faults and carve up the grid into smaller, autonomous islands. From the perspective of the most critical loads, the restoration of power begins at the bottom, with on-site power instantly cutting in to maintain the functionality of command and control systems that are essential in coordinating the step-by-step restoration of the larger whole.

Major utilities establish electric service priority protocols to prioritize power restoration efforts after a major outage, typically targeting hospitals and other emergency services. Such programs acknowledge that certain users and uses are atypically dependent on supplies of electric power and suffer unusually serious consequences when their power fails. Thus, they are given priority over business restoration. All such priorities are swept aside, however, when a high-level failure cuts off power to an entire region. Then, the focus is almost entirely on the systematic restoration of power from the top down, beginning with the highest-power stations, trunks, and switching centers, in a process structured largely to minimize damage to the utility's own, most essential equipment.[5] It is thus likely in both cases of isolated and of widespread, extended outages that financial facilities will remain relatively low in the restoration priority hierarchy. Ensuring the security and continuity of critical power will increasingly require facility-specific planning and systems, with relevant coordination with, and occasionally engineering help from, local utilities.

[5] See, for example, *Blackstart Regional Restoration Plan*, Southeastern Electric Reliability Council (March 14, 2003).

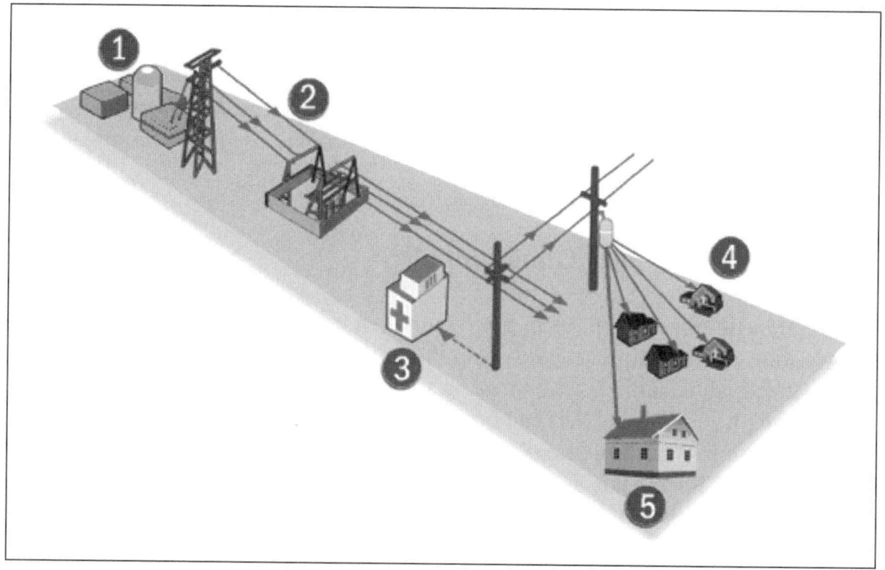

Power restoration process.

While each utility has its own restoration plan, there are elements that will likely be common. The following plan is duplicated here courtesy of JEA, which owns, operates, and manages the electric system established by the City of Jacksonville in 1895. JEA is the largest community-owned utility in Florida and the eighth largest in the United States.

1. The first step is damage assessment, which includes physical inspections of facilities and plants. Once damage assessments have been made, repairs begin.
2. Repairs begin at generating facilities and transmission lines from plants, and to water and wastewater treatment facilities.
3. Next, main line repairs begin on electric circuits, water systems, and sewer systems that serve critical facilities such as hospitals, police stations, and fire stations.
4. The goal is to restore services to the greatest number of customers as soon as possible.
5. Once the large impact areas have had power restored, restoration of power begins for small pockets or individuals still without power.

NERC published Standard EOP-005-0[6] to "ensure plans, procedures, and resources are available to restore the electric system to a normal condition in the

[6] ftp://www.nerc.com/pub/sys/all_updl/standards/rs/EOP-005-0.pdf

event of a partial or total shut down of the system." As mentioned earlier, each utility may have a predetermined order in which it restores power. It is imperative that financial institutions understand "if" their utility has a prioritization order; "what" the prioritization order is; and "if and where" within that order the financial services industry falls. Specific questions are outlined in the section that immediately follows.

Electric Power/Utility Questionnaire[7]

Financial institutions should be in touch with their electric power utility to establish the working relationship needed to help through real or potential service interruptions. Check with facilities management and/or accounts payables offices as staff may already have contacts that can be used to establish a business continuity planning partnership. Communicate directly with the provider and ask to speak with the person responsible for business continuity, crisis or emergency management. The following guidelines may help as you try to ascertain information:

- Explain that you need information to help in developing a plan.
- Refer to regulatory guidelines or business continuity requirements that necessitate planning with critical infrastructure providers.
- Seek to establish an ongoing relationship with your counterpart. Such a relationship can be the difference between getting access to good information at the time of disaster and getting only that information available to the general public.
- Get involved in crisis management testing performed by the utility.
- Understand the utility's operating strategy and position within any larger regional or national infrastructure.

Questions

Item	Description
	General
1.	Do you have a working and ongoing relationship with your electric power utility?
2.	Do you know who in your financial institution currently has a relationship with your electric power utility—i.e., facilities management or accounts payable?
3.	Do you understand your electric power utility's electric service priority" protocols?
4.	Do you understand your electric power utility's restoration plan?
5.	Are you involved with your electric power utility's crisis management/disaster recovery tests?
6.	Have you identified regulatory guidelines or business continuity requirements that necessitate planning with your electric power utility?

[7] This section is excerpted from and reprinted with the permission of the Securities Industry Association Business Continuity Planning Committee's "Critical Infrastructure Guide" published in January 2005.

Specifically for an Electric Power Utility

7. What is the relationship between the regional source power grid and the local distribution systems?
8. What are the redundancies and the related recovery capacity for both the source grid and local distribution networks?
9. What is the process of restoration for source grid outages?
10. What is the process of restoration for local network distribution outages?
11. How many network areas are there in (specify city)?
12. What are the interrelationships between each network segment and the source feeds?
13. Does your infrastructure meet basic standard contingency requirements for route grid design?
14. What are the recovery time objectives for restoring impacted operations in any given area?
15. What are recovery time objectives for restoring impacted operations in any given network?
16. What are the restoration priorities to customers—both business and residential?
17. What are the criteria for rating in terms of service restoration?
18. Where does the financial services industry rank in the priority restoration scheme?
19. How do you currently inform clients of a service interruption and the estimated time for restoration?
20. What are the types of service disruptions, planned or unplanned, that (specify city) could possibly experience?
21. Could you provide a list of outages, type of outage and length of disruption that have affected (specify city) during the last 12 months?
22. What are the reliability indices, and who uses them?
23. During an outage, would you be willing to pass along information regarding the scope of interruptions to a central industry source—for example, a financial services industry business continuity command center?
24. Are the local and regional power utilities cooperating in terms of providing emergency service? If so, in what way? If not, what are the concerns surrounding the lack of cooperation?
25. Would you be willing to provide schematics to select individuals and/or organizations on a non-disclosure basis?
26. Could you share your lessons learned from the events of 9/11 and the regional outage of 8/14/03?
27. Are you familiar with the "Critical Infrastructure Assurance Guidelines for Municipal Governments" document written by the Washington Military Department Emergency Management Division? Is so, would you describe where (specify city) stands in regard to the guidelines set forth in that document?
28. Independent of the utility's capability to restore power to its customers, can you summarize your internal business continuity plans, including preparedness for natural and manmade disasters (including, but not limited to, weather-related events, pandemics, and terrorism)?

III NEEDS ANALYSIS/RISK ASSESSMENT

Evolving Risk Landscape

Planning rationally for infrequent but grave contingencies is inherently difficult. Financial institutions that have prepared properly for yesterday's risk profiles may be unprepared for tomorrow's. The risk-of-failure profiles of the past reflect the relatively benign threats of the past—routine equipment failures, lightning strikes on power lines, and such small-scale hazards as squirrels chewing through insulators or cars colliding with utility poles. Now, in addition to concerns about weather-related outages (hurricanes and ice storms in particular), as well as recent experiences underscoring the possibility of widespread operational outages, there is as well the possibility of deliberate attack on the grid. The latter changes the risk profile fundamentally—that possibility poses the risk of outages that last a long time and that extend over wide areas. The planning challenge now shifts from issues of power *quality* or *reliability* to issues of business *sustainability*. Planning must now take into account outages that last not for seconds, or for a single hour, but for days.

Events such as the terrorist attacks of September 11th, the Northeast Blackout of 2003, and the 2006 Hurricane season have emphasized our interdependencies with other critical infrastructures—most notably power and telecommunications.[8] There are numerous national strategies and sector specific plans—all of which highlight the responsibility of the private sector for "building in increased resiliency and redundancy into business processes and systems."[9] These events have also prompted the promulgation and revision of laws, regulations, and policies governing reliability and resiliency of the power industry. Some of these measures also delineate controls required of some critical infrastructure sectors to maintain business-critical operations during a critical event. An FI's ability to comply with these legislative and regulatory measures is therefore impacted by its success in providing its key facilities with critical power. A more detailed discussion of the regulatory environment can be found later in this section.

The need to provide continuous operation under all foreseeable risks of failure, such as power outages, equipment breakdown, internal fires, natural phenomena, and terrorist attacks, requires use of many techniques to enhance reliability. These techniques include redundant systems and components, standby power generation and UPS systems, automatic transfer and static switches, and the use of probability risk analysis modeling software to predict potential future outages and develop maintenance and upgrade action plans for all major systems.

[8] For a review of steps financial institutions can take to enhance resiliency and business continuity specific to telecommunications see the *BITS Guide to Business-Critical Telecommunications Services*, published in 2004.

[9] National Infrastructure Protection Plan, Draft Version 6.0, January 2006.

Critical Power Options

Financial institutions require highly sophisticated uninterruptible power systems, generators, HVAC systems, transfer switches, and other high-voltage electrical and mechanical systems to ensure fail safe power delivery and operation. The first step in the needs analysis process is to determine the level of criticality. Individual institutions establish their level of criticality based on their own criteria and their customers' criteria. The primary criterion is availability. What is the impact if a specific facility is lost for a day, an hour, or three minutes? Is failure an option?

Traditionally, when discussing critical facilities, there is an assumption that those facilities are data centers. This *Guide* broadens the definition of "critical facility" to encompass any location where a critical operation is performed. So, a critical facility can include, but is not limited to, all work area environments such as branch backroom operations facilities, headquarters, or data centers. The critical level of an operation dictates the reliability and the security of the facility in which the operation is performed. Because the cost of these facilities is driven by reliability requirements, it is critical to calculate the reliability of various design options and budget requirements. Developing innovative design strategies and, subsequently, state-of-the-art critical facilities requires a deep understanding of both design and operational issues. Clearly, design impacts operation, and operation impacts design. Redundancy adds considerable cost to the construction and operation of a mission critical facility. It is critical to optimize engineering design for high reliability and performance versus cost.

Reliability is one of the major cost drivers associated with construction and operations. The availability metric is commonly used for measuring reliability of critical facilities and is usually expressed as a percentage. For modern critical facilities, the benchmark availability is in the range of 99.999% ("five nines") to 99.9999% ("six nines"). To achieve six nines availability the engineered systems will have to incorporate designs that include system + system $[2(N+1)]$ redundancy. It is worth nothing that engineered systems in a critical facility are often overdesigned to include too much redundancy. That is, systems become more complex than they need to be, which leads to decreased reliability.

Defining the Four Tiers of Critical Facilities

Critical facilities can be generally classified by Tiers, with Tier 1 being the most basic and Tier 4 being the most reliable facility. Critical functions of financial institutions will usually require a facility in the Tier III to Tier IV range. The commercial/financial industry definitions and characteristics for the four tiers are as follows:

Tier I—Basic, Nonredundant. A Tier I facility is susceptible to disruptions from both planned and unplanned activities. Of the four tiers, Tier I represents minimum attention to availability and criticality of up-time operations. It is rare that any facility, even noncritical departmental data depositories, is supported by a Tier I or even Tier II configuration since both lack backup and redundancy support. A basic nonredundant facility must be shut down completely on an annual basis in order to perform any maintenance and repair work. Urgent situations may require shutdowns on a more frequent basis. Errors in operation or spontaneous failures of site infrastructure components or distribution paths will cause disruption. Typical equipment configurations for electrical and mechanical support of the facility are represented by a single '*N*' path and backup configuration. '*N*' in such configurations represents the minimum number of systems or component (also called paths) required to operate.

Tier II—Basic, Redundant. This tier of facility is also susceptible to disruptions from both planned and unplanned activity. Except for maintenance of uninterruptible power supply (UPS) modules and other redundant capacity delivery components, a basic redundant facility must be shut down completely on an annual basis to perform maintenance and repair work to the distribution systems. Urgent situations may require shutdowns on a more frequent basis. Errors in operation or spontaneous failures of site infrastructure distribution paths will cause a facility disruption. Unexpected failures of capacity components may cause a facility disruption.

Tier III—Concurrently Maintainable. This tier provides for any planned activities to be conducted without disrupting the computer hardware operation in any way. Planned activities include preventative and programmable maintenance, repair and replacement of components at the end of their lives, addition or removal of capacity components, testing of components and systems, or reliability-centric maintenance (RCM). This means that planned activities must never remove from service more than the redundant component or distribution path. Every component must be able to be removed from service on a planned basis without disruption to the load. This requires sufficient capacity to carry the full load on one path while performing maintenance or testing on the other path. Unplanned activities such as errors in operation or spontaneous failures of facility infrastructure distribution paths will cause a facility disruption. Unexpected failures of components may cause a facility disruption.

Tier IV—Fault and Failure Tolerant. This tier facility can withstand at least one unplanned failure, error, or event with no critical load impact. Fault-tolerant functionality also provides the ability to permit the conduct of any planned activity without disrupting the critical load in any way. With this level of capability, the site infrastructure generally has the same degree of flexibility and redundancy as the information technology architecture. At this level, any component must be able to fail without disruption to the load.

[10] "Tier Classifications Define Site Infrastructure Performance," A white paper by W. Pitt Turner, IV., John H. Seader and Kenneth G. Brill, Uptime Institute, www.uptimeinstitute.org.

Load Classifications

Critical, Essential or Discretionary Load

The key to overall effectiveness is the ability to completely and accurately fulfill missions within the critical facility and within the network of facilities. The backup power system typically does not provide power to all building systems. The first step in the design process is to identify risks—that is, the impact that a power outage would have on the facility and the operations. A key component of this risk assessment is determining which loads need to be connected to the back up power system. The classifications of loads that support critical facilities are referred to as critical, essential or discretionary.

Critical Load. That portion of the load requiring 100% continuity in power service. This equipment must have an uninterrupted power input to prevent damage or loss to a facility (including its information technology and infrastructure), to its specific business functions, or to prevent danger of injury to personnel.

Essential Load. That portion of the load that directly supports routine accomplishment of site operations. Generally, these are loads that can tolerate power outages without loss of data and without adversely affecting the vital mission.

Discretionary Load. That portion of the load that indirectly supports the operations at the facility. This generally means loads associated with administration and office functions. This load is often referred to as "sheddable" and can be curtailed without overall adverse impact on the facility or the business.

Systems Requirements

There are further detailed requirements for each discipline (e.g., electrical, mechanical, architectural), subsystem (e.g., AC power, DC power, backup power, emergency power), and element of the subsystem. This detail is often manifest in checklists with which individual facility and IT engineers can assess their situation, measure the reality of their hardware and software against industry standards, and plan and budget for achieving compliance.

These requirements could also map into a compliance matrix of assessment checklists and tier rankings. Such a matrix would enable those at the highest levels to understand their institution's level of preparedness without having to weed through reports that may or may not accurately state their readiness or vulnerabilities.

For electrical systems, a complete analysis must include on-site power generation, electrical distribution, service entrance (utility feed), uninterruptible power supplies (UPS) and batteries (for critical systems), generators (for essential systems), power distribution units (PDU), branch circuits, fire alarm systems, grounding systems, and conduit and static transfer switches to switch between power sources in response to interrupted power. Although a totally critical system, the electrical system is not the only system that requires evaluation.

Mechanical system assessment must include the mechanical plant (e.g., control plant chillers, cooling towers, pumps, piping, fluid coolers, air compressors), strategies for airflow and rack cooling (including high density racks), fuel oil and other fluid storage and distribution, generator exhaust, air conditioning units, direct expansion system, and humidification. Potential failure modes for mechanical systems, such as floods, very low temperatures affecting outdoor mechanical equipment, and makeup water for cooling towers must be addressed.

Architectural issues include space adjacencies, physical separation of systems, minimum seismic requirements (tailored locally and approved regionally), raised floor issues (bolted stringers, colored aisle tiles, field tile loading capacity, pedestal and base anchorage), natural disaster planning and avoidance, physical security, and managing vendor activities during service and maintenance events.

Reliability Modeling and Probabilistic Risk Assessment

Once these determinations are made, the process can begin to develop a complete facility assessment to determine needs and budget constraints. Recognizing that reliability is a major cost-driver for these facilities, there are some leading-edge approaches that can be applied to optimize the ratio between reliability and budget requirements. These approaches should follow established guidelines for designing reliability and maintainability into the facility. This methodology needs to be included in the programming/conceptual phase of the design. Each design alternative needs to quantify performance (reliability and availability) using Probabilistic Risk Assessment Analysis against cost analysis, in order to optimize the main decisions in the initial phase of the project.

Probabilistic Risk Assessment is a decision tool that has been gaining utility in the design and construction industry, especially for facilities designed with various system redundancies and with millions of dollars of assets at risk. This technique incorporates reliability models and provides for quantitative assessment of decision making. Traditionally used extensively in the aviation, nuclear, and chemical industry, the analysis of these types of redundancies requires the use of probability simulations to calculate failure rate, reliability, availability, unreliability, and unavailability.

Reliability modeling quantifies the resilience of critical nodes in the facility using matrices such as probability of failure or availability. The reliability evaluation process includes the following steps:

- Analyze existing systems and calculate reliability of the present environment.
- Develop recommendations and solutions for improving system resiliency.
- Calculate reliability of the upgraded configuration.
- Estimate costs of the upgrades and improvements.

Reliability predictions are only as good as the ability to model the actual system. In past reliability studies, major insight was gained for various topologies of the electrical distribution system using standards from IEEE Standard 493-1997, Recommended Practice for the Design of Reliable Industrial and Commercial Power Systems. However, aspects of the electrical distribution system for a critical facility differ from other industrial and commercial facilities and reliability modeling should include other sources of information (e.g., from Telcordia, DOD military handbooks and technical manuals, and other real world data).

Regulatory Environment

The following are some of the most notable laws and guidance affecting the financial services industry, compliance with which might be impacted by success in providing its key facilities with critical power.

Sound Practices to Strengthen the Resilience of the U.S. Financial System

The Federal Reserve, the Office of the Comptroller of the Currency, and the Securities and Exchange Commission issued the *Interagency Paper on Sound Practices to Strengthen the Resilience of the U.S. Financial System* in September 2002. The purpose of the paper was to advise financial institutions of steps essential to increase financial services resilience and business continuity. The paper identifies three new business continuity objectives that are applicable to all financial services institutions. These objectives aim to relieve immediate pressure placed on the financial system by a wide-scale disruption of critical financial markets.

These objectives are: the rapid recovery and timely resumption of critical operations following a wide-scale disruption; rapid recovery and timely resumption of critical operations following the loss of inaccessibility of staff in at least one major operating location; and a high level of confidence, through ongoing use or robust testing, that critical internal and external continuity arrangements are effective and compatible. These business continuity objectives should be pursued by all financial institutions because of the high degree of interdependency in the financial services sector.

In addition to these three new business continuity objectives, the paper details four broad sound practices. Unlike the business continuity objectives, the sound practices have been identified specifically for core clearing and settlement organizations and for firms that play significant roles in critical financial markets. The paper identifies timelines for the implementation of these sound practices by such organizations.

The first sound practice is to identify clearing and settlement activities in support of critical financial markets. Second, organizations should determine appropriate recovery and resumption objectives for clearing and settlement activities in support of critical markets. Here, the paper specifies that core clearing and settlement organizations should be able to recover and resume clearing and settlement on the business day of the disruption, with a two-hour goal for recovery. Although firms that play significant roles in critical financial markets are also expected to recover in the same business day, their goal is to recover within four hours of the disruption.

The third sound practice is to maintain sufficient geographically dispersed resources to meet recovery and resumption activities. The paper advises careful examination of the geographic diversity between primary and backup sites, including diversity of transportation, telecommunications, and power infrastructure. In addition, organizations should be sure that there is diversity in the labor pool of the primary and backup sites, such that a wide-scale event would not simultaneously affect the labor pool of both sites. Appropriate diversity should be confirmed through testing. The fourth sound practice is to routinely use or test recovery and resumption arrangements. Here the paper stresses the importance of testing backup arrangements with third-party service providers, major counterparties, and even customers. Backup connectivity, capacity, and data integrity should be tested, and scenarios should include wide-scale disruptions.

FFIEC Information Technology Booklets

The Federal Financial Institutions Examination Council[11] (FFIEC) issued an Information Technology Booklet on Business Continuity Planning in May 2003. The booklet outlines broad expectations for financial institutions and includes the procedures by which federal examiners evaluate the adequacy of a financial institution's business continuity plan and information security program. The booklet includes references to power issues including, for example, the expectation that financial institutions conduct comprehensive business impact analyses (BIA) and risk assessments. The BIA should identify and prioritize business functions, and it should state the maximum allowable downtime for critical business functions. It should also estimate data loss and transaction backlog that may result from critical business function downtime. Examiners review, among other things, a financial institution's risk assessment to ensure that it includes disruption scenarios and the likelihood of disruption affecting information services, technology, personnel, facilities, and service providers. Such disruption scenarios should include both internal and external sources, such as natural events (e.g., fires, floods, severe weather), technical events (e.g., communication failure, power outages, equipment and software failure), and malicious activity (e.g., network security attacks, fraud, terrorism).

Basel II Accord

The globalization of financial services firms and the rise of sophisticated information technology has made banking become more diverse and complex. The Basel

[11] The FFIEC is composed of the Board of Governors of the Federal Reserve System (FRB), the Federal Deposit Insurance Corporation (FDIC), the National Credit Union Administration (NCUA), the Office of the Comptroller of the Currency (OCC), and the Office of Thrift Supervision (OTS).

Capital Accord was first introduced in 1988 by the Bank for International Settlements (BIS) and was then updated in 2004 as the Basel II Accord. Basel II provides a regulatory framework that requires all internationally active banks to adopt similar or consistent risk-management practices for tracking and publicly reporting their operational, credit, and market risks. Basel II's risk management guidelines implicate the collection, storage, and processing of data. Financial organizations also must implement operational controls, such as power protection strategies to ensure reliability, availability, and security of their data and business systems, to minimize operational risk.

National Fire Protection Association (NFPA) 1600–Standard on Disaster/Emergency Management and Business Continuity Programs—2004 Edition[12]

NFPA 1600 provides a standardized basis for disaster/emergency management planning and business continuity programs in private and public sectors by providing common program elements, techniques, and processes.

Other noteworthy laws and/or guidance with less specific applicability to financial services are:

- **U.S. PATRIOT Act of 2001**. Designed to support counter terrorism activities, threat assessment, and risk mitigation; to support critical infrastructure protection and continuity; to enhance law enforcement investigative tools; and to promote partnerships between government and industry.
- **Sarbanes–Oxley Act of 2002 (SOX)**. Enacted in response to major instances of corporate fraud, abuse of power, and mismanagement. SOX required major changes in securities law, reviews of policies and procedures with attention to data standards and IT processes. A key part of the review process is how vital corporate data are stored, managed, and protected. From a power protection standpoint, the focus is on availability, integrity, and accountability. Critical data must not be lost or corrupted, tampered with, or rendered unavailable.
- **High-Level Principles for Business Continuity**. In December 2005. The Joint Forum[13] released High-Level Principles for Business Continuity. The paper proposes to set out a framework for international standard setting organizations and national financial authorities. These organizations identify seven of high-level business continuity principles to increase resiliency to a major operational disruption. The principles are
    - Assigning responsibility for business continuity to an institution's board of directors and senior management.

[12] NFPA 1600 Standard on Disaster/Emergency Management and Business Continuity Programs—2004 Edition, Copyright © 2004, National Fire Protection Association, All Rights Reserved.
[13] The Joint Forum is composed of the Basel Committee on Banking Supervision, the International Organization of Securities Commissions and the International Association of Insurance Supervisors.

- Planning for the occurrence of and recovery from major operational disruptions.
- Establishing recovery objectives proportionate to the risk posed by a major operational disruption.
- Including organizational and external communications in business continuity plans.
- Including cross-border communication components in business continuity plans.
- Testing of business continuity plans. Evaluation and, if necessary, updating should follow continuity testing.
- Reviewing of institutional business continuity plans by financial authorities.

Questions

Item	Description
29.	How much does each minute, hour, or day of operational downtime cost your company if a specific facility is lost?
30.	Have you determined your recovery time objectives for each of your business processes?
31.	Does your financial institution conduct comprehensive business impact analyses (BIA) and risk assessments?
32.	Have you considered disruption scenarios and the likelihood of disruption affecting information services, technology, personnel, facilities, and service providers in your risk assessments?
33.	Have your disruption scenarios included both internal and external sources, such as natural events (e.g., fires, floods, severe weather), technical events (e.g., communication failure, power outages, equipment and software failure), and malicious activity (e.g., network security attacks, fraud, terrorism)?
34.	Does this BIA identify and prioritize business functions and state the maximum allowable downtime for critical business functions?
35.	Does the BIA estimate data loss and transaction backlog that may result from critical business function downtime?
36.	Have you prepared a list of "critical facilities" to include any location where a critical operation is performed including all work area environments such as branch backroom operations facilities, headquarters or data centers?
37.	Have you classified each critical facility using a critical facility ranking/rating system such as the Tier I, II, III, IV rating categories?
38.	Has a condition assessment been performed on each critical facility?
39.	Has a facility risk assessment been conducted for each of your key critical facilities?
40.	Do you know the critical, essential, and discretionary loads in each critical facility?
41.	Must you comply with the regulatory requirements and guidelines discussed in this chapter?
42.	Are any internal corporate risk and compliance policies applicable?
43.	Have you identified business continuity requirements and expectations?

44. Has a gap analysis been performed between the capabilities of each company facility and the corresponding business process recovery time objectives residing in that facility?
45. Based on the gap analysis, have you determined the infrastructure needs for your critical facilities?
46. Have you considered fault tolerance and maintainability in your facility infrastructure requirements?
47. Given your new design requirements, have you applied reliability modeling to optimize a cost effective solution?
48. Have you planned for rapid recovery and timely resumption of critical operations following a wide-scale disruption?
49. Following the loss of inaccessibility of staff in at least one major operating location, how will you recover and timely resume critical operations?
50. Are you highly confident, through ongoing use or robust testing, that critical internal and external continuity arrangements are effective and compatible?
51. Have you identified clearing and settlement activities in support of critical financial markets?
52. Do you employ and maintain sufficient geographically dispersed resources to meet recovery and resumption activities?
53. Is your organization sure that there is diversity in the labor pool of the primary and backup sites, such that a wide-scale event would not simultaneously affect the labor pool of both sites?
54. Do you routinely use or test recovery and resumption arrangements?
55. Are you familiar with National Fire Protection Association (NFPA) 1600—Standard on Disaster/Emergency Management and Business Continuity Programs, which provides a standardized basis for disaster/emergency management planning and business continuity programs in private and public sectors by providing common program elements, techniques, and processes?

IV INSTALLATION

Critical power is protected and secured by deployment of specialized infrastructure equipment designed to provide a continuous facility power stream and immunity from public power "grid" disturbances.

Based on the results of the Needs Analysis and Risk Assessment process, decisions will result authorizing projects to install or upgrade critical power facility infrastructure at your financial institution.

Depending on the size and scope of the project, a range of electrical, mechanical, security, and fire protection infrastructure will be impacted even though the project purpose is improvement of the critical power. The range of infrastructure typically included in a project scope can include:

Electrical Systems

- Transformers
- UPS systems

- Emergency generators and controls
- Utility switchboards
- Automatic transfer switchgear
- Auto-static transfer switches
- Power distribution units
- Transient voltage protection

Mechanical Systems

- Computer room air handling units (CRAH)
- Water chillers, pumps
- Cooling towers
- Water storage tanks (chilled, makeup)
- Automatic controls (BMS, BAS)
- Fuel oil systems for diesel generators
- Fire protection
- Security

The five typical phases of a project to provide critical power are design, procurement, construction, commissioning/acceptance testing, and transition to operations.

Design

The design phase of the project involves a comprehensive review and identification of the project purpose resulting in issuance of detailed plans and specifications that will be used by the contractors to install and by the operations staff to understand the project intent. The design phase should include a qualified architect, engineering, and design firms, and the design team members should include representation from the end-user, facility management, and IT staff. It is during the design phase that environmental issues should be considered. These issues can range from avoiding placement of equipment in flood prone areas to ensuring compliance with local ordinances regarding noise abatement.

Procurement

The resulting design documents can be sent to qualified contractors for bidding or pricing. Depending on the institution, the procurement staff may be facilities, project, or dedicated purchasing/procurement staff. Appropriate contractors and equipment vendors are then chosen.

Construction

Entering the construction phase, the work flow should shift to construction and to project management professionals who are experienced in this type of mission critical infrastructure. The construction team should include financial institution facilities operations staff who will gain invaluable exposure to the project and have a vested interest in project success because of their operational role once the infrastructure is in service.

Commissioning/Acceptance Testing

Before a new facility or new infrastructure in an existing building goes on-line, it is crucial to resolve all potential equipment problems during a commissioning and acceptance phase. Commissioning is a systematic process of ensuring, through documented verification, that all building systems perform according to the documented design intent, and to the owner's operational needs. The goal is to provide the owner with a safe, reliable installation. Commissioning specifications should be written and included in bid documents

Acceptance testing will be the construction team's sole opportunity to integrate and commission all the systems, given the facility's 24/7 mission critical status. There is no "one size fits all" formula and the commissioning agent facilitates a highly interactive process through coordination with the owner, design team, construction team, and vendors during the various phases of the project. Prior to installation at the site, all systems should be tested at the factory and witnessed by an independent test engineer familiar with the equipment. However, reliance on the factory testing and competence of independent test engineers is insufficient.

Once the equipment is delivered, placed, and wired, the second phase of certified testing and integration begins. The goal of this phase is to verify and certify all components work together while fine tuning, calibrating, and integrating all systems. A tremendous amount of coordination is required during this phase. The facilities engineer and commissioning team work with the factory, field engineers, and independent test consultants to coordinate testing and calibration. Critical circuit breakers must be tested and calibrated prior to placing any critical electrical load on them. After all tests are completed, results must be compiled and the certified test reports prepared, which will establish a benchmark for all future testing. Steps to educate staff regarding each major system and piece of equipment should be included in the construction process. This training phase is an ongoing process that begins during construction and continues over the life of the facility.

Transition to Operation

Transition to operations is the final phase of the project. This phase ensures that the design and construction team smoothly hands off the new infrastructure or facility to a confident, fully trained operations team that is charged with its reliable and safe operation. Plans for this transition should be discussed during the design concept phase. Steps should be included to ensure that accurate documentation, detailed tracking/resolution of open items and maximization of training opportunities are incorporated into all project phases.

Security Considerations

It is important to address physical and cyber security needs of critical infrastructure since it includes systems, facilities, and assets so vital that if they are destroyed or incapacitated could disrupt the safety and the economic condition of the FI. Operation of these systems typically utilizes environmentally hazardous diesel fuel oil and lead acid batteries which require special accommodations. Security requirements may include capabilities to prevent and protect against intrusion, hazards, threats, and incidents and to expeditiously recover and reconstitute critical services.

A sensible and cost-effective security approach can provide a protection level achieved through design, construction, and operation that mitigates adverse impact to systems, facilities, and assets. This can include vulnerability and risk assessment methodologies which determine prevention, protection, monitoring, detection, and sensor systems to be deployed in the design. It is essential to include representation from the financial institution security department and from the engineering/design firm from the onset of the initial design.

The increased use of and advances in information technology, coupled with the prevalence of hacking of and unauthorized access to electronic networks, requires physical security to be complemented by cyber security considerations. Before enabling remote access for monitoring and/or control of critical infrastructure systems, cyber security protection must be assured.

Questions

Item	Description
	Design
56.	Has the owner, working with an engineering professional, developed a Design Intent Document to clearly identify quantifiable requirements?
57.	Have you prepared a Basis of Design document that memorializes in a narrative form, the project intent, future expansion options, types of infrastructure systems to be utilized, applicable codes and standards to be followed, design assumptions, and project team decisions and understandings?
58.	Will you provide the opportunity to update the Basis of Design to reflect changes made during the construction and commissioning process?
59.	Are the criteria for testing all systems and outlines of the commissioning process identified and incorporated into the design documents?
60.	Have you identified a qualified engineering and design firm to conduct a peer project design review?
61.	Have you considered directly hiring the commissioning agent to provide true independence?
62.	Have you discussed and agreed to a division of responsibilities between the construction manager and the commissioning agent?
63.	Do you plan to hire the ultimate operating staff ahead of the actual turnover to operations so they will benefit from participation in the design, construction and commissioning of the facility?
64.	Have you made a decision on the commissioning agent early enough in the process to allow participation and input on commissioning issues and design review by the selected agent?

65. Is the proper level of fire protection in place?
66. Is the equipment (UPS or generator) being placed in a location prone to flooding or other water damage?
67. Do the generator day tanks or underground fuel cells meet local environmental rules?
68. Does the battery room have proper ventilation?
69. Has adequate cooling or heating been specified for the UPS, switchgear or generator room?
70. Are the heating and cooling for the mechanical rooms on the power protection system?
71. Have local noise ordinances been reviewed and does all the equipment comply with the ordinances?
72. Are the posting and enforcement of no-smoking bans adequate, specifying, for example, no smoking within 100 ft?
73. Are water detection devices used to alert building management of flooding issues?

Procurement

74. Is there a benefit to using an existing vendor or supplier for standardization of process, common spare parts, or confidence in service response?
75. Have the commissioning, factory, and site testing requirement specifications been included in the bid documentation?
76. If the project is bid, have you conducted a technical compliance review which identifies exceptions, alternatives, substitutions, or non-compliance to the specifications?
77. Are the procurement team members versed in the technical nuances and terminology of the job?
78. If delivery time is critical to the project, have you considered adding late penalty clauses to the installation or equipment contracts?
79. Have you included a bonus for early completion of project?
80. Have you obtained unit rates for potential change orders?
81. Have you obtained a GMP (guaranteed maximum price) from contractors?
82. Have you discussed preferential pricing discounts that may be available if your institution or your engineer and contractors have other similar large purchases occurring?

Procurement Construction

83. Do you intend to create and maintain a list of observations and concerns that will serve as a check list during the acceptance process to ensure that these items are not overlooked?
84. Will members of the design, construction, commissioning agent, and operations team attend the factory acceptance tests for major components and systems such as UPS, generators, batteries, switchgear and chillers?
85. During the construction phase, do you expect to develop and circulate for comment the start up plans, documentation formats and pre-functional checklists that will be used during startup and acceptance testing?
86. Since interaction between the construction manager and the commissioning agent is key, will you encourage attendance at the weekly construction status meetings by the commissioning team?
87. Will an independent commissioning and acceptance meeting be run by the commissioning agent ensuring that everything needed for that process is on target?
88. Will you encourage the construction, commissioning, and operations staff to walk the job site regularly to identify access and maintainability issues?
89. If the job site is an operating critical site, do you have a risk assessment and change control mechanism in place to ensure reliability?
90. Have you established a process to have independent verification that labeling on equipment and power circuits is correct?

(*continued*)

(continued)

Item	Description
	Commissioning and Acceptance
91.	Do testing data result sheets identify expected acceptable result ranges?
92.	Are control sequences, check lists, and procedures written in plain language, not technical jargon that is easily misunderstood?
93.	Have all instrumentation, test equipment, actuators and sensing devices been checked and calibrated?
94.	Is system acceptance testing scheduled after balancing of mechanical systems and electrical cable/breaker testing are complete?
95.	Have you listed the systems and components to be commissioned?
96.	Has a detailed script sequencing all activities been developed?
97.	Are all participants aware of their responsibilities and the protocols to be followed?
98.	Does a team directory with all contact information exist and is it available to all involved parties?
99.	Have you planned an "all hands on deck" meeting to walk through and finalize the commissioning schedule and scripted activities?
100.	Have the format and content of the final report been determined in advance to ensure that all needed data is recorded and activities are scheduled?
101.	Have you arranged for the future facility operations staff to witness and participate in the commissioning and testing efforts?
102.	Who is responsible for ensuring that all appropriate safety methods and procedures are deployed during the testing process?
103.	Is there a process in place that ensures training records are maintained and are updated?
104.	Who is coordinating training and ensuring that all prescribed training takes place?
105.	Will you videotape training sessions to capture key points and for use as refresh training?
106.	Is the training you provide both general systems training as well as specifically targeted to types of infrastructure within the facility?
107.	Have all vendors performed component level verification and completed pre-functional check lists prior to system level testing?
108.	Has all system level acceptance testing been completed prior to commencing the full system integration testing and "pull the plug" power failure scenario?
109.	Is a process developed to capture all changes made and to ensure that these changes are captured on the appropriate as built drawings, procedures, and design documents?
110.	Do you plan to re-perform acceptance testing if a failure or anomalies occur during commissioning and testing?
111.	Who will maintain the running punch list of incomplete items and track resolution status?
	Transition to Operation
112.	Have you established specific Operations Planning meetings to discuss logistics of transferring newly constructed systems to the facility operations staff?
113.	Is all as-built documentation, such as drawings, specifications, and technical manuals complete and has it been turned over to operations staff?
114.	Have position descriptions been prepared that clearly define roles and responsibilities of the facility staff?

115. Are Standard Operating Procedures (SOP), Emergency Action Procedures (EAP), updated policies, and change control processes in place to govern the newly installed systems?
116. Has the facility operations staff been provided with warranty, maintenance, repair, and supplier contact information?
117. Have spare parts lists, setpoint schedules after Cx is complete, TAB report, and recommissioning manuals been given to operations staff?
118. Are the warranty start and expiration dates identified?
119. Have maintenance and repair contracts been executed and put into place for the equipment?
120. Have minimum response times for service, distance to travel, and emergency 24/7 spare stock locations been identified?

Security Considerations

121. Have you addressed physical security concerns?
122. Have all infrastructures been evaluated for type of security protection needed (e.g., card control, camera recording, key control)?
123. Are the diesel oil tank and oil fill pipe in a secure location?
124. If remote dial in or Internet access is provided to any infrastructure system, have you safeguarded against hacking or do you permit read only functionality?
125. How frequently do you review and update access permission authorization lists?
126. Are critical locations included in security inspection rounds?

V MAINTENANCE AND TESTING

Electrical maintenance is a necessity not a luxury. Understanding the risk and sensitivity of mission critical sites affords a financial institution mitigation of downtime with regard to a range of mission critical engineering services.

An effective maintenance and testing program for a mission critical electrical load is key to protecting the investment by safeguarding against power failures. Maintenance procedures and schedules must be developed, staff properly trained, spare parts provisioned, and mission critical electrical equipment performance tested and evaluated regularly.

There are various approaches to establish a maintenance program. In most cases, a program will include a blend of the strategies listed below:

- Preventive maintenance (PM) is the completion of tasks performed on defined schedule. The purpose of PM is to extend the life of equipment and detect wear as an indicator of pending failure. Tasks describing the maintenance procedures are fundamental to a PM program. They instruct the technician on what to do, what tools and equipment to use, what to look for, how to do it, and when to do it. Tasks can be created for routine maintenance items or for breakdown repairs.
- Predictive maintenance uses instrumentation to detect the condition of equipment and to identify pending failures. A predictive maintenance program uses these equipment condition indices for the purpose of scheduling maintenance tasks

- Reliability-centered maintenance (RCM) is the analytical approach to optimize reliability and maintenance tasks with respect to the operational requirements of the business. Reliability, as it relates to business goals, is the fundamental objective of the RCM process. RCM is not equipment-centric, but business-centric. RCM analyzes each system and how it can functionally fail. The effects of each failure are analyzed and ranked according to their impact on safety, mission, and cost. Those failures which are deemed to have a significant impact, are further explored to determine the root causes. Finally, maintenance is assigned based on effectiveness, with a focus on condition based tasks.

The objective of a maintenance program is to use a blend of predictive, preventative and RCM techniques to reach the optimum point at which the benefits of reliability are maximized while the cost of maintenance is minimized.

The appropriate frequency of electrical maintenance should be driven in part by the level of reliability an institution requires. Specifically, risk tolerance expectations and uptime goals must be weighed. An institution satisfied with 99% reliability, or 87.6 hours of downtime per year, will run a maintenance program every three to five years. However, if 99.999% reliability, or 5.25 minutes of downtime per year is mandatory, then the institution must perform an aggressive preventive maintenance program every six months. The cost of this hard-line maintenance program could range between $300 and $400 annually per kilowatt (kW), not including the staff to manage the program. The Human Resources cost will vary, depending on the location and complexity of the facility.

The other factor affecting maintenance frequency is the state of industry-accepted guidelines. There are several excellent resources available for developing the basis of an electrical testing and maintenance program. For example, the InterNational Electric Testing Association (NETA) publishes the *Maintenance Testing Specifications* that recommend appropriate maintenance test frequencies based on equipment condition and reliability requirements. The National Fire Protection Association's (NFPA) *70B Recommended Practice for Electrical Equipment Maintenance* and *RSMeans Facilities Maintenance and Repair Book* give guidance for testing and maintenance tasks and periodic schedules to incorporate into maintenance programs.

Testing and service individuals should have the highest education, skills, training, and experience available. Their conscientiousness and decision-making abilities are a key to avoiding potential problems with perhaps the most crucial equipment in a facility. Most importantly, learn from previous experiences and from the experiences of others so that operational and maintenance programs continuously improve as knowledge increases. If a task has historically not identified a problem at the scheduled interval, consider adjusting the schedule. Examine maintenance programs on a regular basis and make appropriate adjustments.

Routine shutdowns of a facility should be planned to accommodate preventive maintenance of electrical equipment. Neither senior management nor facility managers should underestimate the cost-effectiveness of a thorough preventative maintenance program.

Initial equipment acceptance testing and ongoing maintenance will not return maximum value unless the test results are evaluated and compared with standards and previous test reports that have established benchmarks. It is imperative to recognize failing equipment and to take appropriate action as soon as possible. All too commonly, maintenance personnel perform maintenance without reviewing prior maintenance records. This approach must be avoided because it defeats the value of benchmarking and trending. By reviewing past maintenance reports, staff can keep maintenance objectives in perspective and rely upon the accuracy of the information contained in these reports when faced with a real emergency.

Every preventative maintenance opportunity should be thorough and complete, especially in mission critical facilities. If they are not, the next opportunity will come at a much higher price: downtime, lost business, and the loss of potential clients. In addition, safety issues arise when technicians rush to repair high voltage equipment.

Questions

Item	Description
	Strategic
127.	Is there a documented maintenance and testing program based on your business risk assessment model?
128.	Is an audit process in place to ensure that this maintenance and testing program is being followed rigorously?
129.	Does the program ensure that maintenance test results are benchmarked and used to update and improve the maintenance program?
130.	Is there a program in place that ensures periodic evaluation of possible equipment replacement?
131.	Is there a process in place that ensures the spare parts inventory is updated when new equipment is installed or other changes are made to the facility?
132.	Have you evaluated the impact of loss of power in your institution and other institutions because of interdependencies?
133.	Has your facility developed Standard Operating Procedures (SOPs), Emergency Action Procedures (EAPs), and Alarm Response Procedures (ARPs)?
134.	Are the SOP, EAP, and ARP readily available and current?
135.	Is your staff familiar with the SOPs, EAPs, and ARPs?
	Planning
136.	Does the system design provide redundancy so all critical equipment can be maintained without a shutdown if required?
137.	Are there adequate work control procedures and is there a change management process to prevent mistakes when work is done on critical systems and equipment?
138.	Are short-circuit and coordination studies up to date?
139.	Do you have a Service Level Agreement (SLA) with your facilities service providers and contractors?
140.	Is there a change management process that communicates maintenance, testing, and repair activities to both end users and business lines?
141.	Do you have standard operating procedures to govern routine facilities functions?
142.	Do you have emergency response and action plans developed for expected failure scenarios?

(*continued*)

(continued)

Item	Description
	Safety
143.	Have you prepared an emergency telephone contact list that includes key service providers and suppliers?
144.	Is there a formal and active program for updating the safety manual?
145.	Are electrical work procedures included in the safety manual?
146.	Has an arc-flash study been performed?
147.	Are specific PPE requirements posted at each panel, switchgear, etc?
148.	Is there a program in place to ensure studies and PPE requirements are updated when system or utility supply changes are made?
149.	Are workers trained regarding safety manual procedures?
150.	Are hazardous areas identified on drawings?
151.	Are hazardous areas physically identified in the facility?
	Testing
152.	Have protective devices been tested or checked to verify performance?
153.	Is a Site Acceptance Test (SAT) and a Factory Acceptance Test (FAT) performed for major new equipment such as UPS systems and standby generators?
154.	Is an annual "pull the plug" test performed to simulate a utility outage ensure that the infrastructure performs as designed?
155.	Is an annual performance and recertification test conducted on key infrastructure systems?
156.	Is there a process in place that ensures personnel have the proper instrumentation and that it is periodically calibrated?
	Maintenance
157.	Does the program identify all critical electrical equipment and components?
158.	Is there a procedure in place that updates the program based on changes to plant equipment or processes?
159.	Does a comprehensive plan exist for thermo infrared (IR) heat scan of critical components? Is an IR scanning test conducted before a scheduled shutdown?
160.	Are adequate spare parts on hand for immediate repair and replacement?
161.	Is your maintenance and testing program based on accepted industry guidelines such as NFPA 70B and on equipment supply recommendations?
162.	Do you incorporate reliability-centered maintenance (RCM) philosophy in your approach to maintenance and testing?
163.	When maintenance and testing is performed do you require preparation of and adherence to detailed work statements and method of procedures (MOPs)?
164.	Do you employ predictive maintenance techniques and programs such as vibration and oil analysis?

VI TRAINING AND DOCUMENTATION

Millions of dollars are invested in the infrastructure supporting 24/7 applications, with major commitments made in design, equipment procurement, and project management. However, investment in documentation, training, and education has been minimal despite the fact that they are essential to achieving and maintaining optimum levels of reliability.

As equipment reliability increases, a larger percentage of downtime results from actions by personnel who are inadequately trained or who lack access to accurate comprehensible data during crisis events. Keeping on-site staff motivated, trained, and ready to respond to emergencies is a challenge. Years ago most organizations relied heavily on their workforce to retain much of the information regarding the mission critical systems. A large body of personnel had a similar level of expertise and remained with their company for decades. Therefore, little emphasis was placed on creating and maintaining a fluid and living document repository for critical infrastructure.

Today's diversity among mission critical systems severely hinders employee ability to fully understand and master all necessary equipment and the information required to keep that equipment running. We can no longer allow engineers and operators to acquire their knowledge of increasingly sophisticated power supply and distribution technology from "on the job training," which proves woefully inadequate in a time of crisis. Instead, a clear plan must be put into place to develop a critical document repository and to continually educate and train employees while enhancing real time experiences.

Elements of a comprehensive training and certification program could include:

- Providing fundamental training on facility electrical, mechanical, and life-safety systems
- Determining staff qualification criteria
- Identifying training topics and developing specific modules
- Creating testing content and certification methods
- Maintaining employee training records and ongoing training requirements

Prudent business practice recognizes the need to plan for employee succession, unexpected staff departure, orientation/training of new employees, as well as education of seasoned employees. An education and training program coupled with a document management system can address these concerns. Site-specific training courses can be developed to target subjects such as UPS switching procedures, emergency generator operation and testing, company policies and procedures, safety, and critical environment work rules.

In the financial services industry, education and training that is standardized, comprehensive, and focused on the job at hand will create a pool of talented individuals possessing the knowledge and information necessary to solve problems during power emergencies. This will lead to shorter and less frequent unplanned downtime. Documentation is essential not only to facilitate the ongoing education and training requirements of a company's personnel, but also to maintain safety and to minimize risk to the company, assuring the integrity of a robust mission critical infrastructure and the institution's bottom line.

A "database" of perpetually refreshed knowledge can be achieved by creating a living document system that provides the level of granularity necessary to operate a mission critical infrastructure. Such a system should be supplemented with a staff training and development program. The living document and training programs may be kept current can each time a capital project is completed or an infrastructure change is made. Accurate and up-to-date information provides first responders with the intelligence and support necessary to make informed decisions during critical events.

Questions

Item	Description
	Documentation
165.	What emergency plans, if any, exist for the facility?
166.	Where are emergency plans documented (including the relevant internal and external contacts for taking action)?
167.	How are contacts reached in the event of an emergency?
168.	How are plans audited and changed over time?
169.	Do you have complete drawings, documentation, and technical specifications of your mission critical infrastructure, including electrical utility, in-facility electrical systems (including power distribution and ATS), gas and steam utility, UPS/battery/generator, HVAC, security, and fire suppression?
170.	What documentation, if any, exists to describe the layout, design, and equipment used in these systems?
171.	How many forms does this documentation require?
172.	How is the documentation stored?
173.	Who has access to this documentation and how do you control access?
174.	How many people have access to this documentation?
175.	How often does the infrastructure change?
176.	Who is responsible for documenting change?
177.	How is the information audited?
178.	Can usage of facility documentation be audited?
179.	Do you keep a historical record of changes to documentation?
180.	Is a formal technical training program in place?
181.	Is there a process in place that ensures personnel have proper instrumentation and that the instrumentation is periodically calibrated?
182.	Are accidents and near-miss incidents documented?
183.	Is there a process in place that ensures action will be taken to update procedures following accidents or near-miss events?
184.	How much space does your physical documentation occupy today?
185.	How quickly can you access the existing documentation?
186.	How do you control access to the documentation?
187.	Can responsibility for changes to documentation be audited and tracked?
188.	If a consultant is used to make changes to documentation, how are consultant deliverables tracked?
189.	Is your organization able to prove what content was live at any given point in time, in the event such information is required for legal purposes?
190.	Is your organization able to quickly provide information to legal authorities including emergency response staff (e.g., fire, police)?
191.	How are designs or other configuration changes to infrastructure approved or disapproved?
192.	How are these approvals communicated to responsible staff?

193. Does workflow documentation exist for answering staff questions about what to do at each stage of documentation development?
194. In the case of multiple facilities, how is documentation from one facility transferred or made available to another?
195. What kind of reporting on facility infrastructure is required for management?
196. What kind of financial reporting is required in terms of facility infrastructure assets?
197. How are costs tracked for facility infrastructure assets?
198. Is facility infrastructure documentation duplicated in multiple locations for restoration in the event of loss?
199. How much time would it take to replace the documentation in the event of loss?
200. How do you track space utilization (including cable management) within the facility?
201. Do you use any change management methodology (i.e., ITIL) in the day-to-day configuration management of the facility?

Staff and Training

202. How many operations and maintenance staff do you have within the building?
203. How many staff members do you consider to be facilities "subject matter experts"?
204. How many staff members manage the operations of the building?
205. Are specific staff members responsible for specific portions of the building infrastructure?
206. What percentage of your building operations staff turns over annually?
207. How long has each of your operations and maintenance staff, on average, been in his or her position?
208. What kind of ongoing training, if any, do you provide for your operations and maintenance staff?
209. Do training records exist?
210. Is there a process in place to ensure that training records are maintained updated?
211. Is there a process in place that identifies an arrangement for training?
212. Is there a process in place that ensures the training program is periodically reviewed and identifies changes required?
213. Is the training you provide general training, or is it specific to an area of infrastructure within the facility?
214. How do you design changes to your facility systems?
215. Do you handle documentation management with separate staff, or do you consider it to be the responsibility of the staff making the change?

Network and Access

216. Do you have a secured network between your facility IT installations?
217. Is this network used for communications between your facility management staff?
218. Do you have an individual on your IT staff responsible for managing the security infrastructure for your data?
219. Do you have an on-line file repository?
220. If so, how is use of the repository monitored, logged, and audited?
221. How is data retrieved from the repository kept secure once it leaves the repository?
222. Is your file repository available through the public Internet?
223. Is your facilities documentation cataloged with a standard format to facilitate location of specific information?
224. What search capabilities, if any, are available on the documentation storage platform?
225. Does your facility documentation reference facility standards (e.g., electrical codes)? If so, how is this information kept up-to-date?

VII CONCLUSION

The following highlights have been distilled from the prior sections of this *Guide*. They are baseline observations and recommendations intended to provide financial institutions with industry business practices for understanding, evaluating, and managing risks associated when the predicted reliability and availability of the electrical system is disrupted.

General Comments

- Financial institutions need to constantly and systematically evaluate their mission critical systems, assessing and reassessing their level of risk tolerance versus the cost of downtime.
- Providing continuous operation under all foreseeable risks of failure is a nontrivial matter and requires a holistic, enterprise approach.
- Communication between managers of business lines, business continuity and facilities is vital.

The Grid and Power Utility Companies

- The grid is inherently subject to external factors that can cause outages, and there is only so much that can be feasibly done to secure it against these possible problems.
- Guaranteeing supplies of critical power at high-demand, high-reliability locations ultimately means adding increasingly sophisticated on-site generating capacity and storage to back up whatever is being done to improve reliability higher in the grid.
- Financial institutions should be in touch with their electric power utility to establish the working relationship needed to help through real or potential service interruptions, including becoming involved in crisis management testing performed by the utility and understanding the power utility's operating strategy and position within any larger regional or national infrastructure.

Needs Analysis/Risk Assessment

- Today's risk profile includes the possibility that outages may last a long time and extend over wide areas. The planning challenge shifts from issues of power *quality* or *reliability* to issues of business *sustainability*. Planning must take into account outages that last not for seconds, or for a single hour, but for days.
- The first step in the needs analysis process is to determine the level of criticality. Individual institutions establish their level of criticality based on their own criteria and their customers' criteria. The primary criterion is availability.

Installation

- Financial institutions require highly sophisticated uninterruptible power systems, generators, HVAC systems, transfer switches, and other high-voltage electrical and mechanical systems to ensure fail safe power delivery and operation.
- Security requirements may include capabilities to prevent and protect against intrusion, hazards, threats, and incidents—both physical and cyber, and to expeditiously recover and reconstitute critical services.
- It is essential to include representation from the financial institution security department from the onset of initial design.

Maintenance and Testing

- An effective maintenance and testing program for a mission critical electrical load is key to protecting the investment by safeguarding against power failures.
- Routine shutdowns of a facility should be planned to accommodate preventive maintenance of electrical equipment. Neither senior management nor facility managers should underestimate the cost-effectiveness of a thorough preventative maintenance program.
- Maintenance procedures and schedules must be developed, staff properly trained, spare parts provisioned, and mission critical electrical equipment performance tested and evaluated regularly.

Training and Documentation

- A large percentage of downtime results from actions by personnel who are inadequately trained or who lack access to accurate comprehensible data during crisis events.
- A clear plan must be put into place to develop a critical document repository and to continually educate and train employees while enhancing real time experiences.

Prudent planning and an appropriate level of investment will help ensure uninterrupted power supply. Minimizing unplanned downtime reduces risk. Financial institutions must consider a proactive rather than a reactive approach. Strategic planning can identify internal risks and provide a prioritized plan for reliability improvements that identify the root causes of failures before they occur. Planning and careful implementation will minimize disruptions while making the business case to fund necessary capital improvements and implement comprehensive maintenance strategies. When the business case reaches the boardroom, the entire organization can be galvanized to prevent catastrophic losses, damage to capital equipment, and physical danger to our employees and customers.

APPENDIX B.1

Acknowledgements and References

The *BITS Guide to Business-Critical Power* was developed by a small, dedicated team of professionals from BITS member organizations, the Critical Power Coalition, Power Management Concepts and BITS staff. It is based on meetings and calls and it draws on the following sources:

- "Maintaining Mission Critical Systems in a 24/7 Environment," Peter M. Curtis
- BITS white paper, "Telecommunications for Critical Infrastructure: Risks and Recommendations" (December 2002)
- BITS Guide to Business Critical Telecommunications Services (2004)
- BITS Forums on Telecommunications Resiliency (June 2002 and June 2004)
- Securities Industry Association Business Continuity Committee Critical Infrastructure Guidelines (May 2004 draft)
- BITS Lessons Learned: Northeast Blackout of 2003 (October 2003)
- Digital Power white paper, "Critical Power" (August 2003)

Peter M. Curtis, Power Management Concepts, and Teresa C. Lindsey, BITS, served as the principal authors of this document. Additionally, the following individuals made significant contributions:

Warren Axelrod, Pershing

Albert Bocchetti, SIA

Mike Carano, LaSalle Bancorporation

Don Donahue, DTCC and FSSCC

Jim Driscoll, Commerce Bank

Patti Harris, Regions

Sue Kerr, Capital One

Paul LaPierre, The Critical Power Coalition

Joe Lee, Wachovia

Mark Mills, The Critical Power Coalition

Melvyn Musson, Edward Jones Investments

Charles Rodger, PNC

Jim Sacks, Fifth Third

Chris Terzich, Wells Fargo

Tom Weingarten, Power Management Concepts

John Carlson, BITS

Cheryl Charles, BITS

John Ingold, BITS

Heather Wyson, BITS

About BITS

BITS was created in 1996 to foster the growth and development of electronic financial services and e-commerce for the benefit of financial institutions and their customers. A nonprofit industry consortium that shares membership with The Financial Services Roundtable, BITS seeks to sustain consumer confidence and trust by ensuring the security, privacy and integrity of financial transactions. BITS works as a strategic brain trust to provide intellectual capital and address emerging issues where financial services, technology and commerce intersect, acting quickly to address problems and galvanize the industry. BITS' activities are driven by the CEOs and their appointees—CIOs, CTOs, Vice Chairmen and Executive Vice Presidents—who make up the BITS Advisory Board and BITS Advisory Council. For more information, go to www.bitsinfo.org.

BITS
1001 Pennsylvania Avenue, NW
Suite 500 South
Washington, DC 20004
202–289–4322
WWW.BITSINFO.ORG

About the Critical Power Coalition

The Critical Power Coalition (CPC) was formed by leading providers, and users, of critical-power products and services. Over the course of the past decade, telecommunications facilities, financial institutions, hospitals, airports, data centers, emergency response centers, manufacturing plants, and other U.S. enterprises and government agencies have invested some $250 billion in hardware, systems, software, and engineering services to ensure the uninterrupted supply of high-quality power to critical facilities and equipment when grid power fails. The pace of investment is rising as government and the private sector grow increasingly dependent on digital equipment, and as concerns rise about the grid's potential vulnerabilities. CPC is focused on the urgent policy, technology, and regulatory issues that must be addressed to ensure the quality, reliability, and continuity of power where it is needed the most.

Critical Power Coalition
1615 M Street NW, Suite 400
Washington, DC 20036
HTTP://WWW.CRITICALPOWERCOALITION.ORG

About Power Management Concepts

Power Management Concepts (PMC) is an engineering and technology company dedicated to preventing costly downtime for clients operating mission critical facilities. "Mission-Critical" is a broad categorization of ultra-high reliability and availability of electrical and mechanical systems that must meet stringent operating criteria to maintain continuous functionality and eliminate costly unscheduled downtime. PMC provides a fully integrated continuum of services including planning, design, project management, preventive maintenance, mission critical technology solutions and training.

Power Management Concepts
20 Crossways Park North
Suite 400
Woodbury, NY 11797
HTTP://WWW.POWERMANAGE.COM

APPENDIX B.2 GLOSSARY[13]

A

AC or ac Abbreviation for alternating current.

Alternating current Electrical current which periodically reverses direction, usually several times per second.

Ampere The measurement unit for electrical current.

Automatic transfer switch A switch that automatically transfers electrical loads to alternate or emergency-standby power sources.

B

Blackout A complete loss of power lasting for more than one cycle. A blackout can damage electronics, corrupt or destroy data, or cause a system shutdown. Blackouts can result from any of a number of problems, ranging from Acts of God (hurricanes or other high winds, ice storms, lightning, trees falling on power lines, floods, geomagnetic storms triggering by sunspots and solar flares, etc.) to situations such as cables being cut during excavation, equipment failures at the utility, vandalism, corrosion, etc. Also known as an outage.

Brownout A prolonged sag, occurring when incoming power is reduced for an extended period. Usually caused when demand is at its peak and the line becomes overloaded.

C

Capacitor Any AC circuit element possessing the property of capacitance (i.e., the ability to store a charge). Normally a capacitor is a dedicated device, designed for the prime purpose of exhibiting the property of capacitance (as opposed to inductive devices, in which

[13]Glossary reproduced courtesy of Liebert. Corporation.

inductance is used by the device to produce other results, such as turning a motor shaft).

Critical load Equipment that must have an uninterrupted power input to prevent damage or loss to a facility or to itself, or to prevent danger of injury to operating personnel.

Current The flow of electricity in a circuit. The term current refers to the quantity, volume, or intensity of electrical flow, as opposed to voltage, which refers to the force or "pressure" causing the current flow. Current may be either direct or alternating. Direct current refers to current whose voltage causes it to flow in only one direction. Common direct current sources are batteries. Alternating current refers to current whose voltage causes it to flow first in one direction, then the other, reversing direction periodically, usually several times a second. A common alternating current source is commercial/household power. This current reverses direction 120 times each second, thus passing through 60 complete cycles each second for a frequency of 60 Hz.

D

Direct current Electrical current which flows consistently in one direction.

E

EMI/RFI Electromagnetic / radio-frequency interference. These high-frequency signals are generally low level (<1 V) and range from 1 MHz up. EMI/RFI filters are generally not suitable for large-amplitude surge suppression.

H

Harmonic distortion A measure of the degree to which the impedance of a UPS affects the shape of the output voltage waveform. Distortion is stated as a percentage and may refer to any single harmonic or to the total waveform, in which case it is referred to as "total harmonic distortion" (THD).

I

Inverter The DC to AC power converter driven by the UPS rectifier-charger or battery via the DC bus. The inverter output drives the critical load.

IEC555 A German standard that requires power factor corrected (PFC) loads.

K

KVA Abbreviation for kilovolt-amperes. (1000 × volt-amperes)

L

Line disturbance analyzer A tool used in analyzing problems in a facility's incoming power. The line disturbance analyzer is connected at the power input to measure and record incoming power, then left in place for long enough to gather data typical of the site.

N

Noise Noise is the result of distortion of the normal line power sine wave by hundreds or thousands of small increases in voltage similar to EMI/RFI, though it encompasses lower frequencies. The amplitude of this type of disturbance is less than a surge but may be as low as EMI/RFI.

Normal line power Commercial electricity supplied by U.S. power utilities is generally delivered as 60-cycle (Hz) alternating current (AC).

O

Overload capacity A UPS's overload capacity is its ability to respond to sudden surges in load current without allowing the output voltage level to decrease.

P

Power conditioning systems A broad class of equipment that includes filters, isolation transformers, and voltage regulators. Generally, these types of equipment offer no protection against power outages.

Power factor corrected (PFC) supply A recently developed type of computer power supply, which exhibits an input power factor equal to one. IEC 555 will force most computers to use a power supply of this type at some point in the future.

Power synthesizer Power synthesizers actually use the incoming utility power as an energy source to create a new sine wave that's free from power disturbances. They can be as much as 99% effective against power disturbances. Types of power synthesizers include magnetic synthesizers (capable of generating a sine wave of the same frequency as the incoming power, 60 Hz), motor generators (which use an electric motor to drive a generator that provides electrical power), and UPSs.

S

Sag A momentary decrease from nominal voltage lasting one or more line cycles. Severe conditions may indicate a need for a UPS or voltage regulator. Also known as a temporary undervoltage (TUV).

Sine wave A periodic oscillation. The fundamental waveform from which other waveforms may be generated by combinations of various group of harmonics. The voltage and current waveforms produced from the power company generators (alternators) are basic sine waves.

Surge A surge is a prolonged overvoltage condition. Surges can damage electronics and corrupt or destroy data.

Spike A spike involves a sudden marked jump in voltage, which can damage electronics and corrupt or destroy data.

Spike/surge protector These products are inexpensive solutions that provide minimal protection against surges, but no protection against sags and outages.

Suppressed voltage ratings Several ranges are assigned by UL for grading transient suppression voltages. For instance, a 400-V rating indicates a maximum peak voltage between 330 and 400 V. These ratings appear between 330 V peak and 6000 V peak.

Swell An increase from nominal voltage lasting one or more line cycles.

T

Transfer time Transfer time can refer to either (a) the speed with which an off-line UPS transfers from utility power to battery power or (b) the speed with which an on-line UPS switches from the inverter to utility power in the event of an inverter failure. In either case, the time involved must be shorter than the length of time that the computer's switching power supply has enough energy to maintain adequate output voltage. This hold-up time may range from 8 to 16 msec, depending on (a) the point in the power supply's recharging cycle that the power outage occurs and (b) the amount of energy storage capacitance within the power supply. A transfer time of 4 msec is most desirable; however, it should be noted that an oversensitive unit may make unnecessary power transfers.

Transient suppression voltage (let-through voltage) The maximum peak voltage occurring within 100 μsec after the test wave.

Transient voltage surge suppressor (TVSS) A device used to reduce voltage surges. Products may be wired in series or in parallel with the AC electrical conductors.

U

UL 1449 United Laboratories, Inc.'s Standards for Safety of Transient Voltage Surge Suppressors (TVSS).

UPS Uninterruptible power supplies (sometimes called uninterruptible power systems). A system designed to protect against short-term power outages.

V

Volt The quantitative unit of measurement of electrical voltage.

Voltage A term referring to the electrical force or potential. A technical synonym for voltage is emf or "electromotive force." Voltage is the parameter of electricity which causes current to flow when a circuit is completed. Voltage is always presented in an energized line, whether or not the circuit is complete (i.e., whether or not current flows).

Voltage regulator A device designed to regulate RMS voltage by removing swells and sags (such as an automatic tap-switching transformer or ferroresonant transformer).

W

Watt The quantitative unit of measurement of actual power. Actual power in an AC circuit is the measurement of the effective energy available for doing work, and it is normally less than apparent power (volt-amperes) because of power factor considerations. Watts may be measured directly, by means of a wattmeter, or may be calculated by multiplying volt-amperes by the power factor of the equipment.

APPENDIX B.3–CASE STUDY—BLACKOUT AUGUST 2003[14]

This case study highlights key lessons learned from the power outage that affected the Northeast from August 14 through 16, 2003. This case study does not represent the efforts of the other financial services industry associations

[14]Compiled by the BITS Crisis Management Coordination and IT Service Providers Working Groups.

and/or coordinating bodies. This only reflects the BITS perspective and lessons learned relevant to its crisis management coordination process and its members' experiences.

On August 14, 2003, a cascading blackout struck a large portion of the northeast United States and eastern Ontario. The outages began just after four o'clock Thursday afternoon and affected financial services institutions through the following day. Although most financial operations suffered only brief interruptions and no data loss, the financial services industry can learn from both the sound planning and from the handful of planning oversights revealed by the blackout.

Most institutions, particularly the exchanges and institutions involved in payment, clearing, and settlement, successfully continued operations utilizing backup power and facilities. Some bank branches and many ATMs (especially stand alone machines) were forced to suspend operations until regular power was restored. However, there were no runs on banks or market panics. Most attribute the calm to timely communication from the government, assuring the public that the blackout was not caused by terrorism.

The blackout did cause some unanticipated problems. Telecommunications problems stemmed from insufficient backup power at the central office switch and internal telecommunication system levels. Despite communication protocols that generally permitted a successful level of communication in the industry and between the industry and government, it is evident that communication systems and protocols warrant closer scrutiny.

One financial institution experienced an unanticipated problem when its steam provider was unable to continue delivery. Because the financial institution relied on steam to power its electronics cooling system, it was forced to install a boiler for cooling and open late on Friday. Institutions should learn from this instance that continuity plans should encompass more remote potentialities and thorough testing should be done before critical events are considered.

In general, the nation's financial services sector withstood the massive power outage with little or no disruption. Verification and notification by Department of Homeland Security (DHS) officials that the power outage was not terrorism-related provided the public with important assurance. Clearly, the nation's power grid and transmission network should be strengthened to prevent power outages of this magnitude. Further research is needed to understand whether software security weaknesses contributed to the outage.

Contingency Planning and Third-Party Providers

- Financial institutions relied on business continuity plans to respond to the power outage and related consequences.
- Data-protection schemes worked almost flawlessly for most large companies affected by the power outage. Recovery planning efforts made by financial institutions since 9/11 enabled them to respond to the crisis effectively.

Recommendations

- Ensure all critical systems are located in facilities with adequate backup power capacity.
- *Evaluate single points of failure, redundancy, and single-provider implications.*
- Evaluate, define, and test procedures for operating and restarting equipment during power failures.
- Ensure that financial institution patch-management programs include software at contingency sites or vendor-controlled sites.
- Validate emergency building access policy/procedures with third-party building management services.
- Establish and maintain strong relationships with critical partners and suppliers such as power, water, and telecommunications providers.
- Maintain quick-ship/contingency agreements with suppliers.

Communication

- Many member organizations have automated notification systems that provided paging services and 800 numbers for associates to use to receive information.
- Alternate communication devices allowed financial institutions to communicate with employees, third parties, customers, and regulators. With limited cell phone service, Blackberries became a primary and important means of communication for many members whose internal communications servers were not disabled.
- Most Government Emergency Telecommunications Services (GETS) Cards worked.
- Some satellite phones did not work in the New York City area because tall buildings and other environmental factors can affect the phones' ability to receive a signal.
- Reported telecommunications problems included inadequate backup power at telecommunications companies and a spike in the volume of calls. (In the hours after the blackout hit, leading wireless carriers reported three to four times the normal volume of calls, a load that virtually guaranteed that many people would hear busy signals and not be able to get through.)
- Many cell tower generators failed due to insufficient fuel to operate and support the increase of wireless communications. Many of the trucks that service these towers depend on commercial power to refuel and encountered roadblocks in their attempts to reach the towers. The National Coordinating Center of the National Communications System coordinated efforts to get the trucks though the roadblocks and helped secure generators for those carriers in need.

- Because so many thousands of servers were effectively "removed" from the Internet so quickly, it caused a sustained surge in BGP (Border Gateway Protocol) traffic to update router tables, effectively blocking other traffic temporarily and slowing the Internet.

Recommendations

- Obtain as many means of communication with key individuals at third-party service providers as possible (including home phones, cell phones, and e-mail addresses).
- Ensure alternative communication channels to communicate with the media, third-party providers, customers, and government agencies.
- Develop an improved system for communicating emergency and building evacuation instructions and employee protocols.

Coordination with Federal, State, and Local Government

- Government officials provided accurate and timely information, which helped to maintain order. Increased presence by public safety officials helped to alleviate fears and minimize looting and civil unrest.
- Overall communication between government officials and the private sector was successful. Officials from the Federal Reserve, DHS, and Treasury were very responsive to BITS' requests for information and coordinated effectively with the Financial Services Sector Coordinating Council for Critical Infrastructure Protection and Homeland Security (FSSCC).
- All of the major cities affected by the blackout had post-9/11 emergency procedures in place. When electric water pumps shut down in Cleveland, authorities tapped private water trucks the city had arranged to be available in emergencies. Communities in suburban Detroit collaborated to evacuate residents living near a potentially dangerous gasoline plant.

Recommendations

- Establish and maintain strong relationships between the industry and federal, state, and local governments.

Transportation, Water, and Fuel

- There was widespread disruption to transportation systems, including trains, subways, and air travel. Limited and often conflicting or inaccurate information was provided to air travel customers.

- Some companies encountered problems with armored car companies or courier services that would not deliver to locations where power had not been restored. Additionally, some couriers could not gain access to areas due to curfews—or were not able to obtain fuel to complete deliveries.
- Some companies reported a shortage of fuel for generators and difficulty in obtaining additional fuel for their generators.
- In some states, water supplies were affected because water is distributed through electric pumps.
- The inability to pump water and use electronic flushing devices rendered many buildings uninhabitable. High-rise buildings were evacuated due to their inability to run fire pumps.

Recommendations

- Ensure that there is adequate food and water at key locations.
- Ensure that ATMs in key locations have an alternative power source in the event of a power failure.
- Ensure that critical business units/facilities functions are adequately protected by standby power generation and/or power protection systems.
- Test power generation and/or power protection systems regularly at full capacity for extended periods of time.

APPENDIX B.4 CAUSE AND EFFECT OF RECENT POWER OUTAGES

APPENDIX B.5

Consolidated List of Key Questions

Below is a consolidation of the questions that appear in the body of this *Guide*. The questions are the starting point for a rigorous examination of a financial institution's critical power environment. The answers will help financial institutions achieve the necessary levels of diversity, recoverability, redundancy, and resiliency.

The questions are presented in a "worksheet" format providing space so that financial institutions:

- can indicate whether the question is applicable; and
- record comments germane to the question.

Location	Recent Power Outages	
	Cause	Effect
Los Angeles	• Massive power outage—Utility worker wiring error (9-12-05)	• Traffic and public transportation problems and fears of a terrorist attack
Gulf Coast (Florida/ New Orleans)	• 2004/05 Hurricanes: Ivan, Charley, Frances, Katrina, etc.	• Millions of customers without power, water, food and shelter, government records lost due to flooding
China	• 20-million kilowatt power shortage—Equivalent to the typical demand in the entire state of New York (Summer 2005)	• Multiple sporadic brownouts • Government shutdown least energy efficient consumers
Greece	• Temperatures near 104 °F • Mismanagement of electric grid (7-12-04)	• Over half of the country left without power
O'Hare Airport	• Electrical explosion (7-12-04)	• Lost power to two terminals • Flight delays over course of a day
Logan Airport	• Electrical substation malfunction (7-5-04)	• Flight delays and security screening shutdown for 4 hours
Italy	• Power line failures • Bad weather (9-29-03)	• Nationwide power outage 57 million people effected
London	• National grid failure (8-29-03)	• Over 250,000 commuters stranded
Northeast, Midwest, and Canada	• Human decisions by various organizations, corporate and industry policy deficiencies, inadequate management (8-14-03)	• 50 Million People effected due to the 61,800 MW of capacity not being available

	QUESTION	APPLICABLE? (Y/N)	COMMENTS
	QUESTIONS 1 THROUGH 28 APPLY TO POWER UTILITIES—SECTION II OF THE GUIDE		
	Internal General Questions Regarding Relationships with Electric Power Utilities		
1.	Do you have a working and ongoing relationship with your electric power utility?		

2.	Do you know who in your financial institution currently has a relationship with your electric power utility—that is, facilities management or accounts payable?		
3.	Do you understand your electric power utility's "Electric Service Priority" (ESP) protocols?		
4.	Do you understand your electric power utility's restoration plan?		
5.	Are you involved with your electric power utility's crisis management/disaster recovery tests?		
6.	Have you identified regulatory guidelines or business continuity requirements that necessitate planning with your electric power utility?		
	Questions to Pose to an Electric Power Utility		
7.	What is the relationship between the regional source power grid and the local distribution systems?		
8.	What are the redundancies and the related recovery capacity for both the source grid and local distribution networks?		
9.	What is the process of restoration for source grid outages?		
10.	What is the process of restoration for local network distribution outages?		
11.	How many network areas are there in (specify city)?		
12.	What are the interrelationships between each network segment and the source feeds?		
13.	Does your infrastructure meet basic standard contingency requirements for route grid design?		
14.	What are the recovery time objectives for restoring impacted operations in any given area?		

(*continued*)

(continued)

	QUESTION	APPLICABLE? (Y/N)	COMMENTS
15.	What are recovery time objectives for restoring impacted operations in any given network?		
16.	What are the restoration priorities to customers—both business and residential?		
17.	What are the criteria for rating in terms of service restoration?		
18.	Where does the financial services industry rank in the priority restoration scheme?		
19.	How do you currently inform clients of a service interruption and the estimated time for restoration?		
20.	What are the types of service disruptions, planned or unplanned, that (specify city) could possibly experience?		
21.	Could you provide a list of outages, type of outage and length of disruption that have affected (specify city) during the last 12 months?		
22.	What are the reliability indices, and who uses them?		
23.	During an outage, would you be willing to pass along information regarding the scope of interruptions to a central industry source—for example, financial services industry business continuity command center?		
24.	Are the local and regional power utilities cooperating in terms of providing emergency service? If so, in what way? If not, what are the concerns surrounding the lack of cooperation?		
25.	Would you be willing to provide schematics to select individuals and/or organizations on a nondisclosure basis?		

26.	Could you share your lessons learned from the events of 9/11 and the regional outage of 8/14/03?		
27.	Are you familiar with the "Critical Infrastructure Assurance Guidelines for Municipal Governments" document written by the Washington Military Department Emergency Management Division? Is so, would you describe where (specify city) stands in regard to the guidelines set forth in that document?		
28.	Independent of the utility's capability to restore power to its customers, can you summarize your internal business continuity plans, including preparedness for natural and manmade disasters (including but not limited to weather-related events, pandemics and terrorism)?		
	QUESTIONS 29 THROUGH 55 APPLY TO NEEDS ANALYSIS/RISK ASSESSMENT—SECTION III OF THE GUIDE		
29.	How much does each minute, hour or day of operational downtime cost your company if a specific facility is lost?		
30.	Have you determined your recovery time objectives for each of your business processes?		
31.	Does your financial institution conduct comprehensive business impact analyses (BIA) and risk assessments?		
32.	Have you considered disruption scenarios and the likelihood of disruption affecting information services, technology, personnel, facilities, and service providers in your risk assessments?		
33.	Have your disruption scenarios included both internal and external sources, such as natural events (e.g., fires, floods, severe weather), technical events (e.g., communication failure, power outages, equipment and software failure), and malicious activity (e.g., network security attacks, fraud, terrorism)?		

(*continued*)

(continued)

	QUESTION	APPLICABLE? (Y/N)	COMMENTS
34.	Does this BIA identify and prioritize business functions and state the maximum allowable downtime for critical business functions?		
35.	Does the BIA estimate data loss and transaction backlog that may result from critical business function downtime?		
36.	Have you prepared a list of "critical facilities" to include any location where a critical operation is performed including all work area environments such as branch backroom operations facilities, headquarters or data centers?		
37.	Have you classified each critical facility using a critical facility ranking/rating system such as the Tier I, II, III, IV rating categories?		
38.	Has a condition assessment been performed on each critical facility?		
39.	Has a facility risk assessment been conducted for each of your key critical facilities?		
40.	Do you know the critical, essential, and discretionary loads in each critical facility?		
41.	Must you comply with the regulatory requirements and guidelines discussed in this chapter?		
42.	Are any internal corporate risk and compliance policies applicable?		
43.	Have you identified business continuity requirements and expectations?		
44.	Has a gap analysis been performed between the capabilities of each company facility and the corresponding business process recovery time objectives residing in that facility?		

45.	Based on the gap analysis, have you determined the infrastructure needs for your critical facilities?		
46.	Have you considered fault tolerance and maintainability in your facility infrastructure requirements?		
47.	Given your new design requirements, have you applied reliability modeling to optimize a cost effective solution?		
48.	Have you planned for rapid recovery and timely resumption of critical operations following a wide-scale disruption?		
49.	Following the loss of inaccessibility of staff in at least one major operating location, how will you recover and timely resume critical operations?		
50.	Are you highly confident, through ongoing use or robust testing, that critical internal and external continuity arrangements are effective and compatible?		
51.	Have you identified clearing and settlement activities in support of critical financial markets?		
52.	Do you employ and maintain sufficient geographically dispersed resources to meet recovery and resumption activities?		
53.	Is your organization sure that there is diversity in the labor pool of the primary and backup sites, such that a wide-scale event would not simultaneously affect the labor pool of both sites?		
54.	Do you routinely use or test recovery and resumption arrangements?		
55.	Are you familiar with National Fire Protection Association (NFPA) 1600—Standard on Disaster/Emergency Management and Business Continuity Programs which provides a standardized basis for disaster/emergency management planning and business continuity programs in private and public sectors by providing common program elements, techniques, and processes?		

(continued)

(continued)

	QUESTION	APPLICABLE? (Y/N)	COMMENTS
	QUESTIONS 56 THROUGH 126 APPLY TO INSTALLATION—SECTION IV OF THE GUIDE		
	Design		
56.	Has the owner, working with an engineering professional, developed a Design Intent Document to clearly identify quantifiable requirements?		
57.	Have you prepared a Basis of Design document that memorializes in a narrative form, the project intent, future expansion options, types of infrastructure systems to be utilized, applicable codes and standards to be followed, design assumptions, and project team decisions and understanding?		
58.	Will you provide the opportunity to update the Basis of Design to reflect changes made during the construction and commissioning process?		
59.	Are the criteria for testing all systems and outlines of the commissioning process identified and incorporated into the design documents?		
60.	Have you identified a qualified engineering and design firm to conduct a peer project design review?		
61.	Have you considered directly hiring the commissioning agent to provide true independence?		
62.	Have you discussed and agreed to a division of responsibilities between the construction manager and the commissioning agent?		

63.	Do you plan to hire the ultimate operating staff ahead of the actual turnover to operations so they will benefit from participation in the design, construction and commissioning of the facility?		
64.	Have you made a decision on the commissioning agent early enough in the process to allow participation and input on commissioning issues and design review by the selected agent?		
65.	Is the proper level of fire protection in place?		
66.	Is the equipment (UPS or generator) being placed in a location prone to flooding or other water damage?		
67.	Do the generator day tanks or underground fuel cells meet local environmental rules?		
68.	Does the battery room have proper ventilation?		
69.	Has adequate cooling or heating been specified for the UPS, switchgear, or generator room?		
70.	Are the heating and cooling for the mechanical rooms on the power protection system?		
71.	Have local noise ordinances been reviewed and does all the equipment comply with the ordinances?		
72.	Are the posting and enforcement of no-smoking bans adequate, specifying, for example, no smoking within 100 ft?		
73.	Are water detection devices used to alert building management of flooding issues?		
	Procurement		
74.	Is there a benefit to using an existing vendor or supplier for standardization of process, common spare parts, or confidence in service response?		
75.	Have the commissioning, factory, and site testing requirement specifications been included in the bid documentation?		
76.	If the project is bid, have you conducted a technical compliance review which identifies exceptions, alternatives, substitutions, or noncompliance to the specifications?		

(*continued*)

(continued)

	QUESTION	APPLICABLE? (Y/N)	COMMENTS
77.	Are the procurement team members versed in the technical nuances and terminology of the job?		
78.	If delivery time is critical to the project, have you considered adding late penalty clauses to the installation or equipment contracts?		
79.	Have you included a bonus for early completion of project?		
80.	Have you obtained unit rates for potential change orders?		
81.	Have you obtained a GMP (guaranteed maximum price) from contractors?		
82.	Have you discussed preferential pricing discounts that may be available if your institution or your engineer and contractors have other similar large purchases occurring?		
	Construction		
83.	Do you intend to create and maintain a list of observations and concerns that will serve as a check list during the acceptance process to ensure that these items are not overlooked?		
84.	Will members of the design, construction, commissioning agent, and operations team attend the factory acceptance tests for major components and systems such as UPS, generators, batteries, switchgear, and chillers?		
85.	During the construction phase, do you expect to develop and circulate for comment the start up plans, documentation formats and prefunctional checklists that will be used during startup and acceptance testing?		

86.	Since interaction between the construction manager and the commissioning agent is key, will you encourage attendance at the weekly construction status meetings by the commissioning team?		
87.	Will an independent commissioning and acceptance meeting be run by the commissioning agent ensuring that everything needed for that process is on target?		
88.	Will you encourage the construction, commissioning, and operations staff to walk the job site regularly to identify access and maintainability issues?		
89.	If the job site is an operating critical site, do you have a risk assessment and change control mechanism in place to ensure reliability?		
90.	Have you established a process to have independent verification that labeling on equipment and power circuits is correct?		
	Commissioning and Acceptance		
91.	Do testing data result sheets identify expected acceptable result ranges?		
92.	Are control sequences, check lists, and procedures written in plain language, not technical jargon that is easily misunderstood?		
93.	Have all instrumentation, test equipment, actuators, and sensing devices been checked and calibrated?		
94.	Is system acceptance testing scheduled after balancing of mechanical systems and electrical cable/breaker testing are complete?		
95.	Have you listed the systems and components to be commissioned?		
96.	Has a detailed script sequencing all activities been developed?		
97.	Are all participants aware of their responsibilities and the protocols to be followed?		

(*continued*)

(continued)

	QUESTION	APPLICABLE? (Y/N)	COMMENTS
98.	Does a team directory with all contact information exist, and is it available to all involved parties?		
99.	Have you planned an “all hands on deck” meeting to walk through and finalize the commissioning schedule and scripted activities?		
100.	Have the format and content of the final report been determined in advance to ensure that all needed data is recorded and activities are scheduled?		
101.	Have you arranged for the future facility operations staff to witness and participate in the commissioning and testing efforts?		
102.	Who is responsible for ensuring that all appropriate safety methods and procedures are deployed during the testing process?		
103.	Is there a process in place that ensures training records are maintained and are updated?		
104.	Who is coordinating training and ensuring that all prescribed training takes place?		
105.	Will you videotape training sessions to capture key points and for use as refresh training?		
106.	Is the training you provide both general systems training as well as specifically targeted to types of infrastructure within the facility?		
107.	Have all vendors performed component level verification and completed prefunctional check lists prior to system level testing?		

108.	Has all system level acceptance testing been completed prior to commencing the full system integration testing and "pull the plug" power failure scenario?		
109.	Is a process developed to capture all changes made and to ensure that these changes are captured on the appropriate as built drawings, procedures, and design documents?		
110.	Do you plan to re-perform acceptance testing if a failure or anomalies occur during commissioning and testing?		
111.	Who will maintain the running punch list of incomplete items and track resolution status?		
	Transition to Operation		
112.	Have you established specific Operations Planning meetings to discuss logistics of transferring newly constructed systems to the facility operations staff?		
113.	Is all as-built documentation, such as drawings, specifications, and technical manuals complete and has it been turned over to operations staff?		
114.	Have position descriptions been prepared that clearly define roles and responsibilities of the facility staff?		
115.	Are Standard Operating Procedures (SOP), Emergency Action Procedures (EAP), updated policies, and change control processes in place to govern the newly installed systems?		
116.	Has the facility operations staff been provided with warranty, maintenance, repair, and supplier contact information?		
117.	Have spare parts lists, setpoint schedules after Cx is complete, TAB report, and re-commissioning manuals been given to operations staff?		
118.	Are the warranty start and expiration dates identified?		
119.	Have maintenance and repair contracts been executed and put into place for the equipment?		

(*continued*)

(continued)

	QUESTION	APPLICABLE? (Y/N)	COMMENTS
120.	Have minimum response times for service, distance to travel, and emergency 24/7 spare stock locations been identified?		
	Security Considerations		
121.	Have you addressed physical security concerns?		
122.	Have all infrastructures been evaluated for type of security protection needed (e.g., card control, camera recording, key control)?		
123.	Are the diesel oil tank and oil fill pipe in a secure location?		
124.	If remote dial in or Internet access is provided to any infrastructure system, have you safeguarded against hacking or do you permit read only functionality?		
125.	How frequently do you review and update access permission authorization lists?		
126.	Are critical locations included in security inspection rounds?		
	QUESTIONS 127 THROUGH 164 DEAL WITH MAINTENANCE AND TESTING—SECTION V OF THE GUIDE		
	Strategic		
127.	Is there a documented maintenance and testing program based on your business risk assessment model?		
128.	Is an audit process in place to ensure that this maintenance and testing program is being followed rigorously?		
129.	Does the program ensure that maintenance test results are benchmarked and used to update and improve the maintenance program?		

130.	Is there a program in place that ensures periodic evaluation of possible equipment replacement?		
131.	Is there a process in place that ensures the spare parts inventory is updated when new equipment is installed or other changes are made to the facility?		
132.	Have you evaluated the impact of loss of power in your institution and other institutions because of interdependencies?		
133.	Has your facility developed Standard Operating Procedures (SOPs), Emergency Action Procedures (EAPs), and Alarm Response Procedures (ARPs)?		
134.	Are the SOP, EAP, and ARP readily available and current?		
135.	Is your staff familiar with the SOPs, EAPs, and ARPs?		
	Planning		
136.	Does the system design provide redundancy so all critical equipment can be maintained without a shutdown if required?		
137.	Are there adequate work control procedures and is there a change management process to prevent mistakes when work is done on critical systems and equipment?		
138.	Are short circuit and coordination studies up to date?		
139.	Do you have a Service Level Agreement (SLA) with your facilities service providers and contractors?		
140.	Is there a change management process that communicates maintenance, testing, and repair activities to both end users and business lines?		
141.	Do you have standard operating procedures to govern routine facilities functions?		
142.	Do you have emergency response and action plans developed for expected failure scenarios?		

(*continued*)

(continued)

	QUESTION	APPLICABLE? (Y/N)	COMMENTS
143.	Have you prepared an emergency telephone contact list that includes key service providers and suppliers?		
	Safety		
144.	Is there a formal and active program for updating the safety manual?		
145.	Are electrical work procedures included in the safety manual?		
146.	Has an arc-flash study been performed?		
147.	Are specific PPE requirements posted at each panel, switchgear, etc?		
148.	Is there a program in place to ensure studies and PPE requirements are updated when system or utility supply changes are made?		
149.	Are workers trained regarding safety manual procedures?		
150.	Are hazardous areas identified on drawings?		
151.	Are hazardous areas physically identified in the facility?		
	Testing		
152.	Have protective devices been tested or checked to verify performance?		
153.	Is a Site Acceptance Test (SAT) and a Factory Acceptance Test (FAT) performed for major new equipment such as UPS systems and standby generators?		
154.	Is an annual "pull the plug" test performed to simulate a utility outage ensure that the infrastructure performs as designed?		
155.	Is an annual performance and recertification test conducted on key infrastructure systems?		

156.	Is there a process in place that ensures personnel have the proper instrumentation and that it is periodically calibrated?		
	Maintenance		
157.	Does the program identify all critical electrical equipment and components?		
158.	Is there a procedure in place that updates the program based on changes to plant equipment or processes?		
159.	Does a comprehensive plan exist for thermo infrared (IR) heat scan of critical components? Is an IR scanning test conducted before a scheduled shutdown?		
160.	Are adequate spare parts on hand for immediate repair and replacement?		
161.	Is your maintenance and testing program based on accepted industry guidelines such as NFPA 70B and on equipment supply recommendations?		
162.	Do you incorporate reliability-centered maintenance (RCM) philosophy in your approach to maintenance and testing?		
163.	When maintenance and testing is performed do you require preparation of and adherence to detailed work statements and method of procedures (MOPs)?		
164.	Do you employ predictive maintenance techniques and programs such as vibration and oil analysis?		
	QUESTIONS 165 THROUGH 225 APPLY TO TRAINING AND DOCUMENTATION—SECTION VI OF THE GUIDE		
	Documentation		
165.	What emergency plans, if any, exist for the facility?		
166.	Where are emergency plans documented (including the relevant internal and external contacts for taking action)?		

(continued)

(continued)

	QUESTION	APPLICABLE? (Y/N)	COMMENTS
167.	How are contacts reached in the event of an emergency?		
168.	How are plans audited and changed over time?		
169.	Do you have complete drawings, documentation, and technical specifications of your mission critical infrastructure including: electrical utility, in-facility electrical systems (including power distribution and ATS), gas and steam utility, UPS/battery/generator, HVAC, security, and fire suppression?		
170.	What documentation, if any, exists to describe the layout, design, and equipment used in these systems?		
171.	How many forms does this documentation require?		
172.	How is the documentation stored?		
173.	Who has access to this documentation and how do you control access?		
174.	How many people have access to this documentation?		
175.	How often does the infrastructure change?		
176.	Who is responsible for documenting change?		
177.	How is the information audited?		
178.	Can usage of facility documentation be audited?		
179.	Do you keep a historical record of changes to documentation?		
180.	Is a formal technical training program in place?		
181.	Is there a process in place that ensures personnel have proper instrumentation and that the instrumentation is periodically calibrated?		

182.	Are accidents and near-miss incidents documented?		
183.	Is there a process in place that ensures action will be taken to update procedures following accidents or near-miss events?		
184.	How much space does your physical documentation occupy today?		
185.	How quickly can you access the existing documentation?		
186.	How do you control access to the documentation?		
187.	Can responsibility for changes to documentation be audited and tracked?		
188.	If a consultant is used to make changes to documentation, how are consultant deliverables tracked?		
189.	Is your organization able to prove what content was live at any given point in time, in the event such information is required for legal purposes?		
190.	Is your organization able to quickly provide information to legal authorities including emergency response staff (e.g., fire, police)?		
191.	How are designs or other configuration changes to infrastructure approved or disapproved?		
192.	How are these approvals communicated to responsible staff?		
193.	Does workflow documentation exist for answering staff questions about what to do at each stage of documentation development?		
194.	In the case of multiple facilities, how is documentation from one facility transferred or made available to another?		
195.	What kind of reporting on facility infrastructure is required for management?		
196.	What kind of financial reporting is required in terms of facility infrastructure assets?		
197.	How are costs tracked for facility infrastructure assets?		

(*continued*)

(continued)

	QUESTION	APPLICABLE? (Y/N)	COMMENTS
198.	Is facility infrastructure documentation duplicated in multiple locations for restoration in the event of loss?		
199.	How much time would it take to replace the documentation in the event of loss?		
200.	How do you track space utilization (including cable management) within the facility?		
201.	Do you use any change management methodology (i.e., ITIL) in the day-to-day configuration management of the facility?		
	Staff and Training		
202.	How many operations and maintenance staff do you have within the building?		
203.	How many of these staff do you consider to be facilities "subject matter experts"?		
204.	How many staff members manage the operations of the building?		
205.	Are specific staff members responsible for specific portions of the building infrastructure?		
206.	What percentage of your building operations staff turns over annually?		
207.	How long has each of your operations and maintenance staff, on average, been in his or her position?		
208.	What kind of ongoing training, if any, do you provide for your operations and maintenance staff?		
209.	Do training records exist?		
210.	Is there a process in place to ensure that training records are maintained updated?		

211.	Is there a process in place that identifies an arrangement for training?		
212.	Is there a process in place that ensures the training program is periodically reviewed and identifies changes required?		
213.	Is the training you provide general training, or is it specific to an area of infrastructure within the facility?		
214.	How do you design changes to your facility systems?		
215.	Do you handle documentation management with separate staff, or do you consider it to be the responsibility of the staff making the change?		
	Network and Access		
216.	Do you have a secured network between your facility IT installations?		
217.	Is this network used for communications between your facility management staff?		
218.	Do you have an individual on your IT staff responsible for managing the security infrastructure for your data?		
219.	Do you have an online file repository?		
220.	If so, how is use of the repository monitored, logged, and audited?		
221.	How is data retrieved from the repository kept secure once it leaves the repository?		
222.	Is your file repository available through the public Internet?		
223.	Is your facilities documentation cataloged with a standard format to facilitate location of specific information?		
224.	What search capabilities, if any, are available on the documentation storage platform?		
225.	Does your facility documentation reference facility standards (e.g., electrical codes)? If so, how is this information kept up-to-date?		

Appendix C

Syska Criticality Levels

SYSKA CRITICALITY LEVEL™ DEFINITIONS

Syska Hennessy Group Critical Facilities are designed and constructed on a scale of 1–10, 10 being the ultimate in "criticality." Most of the facilities we designed and construct (and those we evaluate that are designed or built by others) are ***Criticality Levels***™ 2, 3, 4, or 5, or C2 through C5 facilities.

The term "criticality" follows from Syska's "Critical Facilities" market focus. ***Syska's Criticality Level*** definitions document Syska's approach, which has been developed over many years for major financial and other clients. ***Syska's Criticality Level*** concept is comparable in parts to data center "tier" levels used in the industry. These criticality or tier levels differentiate expected availability and reliability between sites that are designed, constructed, commissioned, maintained, and operated at different budget and priority levels. Whereas many earlier published tier classifications focused on power to the critical load, ***Syska's Criticality Levels*** address everything that is important for availability of a critical facility.

The industry has effectively addressed power to the point that power failures no longer represent the vast majority of critical facility downtime. Also, skyrocketing load and heat densities have made cooling a major area of focus, as we can no longer have an effective backup plan that involves fans and propping open outside doors. In addition, since many of the chronic power and cooling issues of the past have been resolved, issues such as nuisance emergency power off (EPO) activation and cooling control and power failures are entering into the cross hairs of close scrutiny. The business process that the critical facility supports has a direct effect on the need for availability and reliability. A call center or other production facility can potentially recover quickly from an unplanned interruption, often in the range of minutes to hours. The cost of downtime may not range significantly beyond lost hours of production time, added overtime, and some failed IT components. A financial institution that experiences the same unplanned interruption of IT services, however, may take days to weeks to recover. The cost of brief downtime can be staggering for certain business processes. Therefore, ***Syska's Criticality Levels*** addresses the entire facility, operation, and business process in a holistic and balanced approach.

The overall ***Criticality Level*** is based on subjective analysis of all key parameters. This includes not only how robust and redundant the facility is designed, constructed, and commissioned but how it is maintained and operated and how reliably it could be operated. For example, a highly redundant and flexible design that is extremely complex and customized may receive a lower rating than a much simpler design that accomplishes basically the same thing, with fewer, less important features. This is because the maintenance challenges and potential for operator or service technician error is significantly reduced when standardized, straightforward, and simpler designs are used.

Achieving the highest ***Criticality Levels*** (approaching C10) for a critical facility may not be possible with today's technology. And it may require looking beyond the facility. Redundant facilities in diverse locations (maybe even different countries—think Swiss bank accounts) with fully mirrored IT transactions and data bunkering may be required. The overall IT architecture as well as details must be designed and operated at availability levels beyond what may be possible today.

The overall ***Criticality Level*** is subjective and not the result of rigorous mathematical modeling or risk and probability analysis. However, the facility and operations can be broken down into many components or subsystems. These are analyzed based on industry experience and compared to best practices. Each component, configuration, subsystem, or integrated system is evaluated based on where we as an industry are experiencing failures. Some of the components or subsystems can include very objective factors or risk calculations (with the necessary assumptions used). Then the components are assigned weighting factors. The resulting ***Criticality Level*** then becomes relatively consistent between evaluation teams.

A key goal is to achieve a balance of criticality across all parameters. Without balance, for example, a highly robust and redundant (and expensive) MEP support system may suffer the same failure rate as a simple and inexpensive system if the operating staff is not properly trained and armed with good procedures, or budgets do not allow for proper maintenance and re-commissioning.

Important components of critical facilities that do not necessarily affect availability and reliability are not directly included in the ***Criticality Level*** assessment. These components are very important but require discussion and evaluation separate from availability and reliability. These components are indirectly assessed in the process and include the following:

- Capacities including utility, standby, and UPS power and cooling. Capacity or space for critical load increase. A facility with excess installed capacity that can readily accept an increase in critical load can be thought of as more reliable than the same facility that is operated near, at, or in excess of its redundant design load, or where short-term growth plans can be projected to compromise redundancy. These concerns are dealt with by downgrading the facility redundancy in each of the components that apply.

- Expansion capability. If a facility needs or will need to be expanded for space or capacities of support equipment, the facility should be initially designed and constructed with a plan for the ultimate size and capacity. Depending on the ***Criticality Level***, the expansion should have little impact on ongoing IT operations and should have minimal effect on availability. This is a huge challenge because predicting the IT and support needs two years out is very difficult, let alone for 10 years (typical lease term).
- Load power and heat density. Higher density expressed in watts per square foot or watts per rack increases complexity. Raised floor heights and ceiling heights become limitations for high power densities. High load density also makes component failures, especially those involving cooling air distribution less forgiving. For example, the rule-of-thumb for 15 minutes of UPS batteries in case of utility and standby generator failure may no longer apply, as loss of cooling may cause IT thermal shutdown in far less than 15 minutes. As another example, chilled water that is utilized above, alongside, or inside of server racks for high-density loads introduce new failure potential. Leak detection that effectively shuts down all localized cooling and/or shuts down power to a cabinet of servers can be difficult to make truly redundant. These concerns are dealt with in the cooling and other components that apply, making sure the appropriate redundancy and robustness is applied where it is needed.
- Floor loading capacity, ceiling heights, shape of room, columns, and so on. These are limitations and constraints rather than components of availability or reliability. Often minor limitations and constraints in these areas can be overcome by spreading equipment out horizontally across the floor. This gets into the cost of real estate in the particular geographic location.
- Size of the critical portion of the facility. Larger facilities are certainly more complex than smaller ones. However, the ability to scale support equipment is limited, such that very large critical facilities by necessity often involve duplication of smaller support systems. Very large critical facilities are similar in many ways to several small facilities under one roof.

CRITICALITY APPLICATION SUMMARY TABLE

Criticality Level	Facility/Business Process Application	Infrastructure and Redundancy
C0	Not a critical facility or business process.	Corner of office or closet, comfort cooling, plug-strips, plug-in “shoe-box” UPS.
C1	Basic critical facility, or part of facility (closet, network room) supporting office processes that	Dedicated space, 24 × 7 comfort or precision cooling, *N* (nonredundant) UPS (may be

(*continued*)

(continued)

Criticality Level	Facility/Business Process Application	Infrastructure and Redundancy
	are not critical or are backed up regularly. Loss of availability is roughly equivalent to loss of local productivity. Simple and rapid recovery from unplanned downtime.	internally redundant), wet-pipe or dry-pipe sprinklers.
C2	Facility supporting critical business processes that are local and remote. The data/telecom support might be more critical than the process (call center) than it supports, or it might be equal to or less critical than the process it supports (trading floor). Includes call centers, trading floors, telecom/internet facilities, web page and email support, etc. Loss of availability could widely affect productivity, depending on timing. Full recovery after momentary unplanned downtime can take hours. Maintenance downtime can be regularly scheduled weekly or monthly.	N (nonredundant) or $N+1$ redundant precision HVAC & UPS. $2N$ redundant power distribution. UPS maintenance bypass, backup generator, dry-pipe sprinklers, moderate commissioning and customized operations.
C3	Backup corporate facility supporting and/or including critical business processes. Loss of availability widely affects productivity and directly affects customers. Full recovery after momentary unplanned downtime can take hours or days. Maintenance downtime or high-risk windows can be scheduled, quarterly to annually.	$N+1$ or $2N$ redundant HVAC & UPS. Maintenance with load isolated on generator. N, $N+1$ or $2N$ generator. Dry-pipe sprinklers and FM-200 or VESDA. Commissioning with integrated testing. Customized operations and training. Some IT equipment and process redundancy.
C4	Primary corporate facility supporting and/or including) critical business processes. Loss of availability widely affects productivity and directly affects customers. Full recovery after momentary unplanned downtime can take hours or days. On-line maintenance with moderate-risk windows can be scheduled, monthly to annually, with blackout periods. Maintenance shutdown extremely difficult to be scheduled.	Balanced redundancy and robustness. Full $2N$ redundancy throughout all but the most expensive and lowest risk components. Online maintenance capabilities. No significant single point failure potential. Maintenance on one UPS with load supported on other UPS. N, $N+1$ or $2N$ generator. Dry-pipe sprinklers and FM-200 or VESDA. Commissioning with integrated testing. Customized operations and training. Significant IT equipment and process redundancy.

C5	Primary corporate facility supporting and/or including core business processes. Loss of availability directly translates to bottom line. Full recovery after momentary unplanned downtime can take hours to weeks. On-line maintenance with low-risk windows can be scheduled, monthly to annually, with blackout periods. Maintenance shutdown cannot be scheduled.	$2(N+1)$ redundant HVAC & UPS. Maintenance on one UPS system with load supported on other $N+1$ redundant UPS. $N+1$ or $2N$ generator. Dry-pipe sprinklers, FM-200 and VESDA. Full commissioning and regular re-commissioning. High security. Customized operations and training. Significant IT equipment and process redundancy.
C6	Primary, large corporate data center supporting and/or including core business processes. Network of remote data centers that work together. Loss of availability can affect national security, public safety, and so on. Full recovery after momentary downtime can take weeks to months. All maintenance must be online and extremely low risk.	Consider two adjacent, 100% duplicate facilities. $2(N+2)$ redundant HVAC and UPS. Maintenance on one UPS system or pipe loop with load supported on other $N+2$ redundant UPS or loop. $2N$ generator. Dry-pipe sprinklers, FM-200 and VESDA. Full commissioning & regular re-commissioning. Very high security. Customized operations and training. Significant IT equipment and process redundancy.
C7	Future	Future
C8	Future	Future

CRITICALITY FACILITY GLOSSARY

Term	Definition
Single-point failure (SPF) potential	A single component that can break or could be incorrectly operated that has the potential to take down all or a significant portion of the critical facility. An example is a battery failure on a nonredundant UPS system. Sooner or later we expect the utility to fail and if the (nonredundant) battery also fails, it is an SPF. If two unrelated failures are required to take down the facility this is not an SPF. If one UPS system includes two parallel battery strings, each of which could support the load, and one battery in each string fails, the batteries are not an SPF as this requires two unique failures. However lack of maintenance/testing on the batteries could be considered an SPF. If one failure can lead to another failure, then it is an SPF. For example, if a parallel redundant UPS module can fail, and as it is failing it causes the other UPS module to also fail, then the parallel UPS system is an SPF.

(continued)

(continued)

Term	Definition
On-line maintenance	Online maintenance capability requires sufficient redundancy and maintenance bypass flexibility such that the critical load can be supported with a high degree of reliability, while virtually any component or subsystem can be de-energized for service. In a $2N$ redundant chiller system, for example, one chiller can be turned off and isolated for service while the other chiller supports the load. The load is supported during maintenance but redundancy, and therefore availability potential is reduced during the maintenance period. For a high-reliability UPS system, $2(N+1)$ redundancy would allow the load to be supported on one side with $N+1$ redundancy while the other side is maintained.
N redundancy	"N" is the number of building blocks needed to deliver the required capacity. If 1000 tons of chilled water are required to cool the space, then the load is $N = 1000$.
$N+1$ redundancy	If the load is $N = 1000$ and a 1000-ton chiller is used, then to achieve $N+1$ redundancy, two large 1000 ton chillers are required, or $1000+1000$. 1000 additional tons of redundancy is required for 1000 tons of load. If 500-ton chillers are selected, then three smaller 500-ton chillers are required, or $500+500+500$. Only 500 additional tons of redundancy is required for 1000 tons of load.
$N+2$ redundancy	If the load is $N = 1000$ and a 1000-ton chiller is used, then to achieve $N+2$ redundancy, three 1000-ton chillers are required, or $1000+1000+1000$. 2000 additional tons of redundancy is required for 1000 tons of load. If 500-ton chillers are selected, then four 500-ton chillers are required, or $1000+500+500$. Only 1000 additional tons of redundancy is required for 1000 tons of load.
$2N$ redundancy	If the load is $N = 1000$ and a 1000-ton chiller is used, then to achieve $2N$ redundancy, two 1000-ton chillers are required, or 1000 tons on each side. Each chiller side and associated pipes, pumps, and controls are completely isolated from each other such that catastrophic failure of one side cannot cascade into the other side. If 500-ton chillers are selected, then four 500-ton chillers are required, or two 500-ton chillers on each side. For either chiller size, 1000 additional tons of redundancy and duplicative piping, pumps, and so on, is required for 1000 tons of load. However, for the smaller size two additional chillers must be installed and serviced.
$2(N+1)$ redundancy	If the load is $N = 1000$ and a 1000-ton chiller is used, then to achieve $2(N+1)$ redundancy, four 1000-ton chillers are required, or 2000 tons on each side. Each chiller side and associated pipes, pumps, and controls are completely isolated from each other such that catastrophic failure of one side cannot cascade into the other side. 3000 additional tons of redundancy and duplicative piping, pumps, and so on, is required for 1000 tons of load. If 500-ton chillers are selected, then six 500-ton chillers are required, or three 500-ton chillers on each side. If one side of the $2(N+1)$ system is down for maintenance, the other side supports the load at $N+1$ redundancy. Only 1000 additional tons of redundancy is required for 1000 tons of load, however for the smaller size two additional chillers must be installed and serviced.

$2(N+2)$ redundancy	If $N = 1000$ and a 1000-ton chiller is used, then to achieve $2(N+2)$ redundancy, six 1000-ton chillers are required, or 3000 tons on each side. Each chiller side and associated pipes, pumps, and controls are completely isolated from each other such that catastrophic failure of one side cannot cascade into the other side. 5000 additional tons of redundancy and duplicative piping, pumps, etc. is required for 1000 tons of load. If 500-ton chillers are selected, then eight 500-ton chillers are required, or four 500-ton chillers on each side. If one side of the $2(N+2)$ system is down for maintenance, the other side supports the load at $N+2$ redundancy. Only 3000 additional tons of redundancy is required for 1000 tons of load; however, for the smaller size, two additional chillers must be installed and serviced.
Load density	The amount of power load, usually expressed in watts or kW per square foot of critical floor space determines the load density. Because effectively all of the electrical power is transformed into an equivalent amount of heat, a direct calculation of the tons of cooling required to remove the heat can be performed. The problem with the disparity between load density numbers is not with the kW or tons but with the square feet. There is no single industry standard for the aisle and access area around racks and the area required for support equipment such as cooling units or PDUs. Therefore many engineers prefer to express load density in kW per rack or cabinet. Each rack or cabinet may require as little as $12\ ft^2$ or as much as $35\ ft^2$ of critical floor space, including service, aisle, column, and support equipment space.
Reverse transfer UPS	Conventional large UPS systems include a rectifier and inverter (or a motor and generator) that continuously carries 100% of the load. In the event of UPS failure or overload, the static bypass inside the UPS reverse transfers the critical load back to utility bypass. Reverse transfer UPS systems are less efficient and produce more heat. They are generally more reliable because a random failure results in a reverse transfer to bypass. Small UPS units, such as the majority of plug-in, "shoe-box" units and some very-large-capacity UPS designs, operate normally on utility bypass. When power fails they quickly turn on and forward transfer onto backup with stored energy (batteries, flywheels). Forward transfer UPS designs are very efficient and produce little heat. They are generally less reliable because they tend to fail when needed at the same time that utility power fails. Many designs fall somewhere in between these two extremes for efficiency, heat loss, and reliability. These include line-interactive and delta conversion designs.
Precision cooling	Precision cooling is designed for high sensible heat load with bearings, motors, and so on, for continuous 24×7 operation and provides precise control of temperature and humidity.
24×7 cooling	Standard office HVAC equipment is designed with bearings, motors, and so on, for 8 to 12-hour, 5-day-per-week duty cycles. Temperature is not precisely regulated and no humidification, or de-humidification beyond that which occurs from the cooling process, is provided. Operating this equipment for 24×7 high sensible heat loads can require significant additional emergency and preventative service.

SYSKA CRITICALITY LEVEL™ C0–C7 SUMMARY TABLE

Syska's Criticality Levels are summarized in the following table.

Component	C0	C1	C2	C3	C4	C5	C6	C7
Architectural	Standard office	Dedicated room	Dedicated room	Dedicated rooms	Dedicated select rooms	Dedicated facility	Dedicated secure facility	Ultra secure facility
Real estate	Standard office	Standard office	Generator capable	Generator capable	Dedicated utilities	All dedicated, away from threats	All dedicated, away from threats	All dedicated, away from threats
Cooling air distribution	Standard	N, 24×7	$N+1$	$N+1$ or $N+2$	$2N$	$2(N+1)$	$2(N+2)$	$2(N+2)+$
Cooling piping	Standard	Standard	Standard or dedicated	Dedicated, isolation	$N+$ w/isolation or $2N$	$2N$ w/isolation	$2N+$	$2N+$
Cooling heat rejection	Standard	Standard	24×7, N or $N+1$	N or $N+1$	$N+1$ or $2N$	$2N$	$2N+$	$2N++$
Cooling power	Standard	Standard	24×7 on generator	On generator	$2N$ throughout	$2N+$	$2N++$	$2N++$
Cooling controls	Standard	N, 24×7	24×7 on generator	On generator, failsafe	$2N$ throughout	$2N+$	$2N++$	$2N++$
Cooling water pipe leak containment	N/A	N/A	Isolate, contain	Isolate, contain	Isolate, contain	Isolate, contain	Isolate, contain	Isolate, contain
Cooling water treatment	Standard	Standard	Standard	Standard	Premium	Premium	Premium	Premium
Leak detection	N/A	N/A	At units	At units	At units and critical piping	Throughout	Throughout	Throughout

Water storage	N/A	N/A	N/A	For CTs, 24 hours	For CTs, 24 hours	For CTs, fire, 48 hours	For CTs, fire, 72 hours	For CTs, fire, 96 hours
Utility power	Standard	Standard	Standard	Prefer UG, dedicated	Dedicated, prefer 2 xfmr	Premium, 2 xfmr or $2N$	$2N$ or prime generation	Premium $2N$ or prime
Standby power	N/A	N/A	Generator, shared	Dedicated, N, $N+1$ or $2N$	$N+1$ or $2N$	$2N$	$2(N+1)$	$2(N+1)+$
Standby power fuel	N/A	N/A	24–72 hours	24–72 hours	24–72 hours	$2N$, 48–72 hours, on-site polish	$2N$, 72–96 hours, on-site polish	$2N$, 72–96 hours, on-site polish
Switchgear configuration	Standard	Standard	ATS for UPS and HVAC	Dedicated, multiple ATS	$2N$ throughout	$2N$ with DO breakers, many ties	$2N+$, DO breakers, many ties	$2N+$, DO breakers, many ties
Load bank capability	N/A	N/A	Rentals connected to lugs	Rentals with load bank breakers	Rental or permanent, breakers/switchboard	Permanent w/load bank switchboard	Permanent w/load bank switchboard	$2N$ Permanent w/$2N$ load bank switchboard
UPS configuration	Plug-in	N	$N+1$	$N+1$ or $2N$	$2N$	$2(N+1)$	$2(N+2)$	$2(N+2)+$
UPS maintenance bypass	N/A	N/A	Internal or external	External, allows load on generator	N provides bypass for other N	Bypass isolation on all; $N+1$ during maintenance	Bypass isolation on all; $N+2$ during maintenance	Bypass isolation on all; $2(N+2)$ during maintenance

(*continued*)

(continued)

Component	C0	C1	C2	C3	C4	C5	C6	C7
UPS DC energy storage	3-year batteries	3 to 5-year batteries	3 to 5-year batteries	Small—VRLA Large—flooded	Flooded, UPS dedicated	Flooded + alternative	Flooded + alternative	Flooded + alternative
UPS cooling	Standard	Standard	N or $N+1$	N if 2 rooms, $N+1$ if 1 room	$N+1$ or $2N$ per UPS room	$2N$ per UPS room	$2N$ per UPS room	$2(N+1)$ per UPS room
DC power plant	Phone switch-board	Phone switch-board	Phone switchboard, DC plant	Phone switchboard, DC plant	Phone switchboard, DC plant	Phone switchboard, DC plant	Large—48 VDC plant	Large—48 VDC plant
Power distribution configura-tion	Plug strips	PDUs/plug strips	N, $2N$, PDUs, panels	$2N$, PDUs, RDCs, panels	$2N$	$2N$	$2N+$	$2N+$
Lightning prot. and TVSS	N/A	1 stage	1 or 2 stages	2 stages	2 stages	Multi-stage	Extensive	Extensive
Grounding	Standard	Standard	MGB w/home runs	Chemical as needed	Chemical as needed	Chemical as needed	Premium	Premium
Fire suppression	Wet-pipe sprin-klers	Wet-pipe sprin-klers	Pre-action dry pipe	Dry-pipe + gas or VESDA	Dry-pipe + gas or VESDA	Dry-pipe + gas + VESDA	Dry-pipe + gas + VESDA	Sprinkler exemption gas + VESDA
EPO system	N/A	Code min	N plus alarm	$2N$ (A and B stations)	$2N$ (A and B stations)	$2N$ (A and B stations), alarm	$2N$ or exemption	Exemption

Fire alarm	Base building	Base building	Base building, no EPO	Base building, no EPO	Base building, no EPO	Premium, no EPO	Premium, no EPO	Premium +
Monitoring	Base building	Base building, SNMP	Base building, SNMP	BMS, SNMP	NOC	NOC+	2*N* NOC	2*N* NOC+
Security	N/A	Locked room	Room locked and controlled	Room locked and controlled	Controlled, cameras	Controlled, cameras, 24-hour security	Very secure	Ultra secure
Business process	Desktops	IT staff	Network w/IT staff	Network w/IT staff	Corporate network	Major corporate network	Major corporate/gov't network	Hi-level gov't network
IT configuration	Office support	Office support	Network	Network	Redundant components	Redundant processes	Multi-redundant processes	Multi-redundant processes
IT redundancy	N/A	N/A	Minimal process redundancy	Process redundancy, dual-corded	Redundant, replicated	Mirrored	Mirrored +	Mirrored +
Telecom/fiber provider	N/A	1 provider	1 provider	Multiple providers	Multiple providers	Multiple providers	Multiple providers +	Multiple providers +
Telecom/fiber entrance, route, distribution	N/A	1 entry distribution	2 entry, 1 backbone	2 rooms, 2 backbones	2 rooms, 2 distribution	Redundant horizontal cable	Redundant horizontal +	Redundant horizontal +

(continued)

(continued)

Component	C0	C1	C2	C3	C4	C5	C6	C7
Commissioning	N/A	Vendor startup	Performance testing	Performance and IT testing, pull-the-plug	Performance and IT testing, pull-the-plug	Full design, factory, witness, on-site Cx	Full design, factory, witness, on-site Cx +	Full design, factory, witness, on-site Cx +
Maintenance process	Repair/ replace	Repair/ replace	PM, generator exercising	PM, quarterly generator load xfr	PM, quarterly generator load xfr, proactive replacement	PM, quarterly generator load xfr, proactive replacement	PM, quarterly generator load xfr, proactive replacement	PM, quarterly generator load xfr, proactive replace-ment
Maintenance downtime	Required	Required	On utility/ generator	If $N+1$, yes; if $2N$, no	No but higher risk	No but slight risk	Very little risk	Minimized maint risk
Re-commissioning	N/A	N/A	Annual/ triennial	Periodic w, PG & IR	Periodic w, PG & IR	Periodic thorough	Periodic, extraction thorough	Periodic, extraction thorough
Operator staff	N/A	Facility engineer	Trained factory engineer	Trained staff, shifts	Customized training	Psychoanalized staff	Psychoanalized staff	Psychoanalized staff
Operator's documenta-tion	Vendor manuals	Vendor manuals	Standard SOP/EOP	Customized training	Customized training	Customized training	Customized training	Customized training
Change control	N/A	Upgrade drawings	Upkeep of all documents	Document upkeep, new staff training	Document upkeep, new staff training	Document upkeep, new staff training	Document upkeep, new staff training	Document upkeep, new staff training

GENERAL CROSS-REFERENCES TO "TIERS" AND "9'S" COMMONLY USED IN THE INDUSTRY

Syska Criticality	Uptime Institute (rev 0203)	Estimated % Availability	Expected Annual Downtime
C1	Tier I	98	20–40 hours
C2	Tier II	99	10–25 hours
C3	Tier III	99.9	1–15 hours
C4	Tier IV	99.99	0.25–1 hours
C5		99.999	1–20 minutes
C6		99.9999	20–120 seconds
C7		99.99999	
C8		99.999999	
C9		99.9999999	
C10		99.99999999	

Last updated April 29, 2005 Syska Hennessy Group.

Glossary

Automatic transfer switch (ATS) ATS provides an electrical connection between two sources of power, one serving as the general source and the other as a backup component. The normal source of electricity is usually the electric utility, and an emergency source of electricity is usually the standby generator. If the normal source power is not available, the ATS will automatically transfer to the emergency source of power and continue to supply the critical load.

Autotransformer An autotransformer is a transformer in which the primary (input) and the secondary (output) are electrically connected to each other. Because the two windings share turns, there is no isolation.

Availability Availability is expressed as the probability that it will be operational at a randomly selected future instant in time.

Brownout A brownout is an extended low-voltage state. An example of a brownout happens during peak electrical demands in the summer, when utilities cannot meet the requirements and must lower the voltage to limit maximum power. When this happens, computer systems can experience data corruption, data loss, and premature hardware failure.

Circuit Breakers (Low Voltage) Circuit breakers are resettable devices that automatically interrupt power and protect a circuit whenever the electrical load exceeds the manufacturer's specifications. A circuit breaker can also be used as an on–off switch for manual control of a circuit. Circuit breakers fall under the following classifications: magnetic, thermal, and a combination of both.

Closed transition transfer Closed transition transfer momentarily parallels the two power sources during transfer from either direction. It closes to one source before opening to another. In this application, other devices, such as in-phase monitor or an active synchronizer, must protect the transfer

Electrical line noise Electrical line noise is defined as radio-frequency interference (RFI) and electromagnetic interference (EMI) and causes undesirable effects in the circuits of computer systems. Sources of line noise include electric motors, relays, motor control devices, broadcast transmissions, microwave radiation, and distant electrical storms. RFI, EMI, and other frequency problems can cause data error, data loss, storage loss, keyboard, and system lockup.

Emergency systems Emergency systems are generally installed in places of assembly where artificial illumination is required for safe exiting and for panic control in buildings subject to occupancy by large numbers of persons such as hotels, theaters, sports arenas, health care facilities, and similar institutions. Emergency systems may also provide power for such functions as ventilation where essential to maintain life, fire detection and alarm systems, elevators, fire pumps, public safety communications systems, industrial processes where current interruption would produce serious life safety or health hazards, and similar functions.

Exercise clock Exercise clock is a programmable time switch. It automatically initiates starting and stopping of the generator set for scheduled exercise periods. The transfer of load from the normal to alternate source does not need to take place but is recommended in most cases. As mentioned previously, the engine-generator should be exercised for 30 minutes per week.

Flywheel A flywheel is a heavy-rimmed rotating mass used to keep a shaft of a machine turning at a steady speed. A flywheel can supply 15–60 sec of mechanical energy. This mechanical energy is then converted to electrical energy. This system is often interfaced with a UPS system in lieu of batteries.

Frequency variations—usually from backup generators Rare in utility power, frequency variations are most common with backup power systems such as standby generators. Many types of UPS cannot handle frequency fluctuations, which can cause system crashes and equipment damage. Obviously, this can negate the value of having backup capability.

Frequency Variation Frequency variation involves a change in frequency from the normally stable utility frequency of 50 Hz or 60 Hz, depending on the geographic location. This may be caused by erratic operation of emergency generators or unstable frequency power sources. For sensitive electronic equipment, the result can be data corruption, hard drive crash, keyboard lockup, and program failure.

Fuel cell Fuel cells operate like a car battery and make electricity by combining hydrogen ions (drawn from a hydrogen-containing fuel) with oxygen atoms. However, unlike a car battery that must be recharged periodically because the fuel and the oxidizer are powered internally, fuel cells utilize a supply of ingredients from outside the system and produce power continuously as long as the fuel supply is maintained.

Harmonics Harmonics are integral multiples of a periodic wave. Harmonic frequency is always multiples of the fundamental frequency (*example*: first harmonic is 60, second is 120, third is 180). Harmonics display a characteristic sine wave distortion that is visible on a recorded electrical sine wave. Harmonics are generated in many electromagnetic devices and may lead to overheating and a variety of power disruptions.

High voltage spikes High-voltage spikes occur when there is a sudden, rapid voltage peak of up to 6000 V. These spikes are usually the result of nearby lightning strikes, but there can be other causes as well. The effects on vulnerable electronic systems can include loss of data and burned circuit boards.

Infrared scanning or thermographic survey Thermographic testing is a procedure performed by qualified technicians in which an infrared scanning instrument is used to identify elevated temperatures as compared to ambient surroundings. This is a low-cost method for identifying potential crippling electrical problems to critical loads.

In-phase monitors This device monitors the relative voltage and phase angle between the normal and alternate power sources. It permits a transfer from normal to alternate and from alternate to normal only when acceptable values of voltage and phase angle difference are achieved. The advantage of the in-phase transfer is that connected motors continue to operate with little disturbance to the electrical system or to the process.

Isolation transformer This transformer is made of primary (input) and secondary (output) windings that have a complete separation. There is no direct path between the input and output.

Isolated ground receptacle Devices that reduce electrical noise by separating the grounding conductor for the load from the grounding path of the receptacle.

Legally required standby systems Legally required standby systems are typically installed to serve loads, such as heating and refrigeration systems, communications systems, ventilation and smoke removal systems, sewerage disposal, lighting systems, and industrial processes, that, when stopped during any interruption of the normal electrical supply, could create hazards or hamper rescue or firefighting operations.

Line interactive UPSs This type of UPS is also known as "active standby" because it is controlled by a microprocessor that continuously monitors the input line power quality and reacts to variations input. Line filtering attenuates most disturbances. Upon detection of a power failure, the UPS transfers the load to batteries in order to stabilize the power.

Manual transfer switch A manual transfer switch provides that same function, as an ATS except an operator is required to transfer to the alternate power source.

Motor Control Center (MCC) A motor control center is a single enclosure designed with a main disconnect switch. This enclosure houses various disconnect switches, protective devices, variable speed drives (VFD), starters, and other protective electrical control components.

Motor load disconnect device This device initiates the opening of a pilot contact before transfer from either source, and then provides a timed closure after transfer. This will disconnect motors before transfer and reconnect the motors after transfer. It prevents the residual voltage of the motors from causing mechanical shock to motors and driven loads and excessive current surges.

National Electrical Code N.E.C.—The National Electrical Code is the standard of the National Board of Fire Underwriters for electric wiring and apparatus, as recommended by the National Fire Prevention Association and approved by the American Standards Association.

National Electrical Manufacturers Association N.E.M.A.—National Electrical Manufacturers Association, a nonprofit trade association, supported by the manufacturers of electrical apparatus and supplies. N.E.M.A. is engaged in standardization to facilitate understanding between the manufacturers and users of electrical products.

National Fire Prevention Association N.F.P.A.—National Fire Prevention Association publishes the National Electrical Code.

Noise Often generated by normal computer operation, noise triggers these exasperating types of problems: incorrect data transfer, printing errors, keyboard/mouse/monitor lock-ups, program crashes, data corruption, and even damage to computer power supplies.

Offline or standby UPSs Off-line or standby power supply UPSs (sometimes referred to as SPSs) use an inverter, which is parallel-mounted to the load supply line and backs up the critical load. This configuration is usually applicable to small units sized less than 3 kVA. Transfer times can take up to 10 msec, and protection is not included for inrush currents. These systems have the same components as on-line UPS systems but only supply power when the normal source is lost.

Online UPSs This type of UPS includes rectifier and battery banks that are both parallel connected to the input of the unit's inverter without using a transfer switch. The batteries will take over the load if the rectifier fails to deliver power. Normally the rectifier is line-powered and carrying the load while simultaneously charging the battery banks as needed in order to float their charge at the desired value.

Open transition switch Momentarily interrupts load during bypass operation. Load-break type switches bypass either the normal or emergency sources to the load through break-before-make operation. The load will see a momentary power interruption, typically three to five cycles, upon operation of the bypass.

Optional standby systems Optional standby systems are typically installed to provide an alternate source of electric power for facilities such as industrial and commercial buildings, farms, and residences. These systems serve loads such as heating and refrigeration systems, data processing and communications systems, and industrial processes that, when interrupted during a power outage, could cause discomfort, serious interruption of the process, damage to the product or process, or the like.

Outages—blackouts Typically described as a "zero-volt" condition lasting longer than a half-cycle, outages can be caused by utility equipment failure, accidents, Mother Nature, fuses, circuit breakers, and many more. The result for unprotected systems: crashes and hardware damage.

PDU (Power Distribution Unit) PDUs are freestanding electrical distribution centers located within a data center on a raised floor. PDUs contain panel boards for branch circuits and are equipped with a transformer if required to step-down the voltage from 480 V to 120/208 V. A power monitoring system can also be included as an option if required.

Power conditioner A power conditioner is an electrical device that protects a load from more than one power quality problem. Power conditioners generally do not protect against power failure.

Power failure or blackout A blackout is defined as a zero-voltage condition that lasts for more than two cycles. The tripping of a circuit breaker, power distribution failure, or utility power failure may cause a blackout. This condition can lead to data damage, data loss, file corruption, and hardware damage.

Power line conditioners Power line conditioners are connected to a electrical hardware to enhance the power quality. Isolation transformers and capacitive couplings are examples of devices that condition power. They regulate over a certain bandwidth of voltage fluctuations.

Power sags Power sags involve voltages 80–85% below normal for fewer than 3 μsec. Possible causes are heavy equipment being energized, large electrical motors being started, and the switching of power mains (internal or utility). Power sags can have effects similar to those of a power surge, such as memory loss, data errors, flickering lights and equipment shutoff. When a power sag occurs for longer than 1 minute, it is called an under voltage condition.

Power surge A power surge takes place when the voltage is 110% above normal for fewer than 3 μsec. The most common cause is heavy electrical equipment being turned off. Under this condition, computer systems may experience memory loss, data errors, flickering lights and equipment shutoff. When a power surge occurs for longer than 1 minute, it is considered an over voltage condition

Premium system ($N+2$) The premium system meets the criteria of an ($N+2$) design by providing an essential component plus two components for backup.

Preventative maintenance (PM) For continued reliability and integrity, all systems are dependent upon an established program of routine maintenance and operational testing. The routine maintenance and operational testing program shall be based on the manufacturer's recommendations, operations manuals, and expert consultants experienced with that equipment.

Program delayed transition Program (delay) transition extends the operation time of the transfer switch mechanism. A field-adjustable setting can add an appropriate time delay that will momentarily stop the switch in a neutral position before transferring to the energized source. This will allow the motors to coast down and the transformer fields to collapse before connection to the oncoming source.

Reliability Reliability is classically defined as the probability that some item will perform satisfactorily for a specified period of time under a stated set of conditions.

RMS (root mean square) Root mean square is a mathematical concept that is used to determine the equivalent heating value of an electrical sinusoidal waveform. Root mean square is the square of the average of the squares of a set of electrical amplitudes.

Sags—short term, low voltage (brownouts) Sags are the opposite of surges, with voltages falling below 90% of average for at least one half-cycle to one minute. These are triggered by the startup of large loads, utility switching, utility equipment failure, lightning, and power service that's too small for the building demand. Sags can damage hardware and result in system crashes.

Specific gravity (SG) In a lead–acid cell, the electrolyte is a dilute solution of water and sulfuric acid. Specific gravity is a measure of the weight of acid in the electrolyte as compared to an equivalent volume of water. For example, an electrolyte with a specific gravity of 1.215 means that it is 1.215 times heavier than an equivalent volume of water that has a specific gravity of 1.000.

Spikes/Transients—instantaneous, very high voltages Spikes are a brief, intense surge, often lasting no more than a half-cycle, but with voltages 100% or more above normal. Transients cause data corruption, processing errors, incorrect data transfer, keyboard/mouse/monitor lock-ups, and hardware damage.

Standard system ($N+1$) A standard system establishes ($N+1$) reliability by supplying an essential component, such as a UPS, plus one additional component for backup. These components are located between the main power supply and the critical load.

Standby generator A standby generator is an independent reserve source of electric energy. When there is a loss of the normal electric supply, this independent reserve source of power will supply electricity to the user's facility.

Static Transfer Switch (STS) A static transfer switch selects between two or more sources of power and provides the best available power to the critical load. The STS is a solid-sate device based on silicon-controlled rectifier (SCR)

technology. The transfer from the preferred to the alternate source occurs in less than a quarter-cycle.

Storage battery A storage battery is a simple piece of equipment that converts chemical energy into electrical energy. Batteries are used as either a primary or backup source of direct current power.

Surge Suppressors A surge suppressor is an electrical device that provides protection from high-level transients. These should be installed at the service entrance panels, distribution panels, and individual critical loads.

Surges/swells—short term, high voltage Surges occur when voltages increase above 110% of normal for one half-cycle to as much as a minute. Surges can be triggered by rapid reduction in power loads, caused by heavy equipment being turned off, or by utility switching. The result: potential hardware damage.

Switching transients Switching transients take place when there is a rapid voltage peak of up to 20,000 V with a duration of 10–100 μsec. Transients are commonly caused by arcing faults and static discharge. In addition, major power system switching disturbances are initiated by the utility to correct line problems. This may happen several times a day. Effects on computer processing equipment can include memory loss, data error, data loss, and component stress.

Test switch Regular testing is necessary to both maintain the backup power system and to verify its reliability. Regular testing can uncover problems that could cause a malfunction during an actual power failure. A test switch is designed to simulate failure of the normal power source. Activation of the test switch will start the transfer of the load to the alternate power source and then retransfer either (1) when the test switch is returned to normal or (2) after a selected time delay.

Total Harmonic Distortion This term refers to the alteration of a waveshape by the presence of multipliers of the fundamental frequency of a signal.

Transformer A transformer allows high voltage to be disbursed to a usable value. A transformer takes AC voltage and changes its value. Transformers can step-up voltage and can step-down voltage, depending on their ratio setting. Transformers can also be used to isolate critical loads.

Transient Transients are disruptions to a system. Some of the disruptions rise quickly, having a sudden frequency change and then return to normal, these are called impulses. Other transients rise and fall rapidly with a changing frequency that lessens with time. This type of transient is known as oscillatory.

UPS (uninterruptible power supply) A UPS is an electrical device providing an interface between the main power supply and sensitive loads (computer systems, instrumentation, etc.). The UPS supplies sinusoidal AC power free of electrical disturbances and within strict amplitude and frequency specifications. It is generally made up of a rectifier/charger and an inverter together with a battery/charger system for backup power in the event of a main failure. UPS systems offer protection against short-term power failures.

Voltage Regulation Voltage regulation is a specification that states the difference between maximum and minimum steady-state voltage as a percentage of nominal voltage.

Voltage Regulator A voltage regulator is a device that maintains the voltage output of a generator near its nominal value in response to changing load conditions.

Voltage and frequency sensing controls Voltage and frequency sensors, controls, and meters add additional useful features to an ATS. The voltage and frequency sensing controls usually have adjustable settings for "under" and "over" conditions. The meters offer a precise indication of the voltage and frequency, and the sensing controls provide flexibility in various applications.

Waveform distortion The most common waveform distortion is a harmonic, which is a natural multiple of the standard power wave. While harmonics can be triggered by equipment inside the network, utility power may contain harmonics generated hundreds of miles away. Caused by motor speed controllers—even computers themselves—distortions can lead to communications errors and hardware damage.

Index

Maintaining Mission Critical Systems in a 24/7 Environment By Peter M. Curtis